Statistics in Management Science

Arnold Applications of Statistics Series

Series Editor: **BRIAN EVERITT**
Department of Biostatistics and Computing, Institute of Psychiatry, London, UK

This series offers titles which cover the statistical methodology most relevant to particular subject matters. Readers will be assumed to have a basic grasp of the topics covered in most general introductory statistics courses and texts, thus enabling the authors of the books in the series to concentrate on those techniques of most importance in the discipline under discussion. Although not introductory, most publications in the series are applied rather than highly technical, and all contain many detailed examples.

Other titles in the series:

Statistics in Education Ian Plewis

Statistics in Civil Engineering Andrew Metcalfe

Statistics in Human Genetics Pak Sham

Statistics in Finance Edited by David Hand & Saul Jacka

Statistics in Sport Edited by Jay Bennett

Statistics in Society Edited by Daniel Dorling & Stephen Simpson

Statistics in Psychiatry Graham Dunn

Statistics in Management Science

Andrew V. Metcalfe
University of Newcastle upon Tyne, UK

A member of the Hodder Headline Group
LONDON
Copublished in the United States of America by
Oxford University Press Inc., New York

First published in Great Britain in 2000 by Arnold,
a member of the Hodder Headline Group,
338 Euston Road, London NW1 3BH

http://www.arnoldpublishers.com

Copublished in the United States of America by
Oxford University Press Inc.,
198 Madison Avenue, New York, NY10016

Whilst the advice and information in this book are believed to be true and
accurate at the date of going to press, neither the author nor the publisher
can accept any legal responsibility or liability for any errors or omissions
that may be made.

British Library Cataloguing in Publication Data
A catalogue record for this book is available from the British Library

Library of Congress Cataloging-in-Publication Data
A catalog record for this book is available from the Library of Congress

ISBN 0 340 74075 2

1 2 3 4 5 6 7 8 9 10

Commissioning Editor: Liz Gooster
Production Editor: Rada Radojicic
Production Controller: Manjit Sihra
Cover Design: Terry Griffiths

Typeset in 10/11 Times by Academic + Technical Typesetting, Bristol
Printed and bound in Great Britain by Redwood Books, Trowbridge

What do you think about this book? Or any other Arnold title?
Please send your comments to feedback.arnold@hodder.co.uk

Contents

Preface

Most business enterprises now trade in an international market, but the tremendous potential advantage of being able to attract customers from around the world may be offset by increased competition. If businesses are to trade successfully they need to ensure their products, which may be services, are reliable, and reasonably priced. It is also necessary to develop existing products and introduce innovative ones if a business is to prosper in the future. All this has to be achieved while keeping costs within strict limits, and is applicable to manufacturers and service providers, whether in public ownership or the private sector. Statistical methods and operations research techniques offer an opportunity to meet the challenge in an interesting and imaginative fashion. The emphasis is on obtaining accurate market information and realistic forecasts, reducing undesirable variation in processes and improving their performance, and efficient use of resources for research and development. Employees at all levels can contribute, their work becomes more rewarding, waste is minimised, customers are retained and new ones attracted.

There are many introductory texts written for managers and students on management courses. This text is at an intermediate level, and I hope that it will help readers progress from the material covered in a typical introductory statistics course to applications described in management, and related, journals. It has been my intention to motivate readers by emphasising the applications of the methods described. I have tried to explain the mathematical basis of most of the techniques discussed, in an informal manner. A proper understanding of the basic mathematical principles is valuable for several reasons. The most immediate is that it is needed to interpret the output from computer software. A second is that it is necessary to understand the assumptions behind mathematical models in order to decide whether a particular model is appropriate for a given application. A third reason is that familiarity with the underlying mathematical concepts is a prerequisite for more advanced texts, and for the research literature in quantitative management.

I am grateful to students, friends and colleagues, associated with the University of Newcastle upon Tyne, for encouraging my interest in applications of statistics to management. I wish to thank all those people who have given me permission to use data, and in particular the following who provided material specifically for this book: Zalina Abdul Aziz (Universiti Sains Malaysia), Engr. Khalid Al-Hassan, Marzuki Bakar (Temasek Polytechnic), Paul Benneworth, Alison Davis (Newcastle International Airport Ltd), Tony Greenfield (Greenfield Consultants), Raouf Kattan Safinah Ltd, Susan Marshall (Northumbrian Tourist Board), Francis Ong, David Robins (Bank Bottom Engineering) and John Turcan (University of Glasgow). Rob Hyndman's Time Series Data Library at: http://www.personal.buseco.monash.edu.au/~hyndman/TSDL/ has also been a valuable source of data. Any misinterpretations, or other errors are due to me. I am grateful to Minitab Inc. for giving me a copy of Release 12 of Minitab, which I have used extensively. I have appreciated the efficiency of Kirsty Stroud and members of the production team at Arnold. Finally, I thank Lynn Kelly for typing the text so quickly and accurately, and Christine Lachecki for her extensive editorial advice. This book would not have been completed without their help.

ANDREW METCALFE
Department of Engineering Mathematics
1999

andrew.metcalfe@ncl.ac.uk

Notation

Introduction

This list gives the usual meaning of the most commonly occurring symbols. Some also occur with different meanings (r in particular) in specific applications, but these should be clear from the context. There is no standard for statistical notation so I have tried to keep to the most common conventions. However, books rarely use identical notation and you do have to read the accompanying definitions.

General notation

1. In most places a, b and c are constants.

2. Variables are usually

$$W, X, Y, Z \text{ or } w, x, y, z \quad \text{and} \quad T, t \text{ for time}$$

   It is sometimes useful to distinguish explicitly a random variable from the value it takes in a specific instance. Then the upper-case letter is the random variable and the lower-case is the specific value, as in

$$\Pr(X = x)$$

   Adhering to this rule can become rather awkward, and I have chosen to rely on the context in many places. For example, $\hat{\beta}$ represents both the estimator of a coefficient in a regression model and the estimate for a particular data set.

3. The following are usually integers:

$$i, j, k, l, m, n$$

They always are if they appear as subscripts. In time series t is usually an integer if it appears as a subscript.

4. e is the base of natural logarithms (2.71828...). π is the ratio of the circumference of a circle to its diameter (3.14159...). Both notations for the exponential function are used:

$$\mathrm{e}^x \quad \exp(x)$$

5. The square root of -1 is denoted by i.

6. Multiplication is $\times$ or juxtaposition: e.g. 2×5, and ab or $a \times b$. Natural logarithms are used unless base 10 is stated: $\ln(y) = x$ if and only if $y = \mathrm{e}^x$. Logarithms base 10 are written $\log(\ \ )$.

7. Matrices are in bold italic, superscript ^{T} is transpose, det or $\|\ \|$ stands for the determinant. The latter also denotes absolute value or modulus.

8. $\{x_i\}$ is a sample of size n, and $x_{i:n}$ are the ordered data. That is,

$$x_{1:n} < x_{2:n} < \cdots < x_{n:n}$$

9. $[a, b]$ is the interval on the real number line between a and b, i.e. x such that $a \leq x \leq b$ and (a, b) is the interval such that $a < x < b$.

10. A sum of terms is written

$$\sum x_i = x_1 + \cdots + x_n$$

A product of terms is written

$$\prod x_i = x_1 \times \cdots \times x_n$$

If there is any doubt about the range of i it is added to $\sum$ or $\prod$.

11. MOM   method of moments
    MOME   method of moments estimation/estimators/estimates
    ML   maximum likelihood
    MLE   maximum likelihood estimation/estimators/estimates
    $\mathscr{L}$ likelihood function

12. $\{x_t\}$ represents a time series.

13. ARIMA(p, d, q) autoregressive integrated moving average model with parameters p, d and q.

14. ∇ is the backward difference operator and B is back shift operator. That is:

$$\nabla x_t = x_t - x_{t-1} \quad \text{and} \quad Bx_t = x_{t-1}$$

15. A dot is used to represent differentiation with respect to time. As in a general linear system

$$\dot{x} = Ax + Bu$$

$$y = Cx$$

Symbols

$\Pr(\ )$	probability of
$\Pr(B\|A)$	probability of B conditional on A
N, n	population, sample size
$n!$	n factorial
$\Gamma(\ )$	gamma function (in particular, $\Gamma(n+1) = n!$)
$_nP_r, \, _nC_r$	permutations (arrangements) and combinations (choices) of r from n
$\mu, \bar{x}$	population, sample mean
σ^2, σ	population variance, standard deviation
s^2, s	sample estimate of population variance, standard deviation
$CV, \widehat{CV}$	population, sample coefficient of variation
$\gamma, \hat{\gamma}$	population, sample skewness
$\kappa, \hat{\kappa}$	population, sample kurtosis
f_k	frequency of event referred to by k
$p, \hat{p}$	population, sample proportion
$P(x)$	probability function (for discrete variable)
$\sim$	distributed as
$\text{Bin}(n, p)$	binomial distribution: n trials, probability of success p
$\text{Poisson }(\lambda t)$	Poisson distribution with rate λ, time interval t
$F(x), \, f(x)(f_X(x))$	cdf, pdf (subscript added if necessary)
$R(t), h(t)$	reliability or survivor function, hazard or failure rate function
x_ε	upper $\varepsilon \times 100\%$ point of distribution of X
$N(\mu, \sigma^2)$	normal distribution with mean μ and variance σ^2
$Z \sim N(0, 1)$	Z-distributed standard normal
$\Phi(z), \phi(z)$	cdf, pdf of standard normal distribution
$z_{\alpha/2}$	percentage points of $N(0, 1)$ for two sided $(1 - \alpha) \times 100\%$ confidence intervals
$U[a, b]$	uniform distribution on the interval $[a, b]$

$M(\lambda)$	exponential distribution with mean λ^{-1}
t_ν	Student's t-distribution with ν degrees of freedom
χ^2_ν	chi-square distribution with ν degrees of freedom
F_{ν_1, ν_2}	F-distribution with ν_1 and ν_2 degrees of freedom for the numerator and denominator respectively
E[ ]	expected value
$\text{var}(x)$ or σ^2_x	variance of x
$\text{stdev}(x)$	standard deviation of x
$P(x, y)$	bivariate probability function
$F(x, y), f(x, y)$	bivariate cdf, pdf
$\text{cov}(X, Y), \widehat{\text{cov}}$	population, sample covariance
ρ, r	population, sample correlation
$P(y\|x)$	conditional probability function
$f(y\|x)$	conditional pdf
$\beta_0, \beta_1, \ldots, \beta_k$	population coefficients of multiple regression model
$\hat{\beta}_0, \hat{\beta}_1, \ldots, \hat{\beta}_k$	least squares estimators, and estimates, of $\beta_0, \ldots, \beta_k$
E_i, r_i	errors about regression relationship and their estimates (residuals)
σ^2, σ	variance, standard deviation of errors
s	the sample estimate of the standard deviation of the errors, known as the residual mean square
R^2	'R squared' (proportion of variance accounted for by regression), formally known as the coefficient of multiple determination
R^2_{adj}	adjusted R^2
$\gamma(k), c(k)$	ensemble autocovariance function (acvf), sample acvf calculated from a time series
$\rho(k), r(k)$	ensemble autocorrelation function (acf), sample acf calculated from a time series
$\gamma_{xy}(\tau), c_{xy}(k)$	cross-covariance function (ccvf) for continuous signals, sample estimate from a time series
$V, \hat{V}$	variance–covariance matrix and its sample estimate

1

Statistics in management

1.1 Introduction

Managers have always had to deal with uncertainty, but they are now expected to do so in more accountable ways. The subject of statistics provides the means. Statistics is the analysis of data, and the subsequent fitting of probability models. Probability theory leads to a mathematical description of random variation, and enables us to make realistic risk assessments. The availability of personal computers and modern software has increased the range of techniques which are feasible to apply, and sophisticated analyses can be completed within minutes, but the principles behind the analyses must be clearly understood if the results are to be used in a beneficial way. The range of applications within a management context is vast, and I have aimed to show this through practical examples.

When the benefits that arise from the use of statistical methods are quantified the results are impressive. Vonderembse and White (1996) give an example of a company, ITT, which produces wiring harnesses for Ford. Wiring harnesses are welded together at certain points, and some of these welds used to fail. Engineers at ITT conducted an experiment to find a way of reducing the variability in weld strengths. The theory of design of experiments was used to set up an efficient sequence of tests, and the problem was solved within five days. The findings were passed on to other suppliers of these harnesses and there was an overall cost saving of one million dollars a year. Another example was the subject of an article in the *Guardian Higher Education Supplement* of 26th January 1999. One of the students taking a Master of Science in our department spent five months on an industrial project at the sweet factory producing Rowntree's Fruit Pastilles. The production manager at the company was aware of an intermittent problem with the machinery packing the sweets, but no employee had time to investigate it. A student project seemed a good way to proceed. It turned out that the problem was a result of the packing machine becoming jammed with overweight sweets, which had been produced due to high variability in the manufacturing process.

Not only were more sweets substantially different from the target value, but the target value itself had to be higher to ensure compliance with the Weights and Measures Act. The variability was reduced by monitoring more carefully the amount of syrup poured into moulds, and the potential savings from reducing the amount of syrup used exceed £200 000 a year. Other benefits include the reduction in lost production due to the packing machine jamming. In both these examples the cost to the companies was minimal, the only outlay being the cost of teaching staff some straightforward statistical techniques.

Managers in companies rely on statistical forecasts for their strategic planning. According to a report in *The Guardian* of 14th September 1999, unlike the European Airbus consortium, Boeing does not intend designing a super-jumbo 650–800-seat aircraft. Randy Baseler, one of Boeing's senior marketing directors, explained that however they looked at the figures there was likely to be a demand for only 350 such aircraft between now and 2018. Mr Baseler did concede that if Boeing's forecast was revised upward to, say, 700, 'then we might well have to sit up and take notice.' Sometimes it is the range of likely forecasts that is crucial, rather than the best estimate. Ranne (1999) describes an investment model for a Finnish pension company which includes inflation, share prices, property prices, government bond yields, premium loan rates, salary increases, money market rates and so on. He simulates many possible scenarios (one thousand for the diagram in the paper) over 20 years and tracks the behaviour of the fund. A particularly important output variable is the working capital, which is the difference between the total assets and the liabilities of the company and which acts as a buffer against variations in the investment results. The company's investment strategy needs to be as robust as possible to different economic conditions whilst providing a high level of return on investments.

The book is structured by area of application, and although I have attempted to arrange the chapters in a logical order, they are fairly self-contained. It is therefore not necessary to have read all the material in the preceding chapters before reading the one most relevant to your needs. However, many of the underlying ideas reappear in different chapters and links can be made between them. This is an intermediate-level text, but a brief review of the material covered in a typical introductory course, together with an informal discussion of the underlying mathematics, is given in Appendix 2. The book website, which is referred to throughout the text and which contains the larger data sets together with answers to selected exercises and other additional material, is at:

www.arnoldpublishers.com/support/metcalfe

1.2   Making decisions

The first topic in this chapter is decision trees. Drawing a tree helps clarify the uncertain events that have to be considered when decisions are made, and the concept of expected utility leads to a rational strategy. Decision trees can take account of the likely behaviour of other parties, for instance the probabilities that a company is successful with a tender will depend on the price quoted.

However, game theory is more suitable for analysing competitive two-party situations. Here the mathematics indicates that cooperation is a sound option, and provides further justification for an enlightened approach to business dealings.

Bayesian statistical methods are introduced in the second half of the chapter. In a Bayesian approach we interpret probability as our belief about the chances of some given event occurring. However, it is assumed that we will use all available data in a rational, and well-defined, manner so the method is quite disciplined. In the Bayesian paradigm, all our knowledge before some study is summarised by a prior probability distribution. We update this distribution, using data from the study, to obtain the posterior distribution. If all our knowledge, in the context of the study, is incorporated in the prior distribution we should be indifferent to the means of obtaining new information. In principle, we would not insist it is from a random sample. However, there are two good reasons for drawing random samples. The first is that all parties can agree that the procedure is fair, and the second is that it helps justify an assumption of random variation. The example given in Section 4.4.4 concerning a bank's management policy, and comparing management by objectives with the bank's usual management style, serves to illustrate some of these issues.

The Bayesian approach seems very suitable for management applications, but it does lead to more complicated calculations. Bayesian statistics software is available, sometimes free with books (e.g. Pole *et al.*, 1994). Three specific applications, for which a Bayesian approach has marked advantages, are covered. The first is an overview study, conducted for a chain of millinery shops, of the changes in sales, in different countries, attributable to the latest fashions commissioned from a new team of designers. The second is an application updating an asset management plan for a water company. A third, included in the chapter on forecasting, is predicting house sales for a builder when we might wish to include information about special events, such as a change in taxation.

1.3 Simulation

The widespread availability of modern PCs makes computer simulation a powerful means of investigating the performance of complex dynamic systems, especially when their performance is subject to random variation. Very many systems involve queueing of people, such as supermarket check-outs and telephone call centres. Others involve queueing of components, in workshop scheduling, or orders in inventory control. Simulation enables the manager to investigate the effects of making changes without disrupting the existing process.

Managers making business decisions cannot rely on a single forecast, and also need an indication of its accuracy. Simulation is used, in conjunction with time-series analysis, for generating sets of plausible future scenarios.

Simulation is also needed to carry out the calculations involved in fitting some probability models to data. This is especially true of Bayesian models, and the Gibbs sampler, essentially a technique for multivariate integration, is now routinely used by statisticians. Yet another use of simulation programs is the various games used for teaching management principles.

Any computer simulation which includes random elements relies on random number generators. These are mathematical algorithms which generate sequences of numbers that appear as if they are random. They are properly referred to as pseudo-random numbers, but the prefix is commonly omitted. Several of the techniques for generating random numbers from various distributions are covered in Appendix 3.

1.4   Obtaining market information

Finding out peoples' opinion of goods or services is essential for successful marketing. Sampling techniques, methods for deciding on a suitable sample size, and the problem of non-response are all covered. Methods of analysis are also included, because only then can the data obtained from the investigation inform management decisions.

1.5   Explanation and prediction

The objective is to find a relationship between some variable we wish to predict in the future and other variables that will be known at the time of prediction, predictor variables. We do this by fitting a regression model to data for which the predictor variables and the associated variables we wish to predict are known. A simple algorithm, available in most spreadsheet software and in all statistical packages, can be used to estimate coefficients of the predictor variables so that a linear combination of them approximates the variable to be predicted. Once fitted, the model can be used to make predictions when values for the predictor variables only are known. For example, a business development manager, in a company which manufactures marine paint, can predict the demand for paint next year from the published information about ships under construction and the price at which the company sells the paint.

There are many variants and extensions of this basic idea, including neural networks, and those which are most relevant for management problems are described.

The other main topic of this chapter is factor analysis, an interesting technique which aims to explain multivariate data in terms of a few common factors. The case study is based on a questionnaire which was used by a consultant whose speciality was advising organisations how best to implement cultural change.

1.6   Control and improvement of the process

Control and improvement of a process, be it manufacturing or some service, reduces to three stages. The first is to find out which aspects of the process cause the product to fall outside its specification. It may be that some control variables are set at incorrect values, or it may be due to excessive variability of certain operations. The second stage is to reset control variables and to try to reduce the variability. The third is to monitor the improved process and try to ensure the improvements are, at least, maintained, and, ideally, pursued further. Statistical process control (SPC) charts are a valuable technique for monitoring and, with human intervention, improving the process. Lockyer and Oakland (1981) give an example of a company in the paint industry, turning over about £35 million a year, which saved more than £250 000 by installing Shewhart mean and range charts to monitor filling processes and ensure conformity with the 1979 Weights and Measures Act. More recently, the Nashua Corporation stated on its packs of ten computer diskettes: 'Literally hundreds of charts, like that shown here [picture of a Shewhart mean chart on box], control the enemy of any manufacturing process – variation.'

Many industrial processes are too sophisticated, or too dangerous, for operator intervention. Automatic process control (APC) is applied in these cases. APC requires a system of sensors that transfer information to a computer controller, and actuators that apply the control action in response to commands from the controller. Although APC is often associated with industrial processes, there is scope for their application in the service sector, automated sorting of mail being one example. The design of APC systems is a challenging area of work that extends into robotics and should reduce the drudgery of some tasks, allowing people to spend time on the more creative aspects of their work.

1.7 Design of experiments

In 1982 a colleague and I conducted a small survey amongst mid-career managers in North-East England (Gallagher and Metcalfe, 1982). Few had ever run a properly designed experiment but a majority thought it would be worthwhile doing so. Taguchi (e.g. Taguchi, 1986, 1987) has done much to make people more aware of the advances that can be made through careful experimentation, which he has promoted as off-line quality control. But, despite the upsurge of enthusiasm engendered by Taguchi's work, designed experiments still do not seem to be carried out very often (Abdul-Aziz *et al.*, 1998).

An argument sometimes made against running a designed experiment is that an archive of data collected over many months has not been looked at. Analysing this data can be a useful preliminary to running a designed experiment. For example, such an analysis can form the basis for discussions with plant managers and operators, make them more amenable to the idea of a designed experiment, and help emphasise the need to adhere to the experimental protocol; it should help with the choice of treatments and ranges of values; it provides information on variability which can be used to determine sample sizes; and it may supply useful information which could not be elicited from a relatively short experiment.

Simple designs offer tremendous advantages over one variable at a time experiments that will result in misleading conclusions if variables interact. Specially designed software, such as DEX (Greenfield and Savas, 1992), removes the need for any detailed statistical experience. A wide range of applications is considered, from how best to improve an airline's service to selection of the best membrane material to use in filters. Iterative experimentation to improve the yield of a chemical process provides a link with APC which is discussed in Chapter 6.

1.8   Forecasting

Forecasting is an essential component of business planning. Unfortunately forecasts are often wrong, but this is not a good reason for giving up the endeavour. A more useful strategy is to consider the forecast within a range of likely scenarios. Methods for doing so are presented and discussed.

1.9   Optimisation

Most operations research (OR) techniques aim to optimise some feature of a complex system which is subject to various constraints. Some very general optimisation methods are given in Appendix 5, but these are quite inadequate when the size of the problem can increase factorially. There are many specialist OR books available, and in this chapter I have attempted simply to demonstrate the principles of stochastic versions of four widely used techniques and to show the links with statistical methods. The last technique described in the book is stochastic dynamic programming, which formalises the ideas of decision trees.

1.10   Computing

Most modern statistical techniques rely on computers for carrying out the calculations. Most people who work in departments of management in universities will have access to a wide range of mathematical and statistical programs, but managers in industry, for example, may not have access to such a variety of software. However, Excel and Lotus contain many statistical functions, including multiple regression. A few of these functions produce some dubious graphical output, such as histogram in Excel 5.0, but others, such as Excel's bubble plot, are very good. A skilled user of Excel could write macros for all the methods described in this book, but it would take some time to do so. The Nag Statistical Add-Ins for Excel offer an alternative.

If it is available, specialist statistical software is far more convenient for statistical analysis than a spreadsheet. Nearly all of the statistical techniques described in this book can be carried out with Minitab Release 12. It may not have such a wide range of routines as some other well-known statistical software, but I find it the easiest to use and I think its graphics and textual output are particularly well designed. Minitab is compatible with the major spreadsheet software, and it is easy to move a worksheet between Minitab and Excel, for example. I chose to use Matlab for writing simulation programs, and for some of the Bayesian examples which need matrix manipulation.

2
Making decisions

2.1 Introduction

Imagine you are a commercial manager in an oil company. You have been offered an opportunity to drill for oil in a specific area of the North Sea. The company engineer estimates that the cost of drilling will be £5 million, and the company geologist has estimated the probability of striking oil in the specified area as 0.4. You estimate that the total revenue at current prices (Appendix 1) would be £20 million. The cost of the concession would be 20% of revenue. Should you take up the offer? You could compare the drilling cost of £5 million with the expected monetary value (EMV) of the revenue. The EMV is the amount of money available multiplied by the probability it accrues to the company, in units of million pounds:

$$0.4 \times (80\% \text{ of } 20) = 6.4$$

This expectation is a mathematical concept which is equivalent to the average amount you would gain per play in very many plays of a game in which you win 16 (80% of 20) with probability 0.4 and receive nothing with probability 0.6. The company's actual revenue would be either 0 or 16.

The EMV of the option is

$$-5 + 0.4 \times 20 \times \frac{80}{100} = 1.4$$

This is positive, and if you are prepared to use an EMV criterion you would drill for oil. This might be a reasonable decision for a large company, but now suppose a loss of £5 million would nearly bankrupt your company. Your colleagues would be most unlikely to approve of your gamble. Application of the minimax principle, which advocates minimising the maximum loss, would lead to the more acceptable decision of turning down the option. The losses are shown in Table 2.1. The minimax principle may help us justify a common sense decision, but it is unlikely to help us arrive at it

Table 2.1   Losses for the different decision and uncertain event combinations, for the oil drilling example. Drilling could lead to a loss of 5 whereas not drilling leads to a loss of 0. The latter decision minimises the maximum loss

	Strike oil	Do not strike oil
Take up option to drill	−1.4	5
Turn down option to drill	0	0

because the minimax principle ignores the chances of uncertain events occurring.

Probability assessments are themselves uncertain but we can always consider the sensitivity of decisions to the assumed probabilities. We need a device that will enable us to model the severity of the loss of £5 million relative to the desired net profit (i.e. revenue if company strikes oil minus the drilling costs) of £11 million. The concept of utility fulfils this role.

2.2   Utility

Lindley (1985) defines utility as a measure of attractiveness of a consequence made on a probability scale. The more desirable consequence in the oil drilling example introduced in Section 2.1 was a net gain to the company of £11 million. The less desirable consequence was a loss of £5 million. Suppose the company has a working capital of £6 million. You could set up a table of utilities relative to some notional maximum amount which you do not envisage exceeding. Suppose this maximum is £20 million. Then utilities can be specified for any amount of money between £0 and £20 million; subject to the utility of 0 being 0, the utility of 20 being 1 and utility increasing as the amount of money increases. To specify the utility of £6 million you have to answer the following question. For what probability of winning £20 million would it be worth risking your entire working capital of £6 million? If you win, you and your partners can enjoy a lavish early retirement in Bermuda, but if you lose the wager the company is finished. Fortunately you are only having to imagine these alternative scenarios for the purpose of setting up a mathematical model rather than being forced to make the decision. Suppose you settle for a probability of 0.75. The EMV of 20 with a probability of 0.75 is 15, which is reasonable as it needs to be much more than 6 to compensate for the risk. So £6 million has a utility of 0.75. Now imagine you are back in the early days of the company when the working capital was only £1 million. For what probability of winning £20 million would it have been worth risking that first million? You might specify a probability of 0.30, and then complete a table, such as Table 2.2, by interpolation. You can now reconsider the decision in terms of utility. If you take up the option, your probability of gaining 11 monetary units of million pounds (mu) is 0.4, and your probability of losing 5 mu is 0.6. Starting from a capital of 6 mu the expected utility (value) (EUV) is:

$$U(6 - 5) \times 0.6 + U(6 + 11) \times 0.4$$

Table 2.2   Utility function of the commercial manager in the oil company (Sections 2.1, 2.2)

Money (x, million pounds)	Utility $U(x)$
0	0.00
1	0.30
2	0.41
3	0.51
4	0.60
5	0.68
6	0.75
7	0.81
8	0.85
9	0.88
10	0.90
11	0.91
12	0.92
13	0.93
14	0.94
15	0.95
16	0.96
17	0.97
18	0.98
19	0.99
20	1.00

From Table 2.2, $U(1)$ is 0.30 and $U(17)$ is 0.97 and the EUV of the option is 0.568. As this is less than the utility of 6 mu, 0.75, you would turn down the option. The relative utilities are independent of the notional maximum in Table 2.2. For example, if 20 was replaced by 100, and $U(20)$ was set as 0.50, all utilities in the table would be scaled by a factor of 0.50.

If you draw a graph of the utility function specified in Table 2.2 you will see that it starts off steeply and then levels off. Any risk-averse utility function exhibits this feature, which is known as concave downwards. A convenient function which gives a plausible shape for utility of money is:

$$U(x) = 1 - e^{-ax} \quad \text{for } 0 \leq x \tag{2.1}$$

The parameter a can be set at any convenient value. In Figure 2.1 a is set at 0.05 and the utility of 100 monetary units (mu) is 0.9933.

Example 2.1
It would cost 80 mu to rebuild your business premises if they were burnt down. Assume the probability of a devastating fire in one year is 0.001, and that the premises represent your entire capital. You are offered insurance for a premium of 0.5 mu/year.

(a) Assuming your utility function has the form

$$U(x) = 1 - e^{-0.05x}$$

   (i)   Is the insurance worth taking out?
   (ii) What is the risk premium?

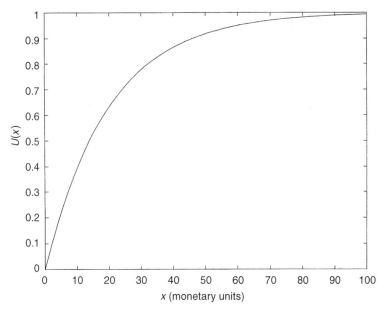

Figure 2.1   Constant risk-averse utility function for money

(b) Assume an insurance company has a utility function of the form $U(x) = 1 - e^{-0.001x}$ and assets of 10 000 mu. Why does the company offer cover for the premium of 0.5 mu?

   (i)   We need to calculate utilities for the monetary value of the premises and for the monetary value of the premises less the premium.

$$U(80) = 1 - e^{-0.05 \times 80} = 0.981\,684\,36$$

$$U(79.5) = 1 - e^{-0.05 \times 79.5} = 0.981\,220\,70$$

Without insurance the expected loss of utility is $U(80) \times 0.001$, which equals 0.9817×10^{-3}. If we insure, the loss of utility is $U(80) - U(79.5)$, which equals 0.4637×10^{-3}. The latter is smaller so it is worth taking out the insurance.

   (ii) The risk premium is defined as the difference between the premium and the EMV of the loss,

$$0.5 - 80 \times 0.001 = 0.42$$

(c) If you accept the insurance company's offer, its expected utility (EUV) is,

$$U(10\,000 - 80 + 0.5) \times 0.001 + U(10\,000 + 0.5) \times (1 - 0.001)$$

So we need to calculate:

$$U(9920.5) = 1 - e^{-0.001 \times 9920.5} = 0.999\,950\,843$$

$$U(10\,000.5) = 1 - e^{-0.001 \times 10000.5} = 0.999\,954\,622$$

and the EUV is 0.999 954 618. This exceeds $U(10\,000)$ which is 0.999 954 600 so the insurance company expects to increase its utility if you buy the insurance.

The utility $U(x)$ given by equation (2.1) is approximately linear for sufficiently small changes in x. To see why this is so, suppose C is the current capital and h is a small change. Then

$$U(C+h) - U(C) = 1 - e^{-a(C+h)} - (1 - e^{-aC})$$

$$= e^{-aC}(1 - e^{-ah})$$

which for small ah is approximately

$$e^{-aC}(1 - (1 - ah)) = ah\,e^{-aC}$$

provided $C + h$ is positive, i.e. $h > -C$. The utility of equation (2.1) also has a property of constant aversion to risk in the following sense. Consider a bet in which you win B, where $B < C$, with probability p and lose B with probability $1 - p$. The value of p which makes this acceptable in utility terms satisfies

$$p \times [U(C+B) - U(C)] = (1 - p) \times [U(C) - U(C - B)]$$

Substitution into equation (2.1) and rearrangement leads to

$$p = \frac{e^{aB} - 1}{e^{aB} - e^{-aB}} \qquad (2.2)$$

The difference between p and one-half, which represents a fair bet in monetary terms, is known as the probability premium, and for this $U(x)$, it does not depend on the current capital. This is unrealistic because most companies would accept a smaller probability premium as the current capital increased. Any function of the form

$$U(x) = 1 - w\,e^{-ax} - (1 - w)\,e^{-bx} \qquad (2.3)$$

with $0 < a, b$ and $0 < w < 1$ gives a decreasingly risk-averse utility (Lindley, 1985). You are asked to investigate a specific case in Exercise 2.1. The probability p in equation (2.2) is only slightly greater than 0.5 for small values of aB, but it increases as B becomes larger. This is realistic.

While the concept of a utility function may not be suitable for deciding many ethical issues, which might include all or nothing risks with a business, it is a good model for less radical business decisions. In the remainder of the book I avoid using any specific utility function by assuming that the company's capital is large enough for the monetary amounts involved in decisions to be proportional to changes in utility.

2.3   Decision trees

2.3.1   Prism Pharmaceuticals

The chief chemist in a small pharmaceuticals company, Prism Pharmaceuticals, has an idea for a new process which might be used to synthesise an expensive drug. If it works, the drug could be synthesised for less than half the current cost. She discusses the proposal with the managing director (MD). She estimates it would cost 15 monetary units (mu) to develop a prototype reactor which would have a 0.4 probability of being successful. If it is successful, it would cost an additional 90 mu to move to full-scale production. Sales can be classed as high, medium or low representing profits of 200, 100 and 50 at current prices. These figures allow for the cost of materials but do not allow for the cost of setting up the production plant.

The chemist and MD are partners in the company, and wonder if they could save money by keeping the process secret instead of taking out a patent. The cost of a patent would be 25 mu. The MD estimates that the probabilities of high, medium and low sales would be 0.7, 0.2 and 0.1 respectively if they have patent protection. If they rely on keeping details of the process secret there is a risk of competitors discovering and copying it. This would reduce sales, and the MD thinks the probabilities of high, medium and low sales should be set at 0.4, 0.2 and 0.4 respectively to take account of the increased risk. Another decision would need to be made about advertising. The sales manager considers that 20 mu of advertising would have a 0.5 probability of increasing low sales to medium, but only a 0.15 probability of increasing medium sales to high.

The decisions and contingent events are shown as a decision tree in Figure 2.2(a), using squares and circles for the decision and chance nodes respectively. Probabilities are added to the branches leaving chance nodes. Payoffs, the overall profit of the enterprise for the different scenarios, are shown at the end of each path through the tree. For example, the payoff for high sales if a patent was taken out is 200 mu less the costs of the prototype and full-scale reactors and the cost of the patent,

$$200 - (15 + 90 + 25) = 70$$

The payoff for low sales, having taken out a patent and advertised unsuccessfully, is

$$50 - (15 + 90 + 25 + 20) = -100$$

The underlying principle for the analysis is to work backwards (rollback) from the ends of the paths, making decisions that maximise the EMV or, more generally, the EUV. If we believe the information on the tree we cannot make a rational decision about building a prototype reactor until we have decided what we would do about advertising and taking out a patent. If sales are low, advertising is worthwhile because the EMV is 5 mu higher, e.g.

$$-50 \times 0.5 + -100 \times 0.5 = -75$$

which is preferable to -80. However, it is not worth advertising if sales are medium because the chance of improving them is not high enough to justify the advertising cost, e.g.

$$50 \times 0.15 + -50 \times 0.85 = -35$$

which is worse than a loss of 30 mu.

We can now eliminate the option to advertise if sales are medium, and eliminate the option to not advertise if sales are low. The EMV of the better decision

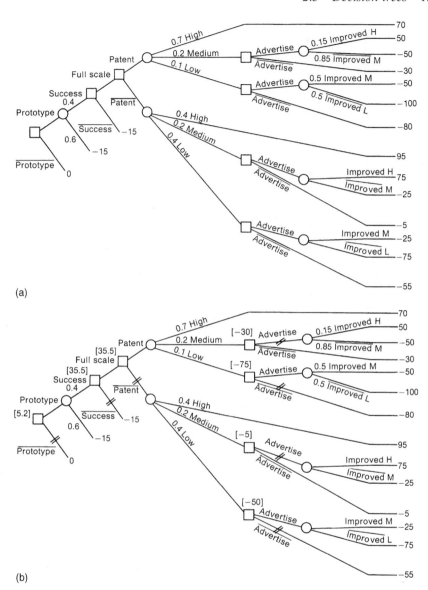

(a)

(b)

Figure 2.2 Decision tree for Prism Pharmaceuticals: (a) decisions, contingencies and costs; (b) with EMVs added. (The overline means 'not'.)

is shown at the decision node, Figure 2.2(b). The next step is to move back to the decision about a patent. The EMV with a patent is

$$70 \times 0.7 - 30 \times 0.2 - 75 \times 0.1 = 35.5$$

whereas the EMV without patent protection is

$$95 \times 0.4 - 5 \times 0.2 - 50 \times 0.4 = 17$$

It would therefore be rash to risk proceeding without a patent, and the decision would be to take out a patent. The option to not take out a patent is eliminated and the EMV of the better option, 35.5, is shown at the decision node. Finally the EMV of developing a prototype reactor is

$$35.5 \times 0.4 - 15 \times 0.6 = 5.2$$

This is positive and the conclusion of the analysis is that the company should develop the idea.

In this case the chemist did not express any doubt about the performance of the full-scale reactor if the prototype was successful. In other cases we might wish to revise probabilities, but still leave some measure of uncertainty, as more information becomes available. Bayes' theorem is used to revise probabilities in the next case.

2.3.2   Swallow Construction

The commercial manager of Swallow Construction has to decide whether to bid for a new overseas airport construction. The cost of preparing a tender would be 20 monetary units (mu). The profit would depend on the ground condition. The manager assesses the probabilities of good (G), fair (F), poor (P) and bad (B) ground conditions as 0.1, 0.5, 0.2 and 0.2 respectively, and would price the tender so that the estimated profit on the construction phase would be 900, 400, 100 and -300 mu respectively. With these prices, she thinks the chance of winning the contract is 10%. Another option to consider is to commission a geotechnical engineer to make a limited survey of the site before preparing a tender. This would cost 10 mu, but the engineer says there would be a 15% chance of such a limited survey being wrong by one classification. For example, if the ground condition is good there is a probability of 0.85 that the survey predicts it is good and a probability of 0.15 that the survey predicts it is fair. If the ground condition is fair there is a probability of 0.15 that it is incorrectly predicted to be good and a probability of 0.15 that it is incorrectly predicted to be poor. The misclassification probabilities for poor and bad ground conditions are similar. The profit estimates have not allowed for the cost of preparing a tender or the cost of a survey, but payments have been discounted (see Appendix 1) to current prices. The decision tree is shown in Figure 2.3. The probabilities of the different ground conditions, given the survey results, are calculated from Bayes' theorem. Let θ_1, θ_2, θ_3 and θ_4 represent good, fair, poor and bad ground conditions, and x_1, x_2, x_3 and x_4 represent the predictions from the survey with the subscripts $1, \ldots, 4$ standing again for good, $\ldots$, bad. Then by the definition of conditional probability,

$$P(\theta_1 | x_1) = \frac{P(\theta_1 \text{ and } x_1)}{P(x_1)}$$

The event x_1 can arise from either good or fair ground conditions, hence,

$$P(x_1) = P(\theta_1 \text{ and } x_1) + P(\theta_2 \text{ and } x_1)$$

The multiplicative rule of probability gives, for example,

$$P(\theta_1 \text{ and } x_1) = P(\theta_1) \times P(x_1 | \theta_1)$$

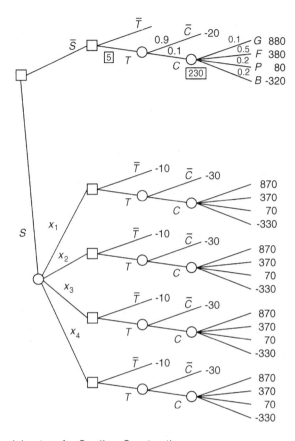

Figure 2.3 Decision tree for Swallow Construction

and the engineer specified the probabilities of predictions given the various ground conditions. Hence

$$P(\theta_1 | x_1) = 0.1 \times 0.85/(0.1 \times 0.85 + 0.5 \times 0.15) = 0.53$$

and it follows that $P(\theta_2 | x_1)$ must equal 0.47. The calculations for the other predictions follow a similar argument, e.g.

$$P(\theta_4 | x_3) = \frac{P(\theta_4 \text{ and } x_3)}{P(x_3)}$$

$$= (0.2 \times 0.15)/(0.5 \times 0.15 + 0.2 \times 0.70 + 0.2 \times 0.15) = 0.122$$

Formally,

$$P(x_i) = \sum_j P(\theta_j) P(x_i | \theta_j)$$

and

$$P(\theta_k | x_i) = P(\theta_k) P(x_i | \theta_k) / P(x_i)$$

Table 2.3 Posterior probabilities of ground conditions given predictions. The subscripts refer to good (1), fair (2), poor (3) and bad (4)

x_i	$\Pr(x_i)$	$\Pr(\theta_1 \mid x_i)$	$\Pr(\theta_2 \mid x_i)$	$\Pr(\theta_3 \mid x_i)$	$\Pr(\theta_4 \mid x_i)$
x_1	0.160	0.531	0.469	0	0
x_2	0.395	0.038	0.886	0.076	0
x_3	0.245	0	0.306	0.571	0.122
x_4	0.200	0	0	0.150	0.850

The results are summarised in Table 2.3. At this stage probabilities can be added to the lower part of the decision tree. The rollback procedure leads to a decision to tender only if the survey predicts good or fair ground conditions, and the EMV is 5.20. This exceeds the EMV of submitting a tender without a preliminary survey, which is 5.0, so the initial decision would be to commission a survey. The manager's tendering problem is an example of Bayesian decision-making (e.g. Aitchison, 1970), and a more formal general algebraic account follows.

2.3.3   Bayesian decision theory

There is a set A of possible actions. Let $U(a_i, \theta_j)$ be the utility of some action a_i if the state of nature is θ_j. We do not know θ_j, but we are prepared to give a probability distribution for all possible states from a set Θ. We then carry out some experiment with a result x_k, which allows us to revise our probabilities for θ_j. The conditional probabilities of x_k given θ_j are assumed to be known. The value of action a_i given data x_k is:

$$V(a_i, x_k) = \sum_j U(a_i, \theta_j) P(\theta_j \mid x_k) \qquad (2.4)$$

The Bayesian action relative to Θ based on data x_k is:

$$V(a^*, x_k) = \max_i V(a_i, x_k) \qquad (2.5)$$

Let d^* represent the vector of a^* corresponding to each x_k. Then the Bayesian worth is

$$B(d^*) = \sum_k V(a^*, x_k) P(x_k) \qquad (2.6)$$

Provided the θ_j form a mutually exclusive and exhaustive set of possible outcomes, i.e. one and only one θ_j is the case,

$$P(x_k) = \sum_j P(\theta_j) P(x_k \mid \theta_j) \qquad (2.7)$$

The conditional probabilities of θ_j given x_k follow from Bayes' theorem

$$P(\theta_j \mid x_k) = P(\theta_j) P(x_k \mid \theta_j) / P(x_k) \qquad (2.8)$$

Example 2.2
Suppose Swallow Construction (Section 2.3.2) has a policy of making a preliminary survey before preparing any tender. There are two possible decisions,

to submit a tender (a_1) or not to submit a tender (a_2), and four possible outcomes from the survey, $x_1, \ldots, x_4$. The array of utilities is calculated:

	a_1	a_2
θ_1	60	−10
θ_2	10	−10
θ_3	−20	−10
θ_4	−60	−10

The probabilities are summarised in Table 2.3, and the values of the two actions for different survey outcomes can be calculated.

$$V(a_1|x_1) = \sum U(a_1|\theta_i)P(\theta_i|x_1)$$

$$= 60 \times 0.531 + 10 \times 0.469 = 36.55$$

$$V(a_1|x_2) = 60 \times 0.038 + 10 \times 0.886 - 20 \times 0.076 = 9.62$$

$$V(a_1|x_3) = 10 \times 0.306 - 20 \times 0.571 - 60 \times 0.122 = -15.68$$

$$V(a_1|x_4) = -20 \times 0.150 - 60 \times 0.850 = -54$$

The value of a_2 is −10 regardless of the state of the ground conditions. Hence, d^* is a_1 if the survey result is x_1 or x_2 and a_2 otherwise.

$$B(d^*) = 36.55 \times 0.160 + 9.62 \times 0.395 - 10 \times 0.245 - 10 \times 0.200 = 5.198$$

This result is the same as we obtained from the decision tree, which is as it should be because the same principles are used.

Continuous form of Bayesian worth

In some applications it is more realistic, or more convenient, to assume the state of nature, the results of an experiment, and the possible actions are defined over a continuum. In the notation of this section the value of an action a is

$$V(a, x) = \int U(a, \theta) f(\theta|x) \, d\theta$$

The Bayesian action a^* is that which maximises $V(a, x)$. The Bayesian worth is

$$B = \int V(a^*, x) f(x) \, dx$$

Decision trees inevitably involve many simplifications and subjective assessments but they do help us appreciate the risks involved in uncertain enterprises. The sensitivity of EMV or EUV decisions to the assumed costs and probabilities can be investigated, and business options can be ranked on the basis of these results. A limitation of decision trees is that the rollback principle is not suitable for analysing situations in which competitors' decisions depend on our actions. The manager in Swallow Construction assumed that the content of the tender would not be divulged to any other companies. Game theory aims to provide some answers to more confrontational situations.

2.4 Game theory

2.4.1 Zero-sum games

A two-person zero-sum game is one in which one person's loss is the other's gain. Two salesmen have been asked to divide a region into two areas so that each has responsibility for sales in one area. Both are concerned that the division is equitable, and neither is willing to accept an arbitrary division of the region into two areas followed by the toss of a coin to decide who is responsible for each area. Instead one will divide the region and the other will choose. The one who divides will make the areas as equal as possible in terms of anticipated sales because he knows the other will choose the area with the higher anticipated sales. The situation is summarised in Table 2.4. The one who divides (D) minimises his maximum loss. More formally, in terms of payment to D, D maximises the row minima (maximin) and the one who chooses (C) minimises the column maxima (minimax). When the maximin and minimax coincide the resulting point is called a saddle point. It is stable because: if C changes strategy D would gain at C's expense, and if D changes strategy C would gain at D's expense. Furthermore, there is no uncertainty over which strategy C and D will choose, and there is no need to keep strategies confidential. However, not all games have saddle points.

The two salesmen, A and B, decide to select who is to do the division, which is the less favourable position, by playing 'Matching Pennies'. Both players simultaneously show one face of a coin. If the coins match A wins, if not B wins. The payoffs are shown in Table 2.5(a). There is now no saddle point, and if B were to know what A would play then B would win. The optimum strategy for both salesmen is to randomly play heads half the time. This is known as a mixed strategy.

Poundstone (1992) introduces a nice variation to 'Matching Pennies' which he calls 'Millionaire Jackpot Matching Pennies'. The payoffs are shown in Table 2.5(b). See Exercise 2.5 for the optimum strategy. John von Neumann had proved his minimax theorem by 1928. The theorem states that there is either a saddle point or an optimum mixed strategy for any finite, two-person, zero-sum game. But these are of limited use as models for business situations. In particular A's loss may not be B's gain.

Table 2.4 Division of region into two areas for sales purposes. Payoffs to divider in which S is half the anticipated sales (mu) for the region, ε is a small number, and L is a large number. $S - \varepsilon$ is the saddle point

		Chooser	
		Choose area with higher anticipated sales	Choose area with lower anticipated sales
Divider	Division as even as possible	$S - \varepsilon$	$S + \varepsilon$
	Division is clearly uneven	$S - L$	$S + L$

Table 2.5   (a) Payoffs to A for 'Matching Pennies'. (b) Payoffs to A for 'Millionaire Jackpot Matching Pennies'

(a)

		B plays	
		Head	Tail
A plays	Head	ε	$-\varepsilon$
	Tail	$-\varepsilon$	ε

(b)

		B plays	
		Head	Tail
A plays	Head	1 million	-1
	Tail	-1	1

2.4.2   The 'Prisoner's Dilemma'

The following model of a business situation is an example of a dilemma which has appeared in many guises over the centuries. It is frequently told in the context of a prisoner being offered a reduced jail term for testifying against his co-accused, and is perhaps best known as the 'Prisoner's Dilemma'. Two companies C and D have developed a novel voice interactive radio. If they both advertise at a modest level (a) they will both achieve an overall profit of 3 monetary units (mu). If they both mount an aggressive advertising campaign (A) they will both achieve an overall profit of only 1 mu. The obvious solution appears to be an agreement to restrict advertising to the modest level. The snag is that if C keeps to the agreement and D reneges on it, then D will make an over-all profit of 5 mu whilst C will be left with no profit at all. These outcomes are summarised in Table 2.6. C reflects that D has recently demonstrated a lack of integrity by enticing some staff with rather specialist knowledge away from C by offering extraordinarily high salaries. Furthermore, if D does keep to the agreement, C could settle the score by reneging. If neither party trusts the other the inevitable outcome is that both companies advertise aggressively. This is a Nash equilibrium (Poundstone, 1992), because C's best strategy is A regardless of what D does and vice versa.

Table 2.6   Payoffs to C and D respectively in monetary units for modest (a) and aggressive (A) advertising

		D	
		a	A
C	a	(3, 3)	(0, 5)
	A	(5, 0)	(1, 1)

If the situation is repeated many times the simple strategy of Tit-for-Tat appears to be near optimum for both players (Axelrod, 1984). The Tit-for-Tat strategy is to cooperate on the first round and then do whatever the other player did in the previous round. A theoretical criticism is that it cannot take advantage of an unresponsive player. A more serious limitation is that if the other player uses a Tit-for-Tat strategy and either starts by defecting, or makes an aberrant defection, the two strategies will alternate defecting and cooperating indefinitely. A neat solution is to play Forgiving Tit-for-Tat in which a defection can be followed by cooperation with a probability p. Values of p as high as $1/3$ work well.

Although this discussion of the 'Prisoner's Dilemma' avoids the restriction to zero-sum games that was made in Section 2.4.1, it is still restricted to two-player games. Much of the work on n-person games rests on an assumption that people will form alliances and that the problem can thereby be reduced to a two-person game. An example is a cartel threatened with state legislation. Both parties might prefer the cooperative solution of industry self-regulation.

2.5   Bayesian statistics

The Bayesian approach to statistics is particularly appealing for management applications because it provides a method for incorporating expert judgement with limited data. This may not be an acceptable way of conducting scientific investigations, but it is a pragmatic approach for making business decisions. Probability is interpreted as our belief about the chance of some event occurring, and we assign probability distributions to unknown population parameters rather than treating them as unknown constants as is done in the frequentist approach. In theory, there is less emphasis on randomisation within the Bayesian approach but this distinction becomes blurred in practice. Frequentist analysis is designed for analysing data that have been obtained through some random sampling scheme. However, we often resort to assuming data were obtained from such a scheme, either because setting up a proper randomisation is prohibitively expensive, as in much market research, or because drawing a random sample is impossible, as is the case if we wish to analyse sales figures over the past years. A corresponding Bayesian analysis might assume exchangeability; that is, the joint distribution remains invariant under any permutation of the suffixes. Although there is no explicit mention of randomisation, taking a simple random sample remains a good justification for assuming exchangeability. Also, exchangeability implies that the variables have the probability structure of a random sample from that distribution. A related concept to exchangeability is mutual independence. Variables can be mutually independent without having identical distributions. A sequence of events $E_1, \ldots, E_n$ are mutually independent if for every finite subset of them

$$\Pr(E_i \text{ and } E_j \text{ and} \ldots E_m) = \Pr(E_i) \times \Pr(E_j) \times \cdots \times \Pr(E_m)$$

Pairwise independence does not imply mutual independence and even

$$\Pr(E_1 \text{ and } E_2 \text{ and} \ldots E_n) = \Pr(E_1) \times \Pr(E_2) \times \cdots \times \Pr(E_n)$$

is not sufficient for mutual independence (Exercise 2.7).

Example 2.3

The following example demonstrates that pairwise independence does not imply mutual independence. A businessman (A) agreed to meet his colleague (B) at a city airport without realising there are two airports serving the city. There is a 50% chance that B flies to the north airport (BN), and hence a 50% chance B flies to the south airport (BS), so A flips a fair coin to decide which airport he will drive to. Let M be the event that they meet.

$$\Pr(AN \text{ and } BN) = \Pr(AN) \times \Pr(BN) = \tfrac{1}{2} \times \tfrac{1}{2} = \tfrac{1}{4}$$

$$\Pr(M) = \Pr(AN \text{ and } BN) + \Pr(AS \text{ and } BS) = \tfrac{1}{4} + \tfrac{1}{4} = \tfrac{1}{2}$$

$$\Pr(AN \text{ and } M) = \Pr(AN) \times \Pr(M \mid AN) = \Pr(AN) \times \Pr(M) = \tfrac{1}{2} \times \tfrac{1}{2} = \tfrac{1}{4}$$

$$\Pr(BN \text{ and } M) = \Pr(BN) \times \Pr(M \mid BN) = \Pr(BN) \times \Pr(M) = \tfrac{1}{2} \times \tfrac{1}{2} = \tfrac{1}{4}$$

Hence the events AN, BN and M are pairwise independent. But

$$\Pr(AN \text{ and } BN \text{ and } M) = \Pr(AN \text{ and } BN) = \tfrac{1}{4}$$

whereas

$$\Pr(AN) \times \Pr(BN) \times \Pr(M) = \tfrac{1}{8}$$

Almond (1995) gives this example in terms of a lighting circuit with two random switches.

2.5.1 Beta distribution

A continuous random variable Y can take values between 0 and 1. The variable Y has a beta distribution with parameters α, β, written as $Y \sim Be(\alpha, \beta)$, if it has a probability density function,

$$f(y) = \frac{1}{B(\alpha, \beta)} y^{\alpha-1}(1-y)^{\beta-1} \quad 0 \le y \le 1 \tag{2.9}$$

where $1/B(\alpha, \beta)$ is the factor which gives an area of 1. The beta function $B(\alpha, \beta)$ occurs in other branches of mathematics and is related to gamma functions (Exercise 2.8 and Table AST.7) by

$$B(\alpha, \beta) = \Gamma(\alpha)\Gamma(\beta)/\Gamma(\alpha + \beta)$$

The mean, mode and variance of the distribution are given by

$$\mu = \alpha/(\alpha + \beta) \tag{2.10a}$$

$$\text{mode} = (\alpha - 1)/(\alpha + \beta - 2) \tag{2.10b}$$

$$\sigma^2 = \frac{\alpha\beta}{(\alpha + \beta)^2(\alpha + \beta + 1)} \tag{2.10c}$$

The parameters can be made the subjects of the formulae, i.e.

$$\alpha = \mu\{[\mu(1 - \mu)/\sigma^2] - 1\} \tag{2.11a}$$

$$\beta = (1 - \mu)\{[\mu(1 - \mu)/\sigma^2] - 1\} \tag{2.11b}$$

Parameter values can be set by substituting the required values for μ and σ^2. If data $\{y_i\}$ from the distribution are available the method of moments estimates of α and β are found by replacing μ and σ by $\bar{y}$ and s_y.

Example 2.4
Loudon and Della Bitta (1993) claim, on the basis of various sources, that brand loyalty for grocery items such as cereal brands is relatively low with only one out of three repurchases of the same brand within a product category. I am prepared to take 1/3 as a best estimate of the probability (p) a shopper buys the same brand next time, but as the original sources go back at least 10 years and are based on surveys in the USA, I have some reservations about the applicability of the estimate for current UK supermarkets. I therefore wish to express my uncertainty about the value of p by specifying a probability distribution for it. A useful interpretation of standard deviation is that a variable has a probability of approximately 2/3 of being within one standard deviation of its mean. I consider that there is a 2/3 chance that p is within 0.15 of its mean, which I assume to be 0.33. The corresponding parameter estimates are obtained by replacing μ and σ, in equations 2.11(a)(b), with 0.33 and 0.15:

$$\alpha = 2.91 \quad \beta = 5.91$$

2.5.2   Estimating a proportion

In the previous example, I represented my opinion about the probability (p) of repurchase of the same brand of cereal by

$$p \sim Be(\alpha, \beta)$$

I shall refer to this as my prior distribution for p because I have just asked a local shopkeeper to tell me how many of the next n customers who buy this type of cereal claim to have bought the same brand last time. I shall revise my opinion about the likely values for p once I have this additional information, and specify a posterior distribution. Let the number making this claim be X out of n. It seems reasonable to assume that the distribution of X for a fixed value of p is $Bin(n, p)$, that is

$$P(x|p) = {}_nC_x p^x (1 - p)^{n-x}$$

This information is used to revise my distribution for p by using Bayes' theorem

$$f(p|x) = \frac{f(p, x)}{P(x)} = \frac{P(x|p) f(p)}{P(x)}$$

We have expressions for $P(x|p)$ and $f(p)$, but not for $P(x)$. However, $1/P(x)$ is just the factor which gives an area of 1 so we can write

$$f(p|x) \propto P(x|p) f(p)$$

and scale to get an area of 1 at the end. The probability density $f(p|x)$ is known as the posterior distribution of p. Once x is known, $P(x|p)$ is more naturally thought of as a function of p known as a likelihood for p (Exercises 2.9 and 2.10). The Bayes' method can be summarised by the statement that the posterior distribution is proportional to the product of the likelihood and the prior distribution. Returning to this specific case

$$f(p|x) \propto p^x(1-p)^{n-x}f(p)$$
$$= p^x(1-p)^{n-x}p^{\alpha-1}(1-p)^{\beta-1}$$
$$= p^{\alpha+x-1}(1-p)^{\beta+n-x-1}$$

So the posterior distribution for p is also a beta distribution,

$$p \sim Be(\alpha+x-1, \beta+n-x-1)$$

Notice that the potentially awkward combinatorial term, $_nC_x$, does not depend on p and is absorbed in the constant of proportionality.

This neat result depends on the choice of a beta prior distribution when the likelihood is binomial. In general a class of prior distributions is conjugate with respect to a form of likelihood if the posterior distribution belongs to the same class as the prior distribution. The posterior distribution would be used as a prior distribution if further survey information became available.

Example 2.5
My prior distribution for the probability (p) a shopper buys the same brand of a type of breakfast cereal next time is

$$p \sim Be(2.91, 5.91)$$

I am now told that 4 out of the last 21 customers to buy this type of cereal claim to have bought the same brand last time. My posterior probability distribution for p is

$$p \sim Be(5.91, 21.91)$$

The prior and posterior density function for p are shown in Figure 2.4.

Highest density region

A $(1-\varepsilon) \times 100\%$ Bayesian confidence interval for p, also known as a highest density region (HDR), is given by $[L, U]$ where

$$Pr(L < p < U) = 1 - \varepsilon$$

and the probability density function of p within the interval is higher than at any point outside the interval. This definition may need modification if the density of p is multi-modal. Lee (1997) gives a construction for the beta distribution, but it relies on a table of HDR points for the logarithmic F-distribution which are slightly different from percentage points of the F-distribution. Other options are to write a program to calculate the HDR directly from the formula, using numerical integration, or to make a rather rough approximation and assume the beta distribution can be approximated by a normal distribution with the same mean and variance.

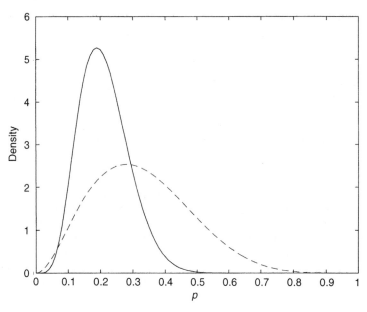

Figure 2.4   Prior (broken line) and posterior densities for probability of repeat purchase of a cereal brand

Example 2.6

Construct an approximate 90% Bayesian confidence interval for p if

$$p \sim Be(5.91, 21.91)$$

The mean and variance of the distribution are given by

$$\mu = 5.91/(5.91 + 21.91) = 0.212$$

$$\sigma^2 = 5.91 \times 21.91/[(27.82)^2 \times (28.82)] = (0.076)^2$$

which leads to an approximate 90% interval

$$0.212 \pm z_{0.05} \times 0.076$$

Since $z_{0.05}$ is given in Table AST.2 as 1.645 this becomes

$$[0.09, 0.34]$$

For comparison, Lee's construction leads to a 90% HDR of [0.12, 0.38].

2.5.3   Predictive distribution

When making business decisions we may require the distribution of some variable x which allows for the uncertainty about the parameter values. This is known as the predictive distribution. Formally, if θ represents the parameters, the predictive distribution is obtained from the joint distribution of x and θ by integrating over θ.

$$f(x) = \int f(x, \theta) \, d\theta$$

It is usually convenient to express the joint distribution as the product of the distribution of θ and the conditional distribution of x given θ. Thus

$$f(x) = \int f(x|\theta) f(\theta) \, d\theta$$

The beta-binomial distribution is the predictive distribution for the number of successes, x, in n independent trials when the probability of success is assumed to have a beta distribution.

$$P(x) = \int P(x|p) f(p) \, dp$$

$$= \int {_nC_x} p^x (1-p)^{n-x} B(\alpha, \beta)^{-1} p^{\alpha-1} (1-p)^{\beta-1} \, dp$$

$$= {_nC_x} B(\alpha, \beta)^{-1} B(\alpha + x, \beta + n - x)$$

Another use of the predictive distribution is to check whether new data are compatible with the current information.

2.5.4 Normal distribution

Suppose a variable X has a normal distribution,

$$X \sim N(\mu, \sigma^2)$$

and we have a prior distribution for the mean μ,

$$\mu \sim N(\mu_0, \sigma_0^2)$$

If we observe a value x of X we can bring our information about μ up to date by constructing the posterior distribution. The principles described in this section lead to

$$\mu \sim N(\mu_1, \sigma_1^2)$$

where

$$\mu_1 = \mu_0 \frac{\sigma_0^{-2}}{\sigma_0^{-2} + \sigma^{-2}} + x \frac{\sigma^{-2}}{\sigma_0^{-2} + \sigma^{-2}} \tag{2.12a}$$

$$\sigma_1^{-2} = \sigma_0^{-2} + \sigma^{-2} \tag{2.12b}$$

If precision is defined as the reciprocal of the variance, these formulae can be expressed succinctly as: posterior mean is the weighted mean of the prior mean and datum, with weights proportional to their precisions; and posterior precision is the sum of the prior precision and the datum precision. The predictive distribution for another observation X is

$$X \sim N(\mu_1, \sigma^2 + \sigma_1^2)$$

The predictive distribution for the original observation would have been $N(\mu_0, \sigma^2 + \sigma_0^2)$. These results are a consequence of variances of statistically independent variables being additive.

Example 2.7

Sales of a certain brand of filter coffee have averaged 1258 boxes per month over the past six months. The standard deviation was 116. A sales manager is prepared to assume market conditions are fairly stable and gives a prior distribution for the mean monthly sales μ as

$$\mu \sim N(1258, (116)^2/6)$$

The distribution of monthly sales X is

$$X \sim N(\mu, (116)^2)$$

and the predictive distribution for next month's sales is

$$N(1258, (116)^2 + (116)^2/6)$$

If the next month's sales are 1023 boxes, is there strong evidence of a decrease in sales? If the manager is willing to assume that market conditions are still reasonably stable, find the posterior distribution of μ.

From the predictive distribution the probability of sales being as low as 1023 boxes is

$$\Pr(Z < (1023 - 1258)/125.3) = \Pr(Z < -1.88) = 0.03$$

This may be slightly worrying, but assuming market conditions are stable the posterior distribution for μ has mean,

$$\mu_1 = 1258 \times \frac{(1/47)^2}{(1/47)^2(1/116)^2} + 1023 \times \frac{(1/116)^2}{(1/47)^2(1/116)^2}$$

$$= 1258 \times 0.859 + 1023 \times 0.141 = 1225$$

and a variance

$$\sigma_1^2 = 1/((1/47)^2 + (1/116)^2) = (44)^2$$

The algebraic details become considerably more complicated if we allow for uncertainty in variances but similar principles hold.

2.5.5   Bayesian hierarchical models

A company owns chains of small millinery shops, which trade under the name of 'March Hare', in several European countries and parts of North America. Sales have been declining and two years ago the managing director commissioned some new designs from several of the graduates showing collections at Graduate Fashion Week which was held in London. Their brief was to produce a range of designs that would appeal to younger customers. The change in sales over the past two years, compared with the preceding two years, has been recorded for all shops. The managing director thinks that the impact of the new collections on sales may not be the same in all countries. She therefore

Table 2.7 Change in sales factors for shops in six countries

Country	y_j	(σ_j^2)	$E[\theta_j \mid \mathbf{y}]$	$(\text{Stdev}(\theta_j \mid \mathbf{y}))^2$
A	1.26	$(0.09)^2$	1.24	$(0.07)^2$
B	1.23	$(0.11)^2$	1.23	$(0.08)^2$
C	1.32	$(0.18)^2$	1.25	$(0.11)^2$
D	0.67	$(0.26)^2$	1.13	$(0.16)^2$
E	1.18	$(0.18)^2$	1.21	$(0.10)^2$
F	1.31	$(0.13)^2$	1.25	$(0.09)^2$

suggests a hierarchical model for the data. Let θ_j be the multiplicative effect on sales for country j, for example a factor (θ_j) of 1.14 corresponds to a 14% increase whereas a factor of 0.92 corresponds to an 8% decrease. The observed factors y_j and their standard deviations σ_j are given for the six countries in Table 2.7. The y_j is the average of the factors for all the shops in the country, and σ_j is the standard deviation of y_j which has been estimated by dividing the standard deviation of the factors for all the shops by the square root of their number. We will overlook the fact that σ_j are estimates. The model is

$$y_j \sim N(\theta_j, \sigma_j^2) \tag{2.13a}$$

with the y_j being mutually independent,

$$\theta_j \sim N(\mu, \tau^2) \tag{2.13b}$$

with the θ_j being mutually independent,

$$\mu \sim \text{suitable prior distribution} \tag{2.13c}$$

$$\tau \sim \text{suitable prior distribution} \tag{2.13d}$$

It is convenient to assume y_j and θ_j are normally distributed, and this assumption will suffice if the factors are fairly close to 1. An alternative would be to work with logarithms of the ratios. The choice of prior distributions for the hyperparameters does not affect the following theoretical development. Let $\boldsymbol{\theta}$ and $\mathbf{y}$ be random vectors $(\theta_1, \ldots, \theta_k)$ and $(y_1, \ldots, y_k)$ respectively. The main objectives are to find

$$f(\theta_j \mid \mathbf{y}) \quad \text{and} \quad f(\mu \mid \mathbf{y})$$

If θ_j is known, knowledge of μ and τ tells us nothing more about the distribution of y_j, i.e.

$$f(\mathbf{y} \mid \boldsymbol{\theta}, \mu, \tau) = f(\mathbf{y} \mid \boldsymbol{\theta}) \tag{2.14}$$

If we can find the joint posterior distribution for all the parameters, given $\mathbf{y}$, we can find the marginal posterior distributions for individual θ_j and μ by integrating over the other parameters. The most direct approach is to write

$$f(\boldsymbol{\theta}, \mu, \tau \mid \mathbf{y}) \propto f(\mu, \tau) f(\boldsymbol{\theta} \mid \mu, \tau) f(\mathbf{y} \mid \boldsymbol{\theta})$$

All three terms in the product on the right-hand side are known, i.e.

$$f(\boldsymbol{\theta} \mid \mu, \tau) = \Pi N(\theta_j \mid \mu, \tau^2)$$

$$f(\mathbf{y} \mid \boldsymbol{\theta}) = \Pi N(y_j \mid \theta_j, \sigma_j^2)$$

and $f(\mu, \tau)$ is whatever we choose to represent our prior belief about μ and τ. However, direct integration of the right-hand side may not be practical, and it is not in a suitable form for a solution by simulation methods because it is not conditional on the data (y).

Gelman *et al.* (1995) advise using the alternative factorisation,

$$f(\boldsymbol{\theta}, \mu, \tau \mid \boldsymbol{y}) \propto f(\tau \mid \boldsymbol{y}) f(\mu \mid \tau, \boldsymbol{y}) f(\boldsymbol{\theta} \mid \mu, \tau, \boldsymbol{y}) \qquad (2.15)$$

to simulate the joint distribution. The third term on the right-hand side, $f(\boldsymbol{\theta} \mid \mu, \tau, \boldsymbol{y})$ follows from equations (2.12a,b) in Section 2.5.4. That is:

$$\theta_j \mid \mu, \tau, y_j \sim N(\hat{\theta}_j, V_j) \qquad (2.16a)$$

where

$$\hat{\theta}_j = (y_j/\sigma_j^2 + \mu/\tau^2)/(1/\sigma_j^2 + 1/\tau^2) \qquad (2.16b)$$

$$V_j = 1/(1/\sigma_j^2 + 1/\tau^2) \qquad (2.16c)$$

These equations are all that are required because

$$f(\boldsymbol{\theta} \mid \mu, \tau, \boldsymbol{y}) = \Pi N(\hat{\theta}_j, V_j)$$

The posterior conditional distribution of μ given τ and $\boldsymbol{y}$, follows from a similar argument. Let $\tilde{\mu}$ be the prior estimate of μ with precision ϕ. Then

$$\mu \mid \tau, \boldsymbol{y} \sim N(\hat{\mu}, V_\mu) \qquad (2.17a)$$

where

$$\hat{\mu} = \left[\sum y_j/(\sigma_j^2 + \tau^2) + \phi\tilde{\mu}\right] / \left[\sum 1/(\sigma_j^2 + \tau^2) + \phi\right] \qquad (2.17b)$$

and

$$V_\mu^{-1} = \sum 1/(\sigma_j^2 + \tau^2) + \phi \qquad (2.17c)$$

The posterior distribution of τ, $f(\tau \mid \boldsymbol{y})$, can be obtained from the following argument (Gelman *et al.*, 1995). First notice that

$$y_j \mid \mu, \tau \sim N(\mu, \sigma_j^2 + \tau^2)$$

Now, it follows from the definitions of conditional distributions that

$$f(\tau \mid \boldsymbol{y}) = f(\mu, \tau \mid \boldsymbol{y})/f(\mu \mid \tau, \boldsymbol{y})$$

$$\propto f(\mu, \tau) f(\boldsymbol{y} \mid \mu, \tau)/f(\mu \mid \tau, \boldsymbol{y})$$

Since $f(\tau \mid \boldsymbol{y})$ does not include μ, the relationship will hold for any choice of distribution for or specific value of μ, and in particular for $\hat{\mu}$. Hence

$$f(\tau \mid \boldsymbol{y}) \propto f(\tau) V_\mu^{1/2} \Pi(\sigma_j^2 + \tau^2)^{-\frac{1}{2}} \exp(-(y_j - \hat{\mu})^2/(2(\sigma_j^2 + \tau^2))) \qquad (2.18)$$

A uniform distribution with some plausible upper bound is a convenient choice for the prior distribution of τ, but the inverse chi-squared family of distributions (χ^{-2}) is often used for variance priors (see Exercise 2.15). The managing director set the prior distribution for μ as uniform over $[0.6, 1.7]$ and the prior distribution for τ as uniform over $[0, 0.4]$. The pdf of the posterior distribution of τ, conditional on $\boldsymbol{y}$, is given by equation (2.18) when the right-hand side of equation (2.17b) is substituted for $\hat{\mu}$. The prior estimate of μ, $\tilde{\mu}$

in equation (2.17b), is equal to the mean of its uniform distribution, which is 1.15, and its precision, ϕ in equation (2.17b), is the reciprocal of the variance of this distribution, $12/(1.7 - 0.6)^2$, which equals 9.92. Both the pdf and the cdf of the posterior distribution of τ can be calculated numerically, and the cdf can be used to generate random numbers from the distribution. The uniform prior distribution for τ restricts the domain of its posterior to $[0, 0.4]$. For each random τ, equations (2.17a, b, c) were used to generate a random μ, conditional on τ and y, after which equations (2.16a, b, c) were used to generate random θ_j, conditional on μ, τ and y. A Matlab program for implementing this algorithm is given on the book website. The results of 10 000 draws gave a mean of 1.22 for μ and a mean of 0.11 for τ. The standard deviations of the μ and τ were 0.08 and 0.09 respectively. Updated estimates of the θ_j are shown in Table 2.7.

2.6   Asset management plans and the Bayes' linear estimator

2.6.1   Background

Water companies in the UK are required to maintain and periodically revise their asset management plans (AMPs), as a condition of their licences. An AMP embodies company strategy for a twenty-year period and estimates the capital investment required to improve, maintain and, where necessary, extend the company's assets. There are many uncertainties associated with such a complex task, for example: the current condition and serviceability of all assets; the scope of work needed to achieve standards; variation in costs for the same nominal work; and trends in demand.

If the AMPs are to be effective they must give a fair reflection of the cost of the work needed, and an indication of the accuracy of the estimate (Metcalfe, 1991, 1994). A single figure for the estimated future cost could be seriously misleading, and contribute to poor decisions. For the purpose of producing an AMP, a company will usually be divided into discrete assets such as trunk mains, treatment works and water towers, and approximately self-contained local distribution areas known as 'zones'. The company will already have considerable information about its assets, partly as a result of previous AMPs, and this can be used to set up a prior model. The prior model can then be regularly updated as new information is obtained. The Bayes' linear estimator is a means for doing this (O'Hagan *et al.*, 1992).

2.6.2   Bayes' linear estimator

Basic Bayes' linear estimator

Let X and Y be two random vectors; denote the prior means by

$$m_x = E[X] \quad m_y = E[Y]$$

the prior variance–covariance matrices by V_x and V_y; and the matrix of covariances between X and Y by C. Then the Bayes' linear estimator (BLE) of Y after observing $X = x$ is

$$\hat{Y}(x) = m_y + C V_x^{-1}(x - m_x) \tag{2.19a}$$

The dispersion matrix is the expected value of the squared error, and is given by:

$$D_y(x) = V_y - CV_x^{-1}C^T \tag{2.19b}$$

The BLE is the linear estimator which minimises the expected squared error. If X and Y have a joint multivariate distribution: $\hat{Y}_{(x)}$ is the expected value of Y given x; the dispersion matrix is the conditional variance of Y; and the BLE is the regression of Y on x, as defined in the population (Appendix 2). The derivation of the BLE (see Appendix 4) makes no assumptions about the distribution of X and Y, but if it is reasonably close to normal we can interpret $\hat{Y}(x)$ and $D_y(x)$ as the posterior mean and variance of Y. The BLE can be compared with making predictions from an empirical regression (Section 5.3). In the latter case the CVx^{-1} of the BLE has been replaced by estimates, $(X^TX)^{-1}X^TY$, from the n data used to fit the regression model. In practice we still have to estimate V_x, C and V_y to use the BLE, but do so in a less formal manner. We can combine information from a variety of sources with engineering judgement. A more sophisticated formulation would allow for uncertainty about these variance–covariance matrices.

In the AMP context, X are the costs for zones for which we have new data. Although we do not have any new data, explicitly, about the costs in zones Y, we know they have similarities with zones X and use the BLE to improve our estimates. However, the basic BLE assumes that we know the values x taken by X and this is not necessarily the case. In general we have an improved, but still imprecise estimate. O'Hagan *et al.* (1992) emphasise that this should be thought of as a posterior estimate of X rather than an observation of X with error and suggest the following modification.

Modified Bayes' linear estimator

Let X be the vector of costs for zones in which further studies have been made. The prior mean and variance–covariance matrix of Y are replaced by posterior values m_y^* and V_y^* according to:

$$m_y^* = m_y + CV_x^{-1}(m_x^* - m_x) \tag{2.20a}$$

$$V_y^* = V_y - CV_x^{-1}C^T + CV_x^{-1}V_x^*V_x^{-1}C^T \tag{2.20b}$$

where m_x^* and V_x^* are the posterior mean and variance of X (see Exercise 2.13). If we have precise estimates, $V_x^* = 0$ and these formulae revert to the basic BLE. In contrast to this, if the posterior information about X is the same as the prior, there is no change in our beliefs about Y. The updating equations 2.20(a) and (b) would not be appropriate if we observed x plus independent error. Then the new observation would be combined with the prior mean rather than be used instead of it. The next example demonstrates the application of the modified BLE to a simplified part of an AMP.

Example 2.8

We have been asked to estimate the cost of mains refurbishment in an area of the local distribution network. Preparatory work has led to the following assumptions.

Table 2.8   (a) Prior data (lengths and unit costs); (b) variance data; (c) update information (for lengths and unit costs)

(a)

Zone	Stratum	Length of main (km)	Unit cost (£k/km)	Total cost (£k)
Zone 1	Rural	1	1	1
Zone 2	Rural	2	1	2
Zone 3	Rural	3	1	3
Zone 4	Urban	2	2	4
Zone 5	Urban	3	2	6

(b)

Level of uncertainty	Base quantity	Percentage standard deviation		
		Overall %	Stratum %	Individual %
Prior	Length	0	0	0
	Unit cost	±30	±80	±50
Detail	Length	0	0	0
	Unit cost	±1	±1	±10

(c)

Zone	Stratum	Length of main (km)	Unit cost (£k/km)	Total cost (£k)
Zone 2	Rural	2	2	4
Zone 5	Urban	3	5	15

(i)   The costs are the products of lengths of main and unit cost of refurbishment.
(ii)   The region can be split into five self-contained zones.
(iii)   The zones can be split into two strata, rural (zones 1, 2 and 3) and urban (zones 4 and 5). Two zones will be surveyed in detail.
(iv)   The equations of the cost model are:

$$\text{total cost} = \sum \text{zone cost}$$

$$\text{zone cost} = \text{length} \times \text{unit cost}$$

(v)   The prior distribution of lengths and costs is shown in Table 2.8.
(vi)   The errors in lengths are negligible. The unit costs are modelled as

$$C = \mu + \nu_i + \varepsilon_{ij}$$

where μ is an overall average, ν_i is an adjustment for stratum i and ε_{ij} is specific to a zone within stratum i. The standard deviations of the three components of prior uncertainty are given in percentage terms in Table 2.8. They have been estimated from components of variance analysis (Section 6.2.2) of ratios of scheme out-turn costs to costs predicted using AMP methods. These costings were made retrospectively by engineers who took no part in the schemes. The finite population correction is ignored. For example, we do not impose a restriction that the sum of the stratum adjustments equals zero. In a real AMP there will be many strata and zones so this simplification has a negligible effect, particularly when compared with the rather rough nature of the estimates of standard deviations. The estimates of the standard deviations after a detailed study are also given in Table 2.8. They must be smaller than the estimates of prior uncertainty because they include all prior information.

It remains to calculate variances and covariances of zone costs. The total error (E) can be split into an error in the overall mean E_0, an error in the stratum adjustment E_s and an individual error (E_I). We assume error components are independent. Then

$$\text{var}(E) = \text{var}(E_0) + \text{var}(E_s) + \text{var}(E_I)$$

Since only the error in unit costs is being modelled, zone cost errors will be the same in relative terms, i.e. let C_i be the cost for zone i

$$\text{var}(C_i) = \text{var}(E) = (0.3)^2 + (0.8)^2 + (0.5)^2 = 0.98$$

Hence the prior standard deviation of a zone is

$$\sigma_E = \sqrt{0.98} = 0.99$$

Now consider another zone in the same stratum. The covariance of costs, C_i and C_j, will be known

$$\text{cov}(C_i, C_j) = \text{E}[(E_0 + E_s + E_{Ii})(E_0 + E_s + E_{Ij})]$$

where E_{Ii} and I_{Ij} are the individual errors. Since

$$\text{E}(E_{Ii}E_{Ij}) = 0$$

the covariance is

$$\text{E}[E_0^2 + E_s^2] = (0.3)^2 + (0.8)^2 = 0.09 + 0.64 = 0.73$$

The correlation between zones in the same stratum is

$$0.73/(0.99)^2 = 0.74$$

A similar argument can be applied to a zone in a different stratum. The covariance of costs, C_i and C_k, will be

$$\text{cov}(C_i, C_k) = \text{E}[(E_0 + E_{si} + E_{Ii})(E_0 + E_{sk} + E_{Ik})]$$

which, assuming independence of the stratum errors,

$$\text{E}[E_0^2] = (0.3)^2 = 0.09$$

Turning to the update information, the total error variance of a zone is

$$(0.01)^2 + (0.01)^2 + (0.1)^2 = 0.0102$$

and the covariance between zone costs in different strata is 0.0001. As there is only one update zone in each stratum, we do not need to calculate the covariances within strata. We now have enough information to use the modified BLE equations 2.20 to calculate posterior estimates for zones for which update information is not available. We first construct the individual matrix elements.

Prior costs   Prior costs   Update costs

$$
m_y = \begin{bmatrix} 1 \\ 3 \\ 4 \end{bmatrix} \begin{matrix} \text{zone 1} \\ \text{zone 3} \\ \text{zone 4} \end{matrix} \qquad
m_x = \begin{bmatrix} 2 \\ 6 \end{bmatrix} \begin{matrix} \text{zone 2} \\ \text{zone 5} \end{matrix} \qquad
m_x^* = \begin{bmatrix} 4 \\ 15 \end{bmatrix} \begin{matrix} \text{zone 2} \\ \text{zone 5} \end{matrix}
$$

The variances and covariances, so far, are in terms of $(\%/100)^2$, and to obtain the variance–covariance matrix they must be multiplied by the estimated mean costs.

The prior variance–covariance matrix between zones where update information will not become available is given by:

$$
V_y = \begin{bmatrix} \text{var(zone1)} & \text{cov(zone1, zone3)} & \text{cov(zone1, zone4)} \\ \text{cov(zone3, zone1)} & \text{var(zone3)} & \text{cov(zone3, zone4)} \\ \text{cov(zone4, zone1)} & \text{cov(zone4, zone3)} & \text{var(zone4)} \end{bmatrix}
$$

Now substituting the numbers i.e. multiplying the covariance (which is in $(\%)^2$ terms) by the product of the mean of the two zones, gives

$$
\begin{bmatrix} 0.98(1 \times 1) & 0.73(1 \times 3) & 0.09(1 \times 4) \\ 0.73(3 \times 1) & 0.98(3 \times 3) & 0.09(3 \times 4) \\ 0.09(4 \times 1) & 0.09(4 \times 3) & 0.98(4 \times 4) \end{bmatrix} = \begin{bmatrix} 0.98 & 2.19 & 0.36 \\ 2.19 & 8.82 & 1.08 \\ 0.36 & 1.08 & 15.68 \end{bmatrix}
$$

The prior variance–covariance matrix between zones where update information will become available is given by:

$$
V_x = \begin{bmatrix} \text{var(zone2)} & \text{cov(zone2, zone5)} \\ \text{cov(zone5, zone2)} & \text{var(zone5)} \end{bmatrix}
$$

$$
= \begin{bmatrix} 0.98(2 \times 2) & 0.09(2 \times 6) \\ 0.09(6 \times 2) & 0.98(6 \times 6) \end{bmatrix} = \begin{bmatrix} 3.92 & 1.08 \\ 1.08 & 35.28 \end{bmatrix}
$$

The update variance–covariance matrix between zones where update information will become available is given by:

$$
V_x^* = \begin{bmatrix} 0.0102(4 \times 4) & 0.0001(4 \times 15) \\ 0.0001(15 \times 4) & 0.0102(15 \times 15) \end{bmatrix} = \begin{bmatrix} 0.1632 & 0.006 \\ 0.006 & 2.295 \end{bmatrix}
$$

Notice that the diagonal elements are the variance of cost for zone i, and the off-diagonal elements are the covariances between zone costs.

We now need the matrix of prior covariances between the zones where update information will not become available and the zones where update information will become available:

$$
C = \begin{bmatrix} \text{cov(zone1, zone2)} & \text{cov(zone1, zone5)} \\ \text{cov(zone3, zone2)} & \text{cov(zone3, zone5)} \\ \text{cov(zone4, zone2)} & \text{cov(zone4, zone5)} \end{bmatrix}
$$

Table 2.9 Summary of analysis using the BLE equations

Zone	Prior			Posterior		
	Zone cost (£k)	Stdev		Zone cost (£k)	Stdev	
Zone 1	1	99%	0.99	1.78	36.2%	0.64
Zone 2	2	99%	1.98	4.00	10%	0.40
Zone 3	3	99%	2.97	5.33	36.2%	1.93
Zone 4	4	99%	3.96	8.55	29.6%	2.53
Zone 5	6	99%	5.94	15.00	10%	1.50
Total	16			34.66		

substituting the numbers gives

$$\begin{bmatrix} 0.73(1 \times 2) & 0.09(1 \times 6) \\ 0.73(3 \times 2) & 0.09(3 \times 6) \\ 0.09(4 \times 2) & 0.73(4 \times 6) \end{bmatrix} = \begin{bmatrix} 1.46 & 0.54 \\ 4.38 & 1.62 \\ 0.72 & 17.52 \end{bmatrix}$$

Substitution into equation (2.20a) gives

$$m_y^* = [1.78 \quad 5.33 \quad 8.55]^T$$

and substitution into equation (2.20b) leads to

$$V_y^* = \begin{bmatrix} 0.41335 & 0.49004 & 0.01524 \\ 0.49004 & 3.72011 & 0.04572 \\ 0.01524 & 0.04572 & 6.40854 \end{bmatrix}$$

The diagonal terms are the posterior variances for zone 1, zone 3 and zone 4. The standard deviations follow by taking square roots.

$$\sigma(\text{zone}1) = \sqrt{0.41335} = 0.643 \quad \text{or in \% terms} \quad \frac{(0.643 \times 100)}{1.778} = 36.2\%$$

$$\sigma(\text{zone}3) = \sqrt{3.72011} = 1.929 \quad \text{or in \% terms} \quad \frac{(1.929 \times 100)}{5.334} = 36.2\%$$

$$\sigma(\text{zone}4) = \sqrt{6.40854} = 2.532 \quad \text{or in \% terms} \quad \frac{(2.532 \times 100)}{8.55} = 29.6\%$$

The results are summarised in Table 2.9. The variance of the posterior total cost is given by the sum of all the elements of the matrix,

$$\begin{pmatrix} V_x^* & C^{*T} \\ C^* & V_y^* \end{pmatrix}$$

where C^* contains the update covariances between zones where update information is available and zones where update information is not available (Exercise 2.14). The variances are on the leading diagonal and the covariances lie above and below it, so adding all the elements gives the sum of the variances and twice the covariances.

The BLE solution is always a compromise between the prior information and the new data. We can see the following.

(i) For rural zones the update represents a 2-fold increase, and we know that the correlation between zones in the same stratum is 0.74, so we would expect the posterior estimates for costs in unsampled zones to go up by a factor rather less than 2. The BLE prediction, for other zones in this stratum, is that prior costs will rise by a factor of 1.78.

(ii) For urban zones the update represents a 2.5-fold increase, and correlation between zones in the same stratum is again 0.74, so we would expect the posterior estimates for unsampled zones to go up by a factor rather less than 2.5. The BLE prediction, for other zones in this stratum, is that prior costs will rise by a factor of 2.15.

(iii) Correlation between zones in different strata is only 0.09, so we would not expect updates in one stratum to have much effect on estimates in the other stratum.

(iv) The standard deviation for costs in unsampled zones has fallen from the prior level of 99% to 36%, a figure much closer to the update value of 10%.

(v) Overall the estimated cost for mains cleaning is increased by a factor of 2.17 (from £16k to £34.66k).

2.7 Gibbs sampling and purchase behaviour

2.7.1 Stochastic model of purchase intention and behaviour

Young *et al.* (1998) presented a stochastic model for the relationship between sociological characteristics of individuals and their purchasing behaviour, which allows for the discrepancy between reported intention and behaviour. Their main example uses data from a study in which members of a large consumer panel were asked whether they intended buying a personal computer for use at home within the next year. The following sociological variables were recorded for each panel member: profess – x_1 coded as 1 if the member was classified as professional and 0 otherwise; student – x_2 coded as 1 if the member was a student and 0 otherwise; use – x_3 coded as 1 if the member used a PC at work or school and 0 otherwise; pacific – x_4 coded as 1 if the house was in the Pacific region of the USA and 0 otherwise; income – x_5 household income expressed in units of US$100 000 to the nearest one decimal place. They modelled the utility of a purchase within a year (u_i on a scale centred at 0) for panel member i as a linear combination of the sociological variables with standard normal random variation added. That is

$$u_i = \boldsymbol{\beta}^T \boldsymbol{x} + z_i \quad \text{for } i = 1, \ldots, n$$

where $\boldsymbol{\beta}^T = (\beta_0, \beta_1, \ldots, \beta_k)$, $\boldsymbol{x}^T = (1, x_{1i}, \ldots, x_{ki})$, $z_i \sim N(0, 1)$, and in general there are k sociological variables and n respondents. The first component of x is 1 to allow for an intercept term (β_0). The purchase behaviour (w_i) is coded as 1 for a purchase and 0 otherwise, and a purchase is made if u_i is positive.

$$w_i = \begin{cases} 1 & \text{if } u_i \geq 0 \\ 0 & \text{if } u_i < 0 \end{cases}$$

This is equivalent to stating that

$$\Pr(w_i = 1) = \Phi(\boldsymbol{\beta}^T \boldsymbol{x}_i)$$

where Φ is the cdf of the standard normal distribution. The stated purchase intention y_i is coded as 1 for yes and 0 for no. Two conditional probabilities are defined, the probabilities of intention given the purchase behaviour,

$$\Pr(y_i = k \mid w_i = j) = p_{jk} \quad \text{for } j, k = 0, 1$$

and the probabilities of purchase given the intention

$$\Pr(w_i = k \mid y_i = j) = q_{jk} \quad \text{for } j, k = 0, 1$$

The survey data are the values of $x_{1i}, \ldots, x_{ki}$ and the declared intention y_i for each respondent. There are published historical values and product class-level estimates for the q_{jk}. The objective is to estimate the β_j which determine the relationship between the sociological variables and the purchase behaviour. Young *et al.* (1998) write down the likelihood function for β in terms of the observable data $(\boldsymbol{x}_i, y_i)$. Rather than multiply this by the prior distribution and attempt a direct numerical integration, they suggest the use of a Markov chain Monte Carlo method known as Gibbs sampling, and provide a particularly clear account of how this can be applied to an interesting estimation problem.

2.7.2   Gibbs sampling

Gibbs sampling is a multivariate random number-generating technique that relies on the fact that knowledge of the full conditional distributions is sufficient to determine a joint distribution, provided it exists. For example, let $f(x, y, z)$ be a trivariate distribution and suppose we know the conditional distributions $f(x \mid y, z), f(y \mid x, z)$ and $f(z \mid x, y)$. The iteration scheme

$$X'_{j+1} \sim f(x \mid y = Y'_j, z = Z'_j) \tag{2.21a}$$

$$Y'_{j+1} \sim f(y \mid x = X'_{j+1}, z = Z'_j) \tag{2.21b}$$

$$Z'_{j+1} \sim f(z \mid x = X'_{j+1}, y = Y'_{j+1}) \tag{2.21c}$$

produces a Gibbs sequence $X'_0, Y'_0, Z'_0, X'_1, Y'_1, Z'_1, X'_2, Y'_2, Z'_2, \ldots$. The effect of the initial values X'_0, Y'_0, Z'_0 dissipates as the sequence progresses, and it is common to ignore the first few hundred triples. The subsequent sequence of X'_i, for example, is a sample from the marginal distribution $f(x)$. Parameters of the distribution can be found, to any desired precision, by calculating the corresponding statistic from the sequence of X'_i. The X'_i are not generated as a sequence of independent variables from $f(x)$, but this does not usually affect parameter estimates if a long sequence is taken. If a random sample is required, for some simulation study perhaps, take every mth triple for some reasonably large m. The only potential drawback of relying on a single long sequence is that in some rather specialist applications the Gibbs sequence may stay in a small subset of the sample space for a long time. Although the construction of the Gibbs sequence may seem intuitively plausible, the proof that it works under reasonably general conditions is quite subtle. However, Casella and George (1992) provide a clear exposition which includes a proof for a simple case and thereby demonstrates the theoretical principles behind the method.

Example 2.9

Let the distribution of X given p be binomial, $\text{Bin}(n, p)$, and suppose p has a beta distribution, $p \sim Be(\alpha, \beta)$. Then the joint distribution is,

$$f(x, y) = {}_nC_x p^{x+\alpha-1}(1-p)^{n-x+\beta-1} \quad \text{for } x = 0, \ldots, n \text{ and } 0 \le p \le 1$$

and the conditional distributions are,

$$P(x|p) \sim \text{Bin}(n, p) \tag{2.22a}$$

$$f(p|x) \sim Be(x + \alpha, n - x + \beta) \tag{2.22b}$$

It was shown in Section 2.5.3 that $P(x)$ is the beta-binomial distribution. Alternatively, the Gibbs sampler can be used to construct the same distribution empirically. In Example 2.5 my posterior probability distribution for p was a beta distribution with α and β equal to 5.91 and 21.91 respectively. I now require the predictive distribution for the number of people making repeat purchases in a sample of 18. I start with p_0' reasonably close to its mean at 0.2. I now generate a random number x_1' from equation (2.22a) and use this to generate a random p_1' from equation (2.22b). I return next to equation (2.22a) and generate another random number x_2' and so on. I run the sequence until I have 2200 pairs (x_j', p_j') and ignore the first 200. The 2000 pairs are a large sample from the bivariate distribution and the marginal distribution of the x_j' is a close approximation to the predictive distribution. I can calculate features of the distribution, such as its mean and variance and lower 1% point from the $\{x_j'\}$.

Estimation of parameters of purchase behaviour model

In this application the unknown parameters are the components of β, and the unobservable u_i which determine the, in general, unobserved w_i. Although we are only interested in β the Gibbs sampler will require both

$$f(\beta|u_i, x_i, y_i) \quad \text{and} \quad f(u_i|\beta, x_i, y_i)$$

Young *et al.* (1998) assume a multivariate normal prior distribution with mean μ and variance–covariance matrix T^{-1} for β and then use the standard results for Bayes' linear models to obtain

$$f(\beta|u_i, x_i, y_i) \sim N((X^TX + T)^{-1}(X^Tu + T\mu), (X^TX + T)^{-1}) \tag{2.23}$$

where $u^T = (u_1, \ldots, u_n)$ and $X^T = (x_1, \ldots, x_n)$. The full conditional distribution for u_i is given by

$$f(u_i|\beta, x_i, y_i) \sim \begin{cases} N(\beta^T x_i, 1)I(u_i \ge 0) & \text{if } w_i = 1 \\ N(\beta^T x_i, 1)I(u_i < 0) & \text{if } w_i = 0 \end{cases} \tag{2.24}$$

where $I(\ )$ is the indicator function which is 1 if its argument is true and 0 otherwise. That is, if w_i equals 1 then u_i must be positive and if w_i equals 0 then u_i must be negative. The y_i conditions the distribution of u_i through the information it provides about w_i through Bayes' theorem.

$$\Pr(w_i = k|\beta, y_i = j, x_i) = \frac{\Pr(y_i = j|w_i = k, \beta, x_i) \times \Pr(w_i = k|\beta, x_i)}{\sum_{l=0}^{1} \Pr(y_i = j|w_i = l, \beta, x_i) \times \Pr(w_i = l|\beta, x_i)}$$

In this expression, the prior distribution, $\Pr(w_i = 1 | \boldsymbol{\beta}, \boldsymbol{x}_i)$, is given by $\Phi(\boldsymbol{\beta}^T \boldsymbol{x}_i)$. The likelihood $\Pr(y_i = l | w_i = 1, \boldsymbol{\beta}, \boldsymbol{x}_i)$ was defined as p_{1l}. In this model $\boldsymbol{\beta}$ and $\boldsymbol{x}_i$ provide no more information about y_i than does w_i, but Young *et al.* (1998) discuss modifications that would allow for utility to provide extra information. Hence

$$\Pr(w_i = 1 | \boldsymbol{\beta}, y_i = 1, \boldsymbol{x}_i) = \frac{p_{11} \Phi(\boldsymbol{\beta}^T \boldsymbol{x}_i)}{p_{11} \Phi(\boldsymbol{\beta}^T \boldsymbol{x}_i) + p_{01}(1 - \Phi(\boldsymbol{\beta}^T \boldsymbol{x}_i))} \qquad (2.25a)$$

$$\Pr(w_i = 1 | \boldsymbol{\beta}, y_i = 0, \boldsymbol{x}_i) = \frac{p_{10} \Phi(\boldsymbol{\beta}^T \boldsymbol{x}_i)}{p_{10} \Phi(\boldsymbol{\beta}^T \boldsymbol{x}_i) + p_{11}(1 - \Phi(\boldsymbol{\beta}^T \boldsymbol{x}_i))} \qquad (2.25b)$$

The corresponding probabilities for w_i equalling 0 are obtained as the complements. A last detail is to express p_{ij} in terms of known quantities. For example:

$$p_{11} = \Pr(y = 1 | w = 1)$$

$$= \frac{\Pr(w = 1 | y = 1) \times \Pr(y = 1)}{\Pr(w = 1 | y = 1) \times \Pr(y = 1) + \Pr(w = 1 | y = 0) \times \Pr(y = 0)}$$

$$= \frac{q_{11} \Pr(y = 1)}{q_{11} \Pr(y = 1) + q_{01} \Pr(y = 0)}$$

The q_{ij} are taken as industry standards and the $\Pr(y = i)$ can be estimated directly from the survey; e.g. $\Pr(y = 1)$ is the proportion of respondents indicating an intention to buy. The Gibbs sampler alternates between the three steps of generating coefficients from equation (2.23), generating w_i from equation (2.25a,b) and hence generating u_i from equation (2.24). Reasonable starting values for the u_i are $+1$ if the panel member intended to buy a personal computer and -1 if he or she did not intend to buy one. The sampler involves generating random numbers from a multivariate normal distribution for $\boldsymbol{\beta}$, and normal distributions for u_i. Procedures for generating random numbers are given in Appendix A3.

In one of their comparative analyses Young *et al.* (1998) assumed values reported in Jamieson and Bass (1989) for purchases of personal computers, that is 42.9% of people who declare an intention to purchase a personal computer within a year actually do so, whereas 3.8% of people who say they do not intend to buy do so. Young *et al.* (1998) ran the Gibbs sampler for 10 000 iterations and used the final 5000 for calculating the mean and variance–covariance matrix for $\boldsymbol{\beta}$. The estimates of the coefficients with their standard deviations in brackets (Young *et al.*, 1998) are:

profess	1.652	(0.145)
student	0.102	(0.185)
use	1.128	(0.140)
pacific	0.472	(0.163)
income	0.753	(0.148)

There is evidence that all the sociological variables, except being a student, are associated with computer purchase. They suggest that the geographic region is associated with sales because a relatively high proportion of people in the Pacific region work in the computer industry, and are therefore more likely to buy a personal computer.

2.8 Software

First Bayes is a teaching package for elementary Bayesian statistics. It is currently offered free, provided it is not used for profit. The internet location is:

http://lib.stat.cmu.edu/DOS/general/first-bayes/1b.html

The rather more specialist WinBUGS (Bayesian inference using Gibbs sampling) is now available free from:

www.mrc-bsu.cam.ac.uk/bugs/

The Bayes Linear Programming Language [B/D] is now available free from:

http://fourier.dur.ac.uk:8000/stats/bd/

Farrow *et al.* (1997) describe the application of a Bayes linear model to help managers with daily planning in a brewery, and the development of their Foresight software. The DPL Decision Analysis software (ADA Decision Systems Staff, 1994) is also useful in this context.

2.9 Summary

1. The expected monetary value (EMV), or more generally expected utility, of some uncertain event is the product of the probability it occurs with the benefit that accrues if it does so.
2. Utility is a measure of attractiveness of a consequence, usually made on a probability scale. If the financial consequences of individual decisions are small in comparison with a company's capital it is reasonable to assume monetary amounts to be proportional to changes in utility.
3. There is either a saddle point or an optimum mixed strategy for any finite, two-person, zero-sum game. However, many two-party business situations are more closely modelled by the 'Prisoner's Dilemma', in which cooperation is to the advantage of both parties.
4. Bayes' theorem for discrete events. If we have k mutually exclusive and exhaustive events $\{A_1, \ldots, A_k\}$, and some observed event B, then

$$\Pr(A_i \mid B) = \frac{\Pr(B \mid A_i) \Pr(A_i)}{\sum_{j=1}^{k} \Pr(B \mid A_j) \Pr(A_j)}$$

5. Bayes' theorem for continuous distributions.

$$f(\theta \mid x) = f(x \mid \theta) f(\theta) / P(x)$$

so

$$f(\theta \mid x) \propto f(x \mid \theta) f(\theta)$$

In words, the posterior distribution for θ given data x is proportional to the product of the prior distribution of θ and the likelihood of θ given the data.

6. A predictive distribution for x is given by:

$$f(x) = \int f(x, \theta)\, d\theta = \int f(x|\theta) f(\theta)\, d\theta$$

7. The beta distribution has the form

$$f(y) = [B(\alpha, \beta)]^{-1} y^{\alpha - 1} (1 - y)^{\beta - 1} \quad 0 \le y \le 1$$

The mean and variance of the distribution are

$$\mu = \alpha/(\alpha + \beta)$$

$$\sigma^2 = \alpha\beta/[(\alpha + \beta)^2 (\alpha + \beta + 1)]$$

Exercises

2.1 Plot the utility function for money (x),

$$U(x) = 1 - [\exp(-x/200) - \exp(-x/20)]/2$$

for x from 0 to 400. Verify that it models a decreasing aversion to risk.

2.2 I have to decide whether to join a roadside rescue motoring organisation (RSR). The probability my car will break down in the next year is 0.4. If it breaks down there is a 10% chance I can persuade a passer-by to help me push it to the nearest garage. A tow would cost 100 monetary units (mu). Alternatively I can join the RSR, which provides either a roadside repair or a tow to the nearest garage, for an annual fee of 38 mu.

(i) Ignore the possibility of more than one breakdown in a year. If I apply an EMV criterion, should I join the RSR?
(ii) If breakdowns have a Poisson distribution with mean 0.4 per year should I join the RSR?

2.3 The manager of Macaw Motors has been told that if Macaw submits a prototype electric motor to a manufacturer of robots, Crow Cybernetics, Macaw may get an order for 1000 motors. Crow will pay 800 mu per motor. The manager has the following additional information.

(i) The cost of producing the prototype is 48 000. This cost must be borne by Macaw whether or not it gets the order.
(ii) The cost of manufacturing the motors using machined parts is 80 000 for tooling and a further 560 per motor.
(iii) It may be possible to manufacture the motors to specification using plastic injection mouldings for some components. The cost of the die for the moulding machine would be 40 000, but the remaining tooling cost would be reduced by 16 000 and the marginal production cost per motor would be reduced to 480.

(iv) The probability that plastic moulded parts will be satisfactory is 0.5. If they are not, the 40 000 for the die will be lost and the full cost of tooling and manufacture would have to be met.

(v) The probability of obtaining the order is 0.4.

Advise the manager, ignoring interest costs.

2.4 Refer to the Swallow Construction decision problem of Section 2.3.2. The limited survey cost 10 mu but could be wrong by one classification. How much would an accurate survey be worth?

2.5 In 'Millionaire Jackpot Matching Pennies' (Table 2.5(b)) B cannot afford to lose M and will almost always play tails. However, B's loss, in EMV terms, can be reduced slightly by playing heads with a small probability q. A will respond by occasionally playing heads with probability p.

(i) Show that the expected payoff (y) to A is given by

$$y = Mpq + 1 - 2p - 2q + 3pq$$

(ii) Differentiate y partially with respect to p and q. Set the partial derivatives equal to 0 and hence show that at a stationary point

$$p = q = 2/(M + 3)$$

Show that this stationary point is a saddle point.

(iii) As a check on the result in part (ii), demonstrate it gives the correct answer when M is 1 rather than 1 million.

(iv) Show that under the optimum strategy A's expected payoff is $1 - 4/(M + 3)$.

2.6 Sherlock Holmes (H) is being pursued by Professor Moriarty (M). H has taken a train from London to Dover, with one stop at Canterbury. M is following on a special train. If H reaches Dover without being met by M he can escape to France. Payoffs to H are shown in the table below.

		M	
		Alight at Canterbury	Alight at Dover
H	Alight at Canterbury	−10	5
	Alight at Dover	20	−10

Advise the two parties.

[Based on an anecdote in Poundstone, 1992.]

2.7 A coin is tossed three times. Let a head (H) represent 1 and a tail (T) represent 0. Then any sequence of 3 flips gives a binary representation of a number between 0 and 7. For example, TTT is 0, THT is 2, HTH is 5, and HHH is 7. Let A be the event that the first flip gives H, and let B be the identical event. Let C be the event that the number is between 1 and 4 inclusive. Show that

$$\Pr(A \text{ and } B \text{ and } C) = \Pr(A) \times \Pr(B) \times \Pr(C)$$

yet A, B and C are not mutually independent (after Lee, 1997).

2.8   The gamma function is defined by

$$\Gamma(\alpha) = \int_0^\infty t^{\alpha-1} e^{-t} \, dt \quad \text{for } 0 < \alpha$$

(i)   Use integration by parts to show that

$$\Gamma(\alpha) = (\alpha - 1)!$$

(ii)   Show that $\Gamma(1) = 1$, and hence justify the statement that $0! = 1$.

(iii)   Use Table AST.7 to find $\Gamma(4.6)$. Check that your answer lies between 4! and 3!. $\Gamma(\alpha)$ is a generalisation of the factorial function $(\alpha - 1)!$ to non-integer values. $\Gamma(\alpha)$ is also defined for negative α but it rarely occurs in statistical applications and has singularities at negative integers.

2.9   Suppose $X \sim \text{Bin}(n, p)$, then

$$P(x|p) = {}_nC_x p^x (1 - p)^{n-x}$$

If x successes have been observed in n trials with unknown p, we can consider $P(x|p)$ as a likelihood function for p. That is

$$\mathscr{L}(p) = {}_nC_x p^x (1 - p)^{n-x}$$

The value of p which maximises $\mathscr{L}(p)$ is known as the maximum likelihood estimate, $\hat{p}$, (MLE) of p. Since $\mathscr{L}(p)$ and $\ln(\mathscr{L}(p))$ have their maxima at the same value of p, maximising $\ln(\mathscr{L}(p))$ is equivalent to maximising $\mathscr{L}(p)$ and the former is often easier. Use calculus to show that $\hat{p}$ is given by x/n.

2.10   Suppose $X_1, \ldots, X_n$ is a random sample from an exponential distribution with mean $1/\lambda$. Then

$$f(x_1, \ldots, x_n) = f(x_1) \ldots f(x_n) = \prod_{i=1}^n \lambda \exp(-\lambda x_i)$$

The likelihood function for λ is

$$\mathscr{L}(\lambda) = \prod_{i=1}^n \lambda \exp(-\lambda x_i)$$

Show that

$$\ln(\mathscr{L}(\lambda)) = n \ln(\lambda) - \lambda \sum x_i$$

and hence verify that

$$\hat{\lambda} = n / \sum x_i$$

2.11   A car manufacturer tests windscreen wiper motors by moving the blade across a dry windscreen. Assume the lifetimes of motors under these conditions (T) have a Weibull distribution with cdf:

$$F(t) = 1 - \exp[-(t/\beta)^\alpha] \quad \text{for } 0 \le t$$

Write down the pdf $f(t)$. Now suppose that a sample of n motors is tested for τ hours, during which m fail at known times $t_1, \ldots, t_m$. The remainder still work at the end of the test. Explain why the likelihood function for α and β is

$$\mathscr{L}(\alpha, \beta) = \prod_{i=1}^{m} f(t_i)(1 - F(\tau))^{n-m}$$

Consider the special case when α equals 1. What is this distribution? Verify that the MLE for β is then

$$\hat{\beta} = \left(\sum t_i + (n - m)\tau\right)/m$$

2.12 Let T be time to failure for some component. Assume T has a cdf $F(t)$, and pdf $f(t)$. The reliability function $R(t)$ is the complement of the cdf, i.e.

$$R(t) = 1 - F(t)$$

(i) Show that

$$\Pr(t < T < t + \delta t \mid T > t)$$

has the form $h(t)\delta t$ where the hazard rate, $h(t)$, is defined by

$$h(t) = f(t)/R(t)$$

(ii) Find $h(t)$ for the Weibull distribution, and the special case of the exponential distribution.

(iii) The mean time before failure MTBF is given by the integral

$$\text{MTBF} = \int_0^\infty t f(t)\, dt$$

Use integration by parts to show that

$$\text{MTBF} = \int_0^\infty R(t)\, dt$$

2.13 Refer to equations 2.20(a) and (b). Suppose x and y are scalars, and take variance of both sides of 2.20(a):

$$V_y^* = V_y + C^2 V_x^{-2}(V_x^* + V_x) - 2CV_x^{-1}C$$

Explain this step, extend the argument to matrices, and hence justify equation 2.20(b).

2.14 Refer to Example 2.8 in Section 2.6.2 and deduce an update covariance matrix C^*. Hence give an approximate 90% prediction interval for the total cost.

2.15 The variable X has an inverse chi-squared distribution with ν degrees of freedom,

$$X \sim \chi_\nu^{-2}$$

if $1/X \sim \chi_\nu^2$. The pdf of the chi-square distribution is:

$$f(x) = x^{\nu/2-1} \exp(-x/2)/(2^{\nu/2}\Gamma(\nu/2)) \quad \text{for } 0 < x$$

Deduce that the pdf of the inverse chi-squared distribution is:

$$f(x) = x^{-\nu/2-1} \exp(-1/(2x))/(2^{\nu/2}\Gamma(\nu/2)) \quad \text{for } 0 < x$$

Show that if $Y = aX$ and $\nu > 4$ then

$$\mu_Y = a/(\nu - 2)$$
$$\sigma_Y^2 = 2a^2/((\nu - 2)^2(\nu - 4))$$
$$\text{mode}(Y) = a/(\nu + 2)$$

3

Monte Carlo methods

3.1 Introduction

An independent general medical practice, the 'Campus Medical Centre' (CMC), is situated on a university campus. Students are required to register with a medical practice, but they can choose any local one that is willing to take them. The CMC's register includes university staff, and other people who live in the neighbourhood, as well as students. Patients are encouraged to make appointments but those with urgent problems, that are not sufficiently serious to be treated as emergencies, can just call in between 9 a.m. and 10 a.m. and wait to be seen by a duty doctor. Many patients are using this facility and the practice manager is considering various options for reducing the queueing time. The feature of this problem that makes it a typical candidate for the Monte Carlo approach is the random nature of arrival patterns and times taken to attend to patients.

Caulcutt (1999) presents a rather different application from his experience of teaching management. His premise is that process management is widely recommended as good practice, and he identifies three essential steps: identify the most important processes that make up the business; measure the performance of these processes; and use the performance measurements to guide management action that will facilitate control and improvement of the process. Many commentators, Seddon (1992) for example, have drawn attention to the sometimes ludicrous results of over reliance on performance indicators. A recent example, reported on television, was of a train company whose drivers improved punctuality of arrivals at a London terminus by missing out a few scheduled stops en route. Deming's (1986) red bead experiment (Exercise 3.1) illustrates the difficulty of making sufficient allowance for natural variability within a process, and Caulcutt (1999) gives an entertaining elaboration of this idea and discusses the consequences of inappropriate action. Participants are put into teams of three or four and are asked to manage four sales processes which are simulated on a computer. They are told that the four processes correspond to four salesmen: Alan,

Brian, Colin and David. In his annual performance appraisal each salesman agreed to sell 15 items, or more, each week. At the end of each week, participants are told how many items each salesman sold and can choose to criticise or praise or take no action by typing *C* or *P* or *N*. The participants are advised to keep a record of sales figures and their actions. The simulation is usually run for 50 sales weeks which takes about half an hour of real time. Alan's sales vary considerably, with a standard deviation of 3.87, about a mean of 15. Most teams criticise Alan after a week in which sales were low, and as sales naturally tend to be higher the next week they infer that criticism improves performance. In fact, the action has no effect on the simulation of Alan's sales. With independent random variation there is a high probability that the sales following a particularly low sales figure will be higher, an effect which is sometimes known as regression towards the mean. Teams also often praise Alan's high sales, which naturally tend to be followed by lower sales. A real Alan would probably act to reduce criticism by holding back orders from good weeks to supplement the shortfall in bad weeks. The effect of this would be to increase the delivery time to customers. Amongst other things, Caulcutt uses the exercise to demonstrate the value of plotting weekly sales as a run chart. It is possible to analyse the sales processes analytically but this would be far less effective as a teaching exercise.

A third example that will be used as a case study is a water supply model WatSup (Tinley Oast Research Ltd, 1998). WatSup can model stochastic water demand in a dendritic system, at the level of detail of supply to a single house, with a one-second time step. This allows a much more realistic assessment of likely peak demands than can be obtained from averaged values. However, the main reason why managers in the water industry now require such comprehensive models is that legislation is forcing water utilities to meet drinking water standards at customers' taps rather than at the treatment plant, and water quality deteriorates while it is in the distribution system (Buchberger and Wu, 1995). Murray and Murray (1999) give other possible applications for slightly modified versions of their program. It can be used to model any queue-based process and examples include computer data transfers, bulk deliveries, plastics manufacture, electricity supply and supermarket checkouts. Other applications include management of inventory.

The Monte Carlo method is a name given to the technique of mimicking systems by setting up computer models in which random events are decided by sequences of random numbers. Applications include: production scheduling, maintenance and repair strategies, inventory control, and control of processes, as well as general queueing problems. There are mathematical models for certain classes of these problems, and corresponding formulae for important properties such as average waiting time, the distribution of queue length and proportion of time the server is busy. Some of this work is covered in Chapter 8. However, the Monte Carlo approach, often referred to as 'stochastic simulation' or just 'simulation', is more versatile. There are several advantages to be gained from using random numbers in simulations rather than past records. The practice manager in the CMC might decide to record times between arrivals and consultancy times over a typical week, and then see what would have happened had a trainee general practitioner been employed to help the duty doctor. However, this approach provides no information about the variability between weeks and no information about the longest queues occurring each month. If the manager fits suitable probability distributions to the week's data, and then uses random numbers from these distributions, many years of queueing situations can be generated in minutes. The parameters of the fitted distributions can be varied to allow for sampling errors in their estimation. In more complicated simulation models it becomes even more important to use random numbers because the Monte Carlo method may expose unfortunate properties of systems which would go unnoticed if the study was limited to a relatively short past record.

The applications mentioned above are referred to as discrete event simulations, because the state of the system is summarised by discrete variables such as the number of people in a queue, the number of items in stock and the number of machines out of service. The behaviour of the system is characterised by times between events that change the system state. In a queueing situation, events include people arriving and servers completing transactions.

3.2 Discrete event simulation

3.2.1 The clock

There are specialist simulation languages, such as SIMULA and GPSS, but general-purpose computer languages can be used. If you have a complex system, or expect to use Monte Carlo techniques frequently, you can probably save a significant amount of time by learning to use a specialist package. The time saved should cover the cost of the software, but you may sometimes need the flexibility of a general-purpose language. A clock is at the centre of any discrete event simulation, and there are two strategies for keeping track of the system over time. The first is to advance time by small constant increments, check whether any events have occurred, and change the system state accordingly. This is known as 'fixed time increment', or 'synchronous', and the time increment is described as the clock period. The second strategy is to increment time until the next event and change the system state appropriately, and it is referred to as 'variable time increment'; or 'asynchronous'. If we are attempting to simulate a system over continuous time, the clock period has to be small in comparison with typical times between events and the variable time increment approach is more efficient. However, we need to compare times until all the events which could logically occur next, and then select the one with the shortest lead time. This may be more awkward to program.

Most computer languages for the simulation of discrete event systems use variable time increments (Mitrani, 1982), but people who write their own code often prefer to work with a fixed time increment. Appleby (1998) uses a fixed time increment for his bus simulation program (BUSTLE available free from www.staff.ncl.ac.uk/john.appleby), which provides a nice demonstration of 'The three-bus syndrome' or 'Why do buses always seem to arrive in threes?' This problem received unexpected publicity in the UK, when a *Guardian* article ('Computer cracks the great bus riddle', 20th May 1999) set off a flurry of correspondence. One of the last of these letters, published under the caption 'There's another theory just behind', and Appleby (1998) refer to an analysis in Eastaway *et al.* (1998). Whichever time increment strategy is used the program needs to keep an account of the number of times different events have occurred, and Ross (1997) gives clear guidelines for doing so.

Example 3.1   Fixed time increments

In Caulcutt's simulation, weekly sales by Alan have a mean of 15 and a standard deviation of 3.87. Sales are constructed from a sequence of pseudo-random standard normal numbers which are treated as a realisation of independent standard normal random variables. Methods for generating sequences of random numbers are given in Appendix 3. The standard normal numbers are multiplied by 3.87, added to 15, and then rounded to the nearest integer.

One of the most striking results from a typical simulation is the effect of Alan's holding orders over from one week to the next. Suppose he is selling reproduction Victorian rocking-horses. At the end of each week he tells the production manager at the small workshop that makes the rocking-horses how many sales he has made. The workshop sends out the horses during the next week, so a customer should receive the goods within two weeks. Now imagine that Alan adopts a strategy for avoiding criticism. If he sells more than 16 rocking-horses he will only declare 16 and hold the remainder until the end of the next week, provided the undeclared sales do not exceed 28. He will only declare sales of less than 14 if he hasn't enough sales from the previous week to make the number up to 14. In this way he should at least reduce the number of criticisms. This simulation is set up with a fixed time increment of one week. A summary, in a suitable form for writing a computer program and a Matlab program, are given on the book website. The results of a typical two-year simulation were that 49%, 24% and 4% of customers had their orders held back one, two and three weeks respectively. Only 23% of customers would have received horses within two weeks.

Example 3.2   Variable time increments

The practice manager in the CMC, described in Section 3.1, intends to investigate the effect of employing a recently qualified doctor (B), who is training to become a general practitioner, to help the duty doctor (A) with morning sessions. She will write a simulation program to keep track of time (t) from 9 a.m. until the last patient has been seen, and the state of the system during this period. The number of patients seen, the average waiting time, maximum number of patients waiting, and time the last patient leaves will be calculated at the end of each day of the simulation. As she simulates more days she will build up a distribution for these variables. The program will include a distribution for the number of patients waiting when the surgeries begin at 9 a.m., and distributions of the times between arrivals and consultation times. In order to choose suitable parameters for these distributions, she has kept a record of arrival patterns and durations of con-sultations over the past three weeks. The distribution of people waiting at 9 a.m. is given in Table 3.1. The mean and standard deviation are 4.056 and 2.689 respectively. The negative binomial distribution (Exercise 3.2) is a two-parameter discrete distribution with probability mass function

$$P(x) = \frac{\Gamma(r+x)}{x!\,\Gamma(r)} p^r (1-p)^x \quad \text{for } 0 < p < 1 \text{ and } x = 0, \ldots$$

Table 3.1 Number of patients waiting in the Campus Medical Centre at 9 a.m. over three weeks (Monday to Saturday)

Number waiting	Frequency (number of days)
0	1
1	2
2	3
3	2
4	4
5	1
6	1
7	2
8	1
9	0
10	1

mean $\mu = r(1 - p)/p$ and variance $\sigma^2 = \mu/p$. The variance is greater than the mean but as p becomes closer to 1 it tends towards a Poisson distribution. The method of moments estimators of the parameters are

$$\hat{p} = \bar{x}/s^2 \quad \text{and} \quad \hat{r} = \bar{x}\hat{p}/(1 - \hat{p})$$

In this case $\hat{p}$ and $\hat{r}$ are 0.561 and 5.183 respectively. The average time between arrivals between 9 a.m. and 10 a.m. was 7.3 minutes, and the manager thinks an exponential distribution will be a reasonable model for these times. Consultation times averaged 12.3 minutes with a standard deviation of 8.4 minutes. A gamma distribution (Exercise 3.3) is bounded below at 0 and can take a wide variety of shapes from highly skewed to near normal. Its pdf is

$$f(x) = \frac{x^{c-1}}{\Gamma(c)\theta^c} \, e^{-x/\theta} \quad \text{for } 0 \leq x$$

and it has mean $\mu = c\theta$ and variance $\sigma^2 = c\theta^2$. The method of moments estimators of θ and c are

$$\hat{\theta} = s^2/\mu \quad \text{and} \quad \hat{c} = \bar{x}/\hat{\theta}$$

respectively, so $\hat{\theta}$ and $\hat{c}$ will be 5.74 and 2.14 for this simulation program. Methods for generating random numbers from the three distributions are described in Examples A3.2, A3.3 and A3.5 respectively. The trainee will be assumed to take twice as long over each consultation. Once times until the next arrival and ends of current consultations have been generated, the program time jumps until the first of these events occurs. The steps involved in a typical day of the simulation are summarised on the book website.

3.2.2 Poisson process

In a point process events occur along a continuous time axis at a discrete set of time instants, the points. These points are the times at which the state of the discrete event simulation changes. The simplest point process is the Poisson process, in which events occur randomly and independently at a constant underlying average rate of λ per unit time. Formally, let $N(t)$ be the number of events that occur in the time interval $[0, t]$. Then the sequence of events is a Poisson process if

(i) $N(0) = 0$.
(ii) The number of events occurring in disjointed time intervals are independent.
(iii) The distribution of the number of events depends only on the length of the interval.
(iv) $\displaystyle\lim_{\delta t \to 0} \frac{\Pr(N(t + \delta t) - N(t) = 1)}{\delta t} = \lambda$

$\displaystyle\lim_{\delta t \to 0} \frac{\Pr(N(t + \delta t) - N(t) > 1)}{\delta t} = 0$

The physical interpretation of condition (iv) is that events cannot occur simultaneously, that the probability of an event occurring in a small length of time δt is approximately $\lambda \delta t$, and that the probability of two or more events occurring in a small length of time is negligible. It follows that the probability of no occurrence in a time interval of length δt is $1 - \lambda \delta t$. We will now show that $N(t)$ has a Poisson distribution with mean λt. We start by dividing the time t into a large number n of equal small time increments δt, that is

$$t = n \delta t$$

Then the number of events has an approximate binomial distribution with n trials and probability of an event occurring on any trial of $\lambda \delta t$.

$$\Pr(N(t) = x) \simeq {}_nC_x(\lambda \delta t)^x (1 - \lambda \delta t)^{n-x}$$

$$= \frac{n \times (n-1) \times \cdots \times (n - x + 1)}{x!} \frac{(\lambda t)^x}{n^x} \left(1 - \frac{\lambda t}{n}\right)^n \left(1 - \frac{\lambda t}{n}\right)^{-x}$$

$$= n \times \left(1 - \frac{1}{n}\right) \times \cdots \times \left(1 - \frac{x-1}{n}\right)(\lambda t)^x \left(1 - \frac{\lambda t}{n}\right)^n \left(1 - \frac{\lambda t}{n}\right)^{-x} \bigg/ x!$$

The approximation becomes exact, as a consequence of assumption (iv) when $n \to \infty$, and hence $\delta t \to 0$. Then

$$\Pr(N(t) = x) = \frac{(\lambda t)^x}{x!} \lim_{n \to \infty} \left(1 - \frac{\lambda t}{n}\right)^n$$

This important limit occurs in other applications, including compound interest, and it equals $e^{-\lambda t}$, as shown in Appendix A1.3. The final result is the Poisson distribution

$$\Pr(N(t) = x) = \frac{e^{-\lambda t}(\lambda t)^x}{x!} \quad \text{for } x = 0, \ldots$$

This result was first obtained by Siméon Poisson in 1837. The mean and variance of $N(t)$ are both equal to λt (Exercise 3.4).

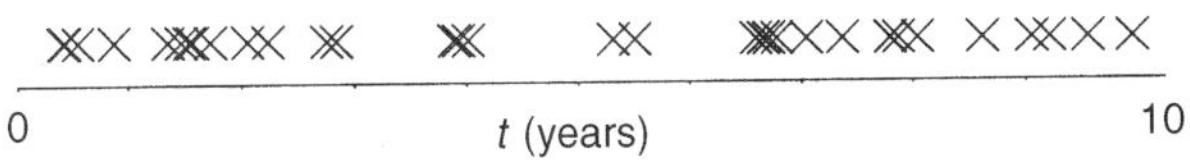

Figure 3.1 Realisation of a Poisson process with rate 3.4/year over 10 years

Example 3.3
A manager in a large chemical manufacturing company keeps a log of any incidents which compromise safety. Over the past few years the average rate of such incidents has been 3.4 per year. Figure 3.1 is a realisation of a Poisson process, with rate 3.4 per year, over ten years.

An alternative description of a Poisson process is in terms of the times between events. Let T be the time from now until the next event in a Poisson process. Assumptions (ii), (iii) and (iv) imply that the probability of no event in a length of time t is the same as the probability of no events occurring in the interval $[0, t]$. Therefore

$$\Pr(T > t) = \Pr(N(t) = 0) = e^{-\lambda t}$$

The $\Pr(T > t)$ is the complement of the cumulative distribution function of T. So

$$F(t) = 1 - e^{-\lambda t} \quad \text{for } 0 \le t$$

and the probability density function is

$$f(t) = \lambda e^{-\lambda t}$$

This is known as the exponential distribution. The mean and variance of T are $1/\lambda$ and $1/\lambda^2$ respectively (Exercise 3.5). Another consequence of assumption (ii) is that $F(t)$ is the distribution of the times between events, because the assumption of independence implies that the fact that an event has just occurred provides no information about the length of time until the next event occurs. It also implies that the length of time we have already waited for an event tells us nothing about how much longer we will wait. It is straightforward to verify that this, Markov property, is the defining feature of the exponential distribution (Exercise 3.9). In particular, if T has an exponential distribution

$$\Pr(T > s + t \mid T > s) = \Pr(T > t)$$

The proof follows from the definition of conditional probability and the fact that if $T > s + t$ then $T > s$.

$$\Pr(T > s + t \mid T > s) = \frac{\Pr(T > s + t \text{ and } T > s)}{\Pr(T > s)}$$

$$= \frac{\Pr(T > s + t)}{\Pr(T > s)} = \frac{e^{-\lambda(s+t)}}{e^{-\lambda s}} = e^{-\lambda t} = \Pr(T > t)$$

The Poisson process is the basis for much of the theoretical work on queueing systems. The simplest models assume customers arrive according to a Poisson process and that service times are exponentially distributed. The assumption about service times is less realistic but the theoretical models nevertheless give useful results and are discussed in Chapter 7. The Poisson process is also assumed for many sub-processes in simulation models. The suitability of a Poisson model can be assessed from available data by, for example, checking whether times between events can reasonably be modelled by an exponential distribution. The rate parameter λ is estimated as the reciprocal of the mean time between events. In a variable time increment simulation, the time until the next event (T) is a random number from the exponential distribution with mean $1/\hat{\lambda}$. If R is from $U[0,1]$ then T is $-\ln(1-R)/\hat{\lambda}$ (Appendix A3.4), or more efficiently $-\ln(R)/\hat{\lambda}$ since R and $1-R$ have identical distributions. In a fixed time increment simulation the probability of an event in the next time interval Δt is $\hat{\lambda}\Delta t$, where it is assumed that Δt is small enough for the probability of two or more events to be negligible. If R is from $U[0,1]$ an event occurs at the next time step if $R < \hat{\lambda}\Delta t$.

The simple Poisson process is rather limited. For example, the manager in the medical centre might expect patients to arrive at a higher underlying average rate at the beginning of the hour set aside for urgent problems than towards the end of the hour. A Poisson process with a time varying rate parameter is known as a time-dependent, or non-stationary, Poisson process.

3.2.3   Time-dependent Poisson process

Suppose point events occur randomly in time and let $N(t)$ denote the number of events that occur by time t. The process $\{N(t), 0 \le t\}$ is a non-homogeneous Poisson process with an intensity function $\lambda(t)$ if

(i) $N(0) = 0$.
(ii)   The number of events that occur in disjointed time intervals are independent.
(iii) $\displaystyle\lim_{\delta t \to 0} \frac{\Pr(N(t + \delta t) - N(t) = 1)}{\delta t} = \lambda(t)$

$\displaystyle\lim_{\delta t \to 0} \frac{\Pr(N(t + \delta t) - N(t) > 1)}{\delta t} = 0$

The time-dependent Poisson process is used in WatSup. Simulating occurrences of events is straightforward with a fixed time increment. The probability of an occurrence in the next small time step Δt is assumed to equal $\lambda(t)\Delta t$. Simulations with variable time increments rely on the following result. Potential events occur according to a Poisson process with rate λ, and are counted as an event with probability $p(t)$, which is independent of any events that have already occurred. The process of counted events is a time-dependent Poisson process with rate $\lambda(t)$ given by $\lambda p(t)$. This result is the basis of the thinning algorithm for generating events up to time t_{end} in a time-dependent Poisson process. The following routine is taken from Ross (1997). First select a λ such that $\lambda(t) < \lambda$ for t from 0 up to t_{end}. The cumulative number of events (i) is counted in I and $S(i)$ is the time at which the ith event occurs.

- *Step 1* $t = 0, I = 0$.
- *Step 2*   Generate random R_1 from $U[0,1]$.
- *Step 3* $t = t - \ln(R)/\lambda$. If $t > t_{\text{end}}$ stop.

- *Step 4* Generate random R_2 from $U[0, 1]$.
- *Step 5* If $R_2 < \lambda(t)/\lambda$ then

$$I = I + 1 \quad \text{and} \quad S(I) = t$$

- *Step 6* Go back to Step 2.

Example 3.4

Murray and Murray (1999) model water use as a customer–server interaction in which customers arrive according to a non-homogeneous Poisson process and engage a water server for a variable period of time and at a variable flow rate. Their WatSup software simulates a dendritic supply network with provision for local storage at each node. The algorithm is made up of four parts: a model of the physical system, system reliability, arrivals of customers with their demands for water, and the supply of the demands. The physical system is made up of modules which include a mains supply, tanks, valves and pumps with variable parameters which enable realistic hydraulic models to be set up. System reliability is investigated by, for example: specifying small probabilities that valves can fail on, fail off, or stick at whatever flow rate pertains at the time of failure; specifying a probability that pipes burst and choosing a distribution for repair times.

The model uses a constant time step (δt) of 1 second. The probability that a person in the system initiates a demand for water in a time step is $\lambda(t)\delta t$, and this probability is independent of the number of preceding time steps in which there has not been a demand. The rate $\lambda(t)$ varies seasonally and throughout the day. The model is approximately a time-dependent Poisson process, but it is not exactly a point process because once begun a demand will persist for several, and possibly many, time steps. Filling a kettle might take 5 seconds whereas running a bath might take 300 seconds. A washing machine cycle will continue for 3600 seconds although the demand will not be continuous. At any one time step the model checks whether any new demands need to be added to the current demands being processed, assesses the reliability of the system, and calculates the flows and volumes by working back from the ends of the system to the mains supply. If the supply is inadequate demands are prioritised and all priority 1 demands are met before priority 2 demands are considered. If it is not possible to satisfy all the priority 1 demands in full they are reduced pro rata.

The original reason for developing WatSup was the statutory need to assess water quality at customers' taps rather than at the treatment works. However, the model enables other questions about the water supply to be investigated. Examples are: the ability of a supply to feed a new housing development or industrial site from an existing source; the effect of changing community social economic habits or the implementation of a water-saving device; the ability to size components within a system with a guaranteed reliability; and the ability to find the states various parts of a system should be at to enable routine maintenance to be carried out with the minimum disruption. The model can also be used to justify borehole abstraction licences.

With a small amount of modification WatSup can be used to model any queue-based process. A concrete products manufacturing plant can serve to demonstrate how the software can be used to simulate and control an industrial process. Concrete products such as paving slabs, bricks and pre-cast building materials typically require sand, aggregate, cement, water and pigment in their manufacture. Each material is used in different quantities and can be treated as a fluid for this application. As each product requires different volumes and proportions of each material in manufacture, and as production quantity of each product may change according to the market-place, materials ordering can be problematic. A model could be set up to examine the materials ordering requirements, dependent upon the products being made and the current state of the supply of the various materials, with the aim of increasing the turnover of the stored materials. This is particularly important when the materials have a short shelf life. The under-standing of such complex industrial processing and the interactions between the processes can lead to benefits in reduced material storage and lower costs.

3.3   Some generalisations of Poisson processes

3.3.1   Renewal processes

A renewal process is a generalisation of a Poisson process. The intervals between events are independently distributed but can have any form of prob-ability density function rather than an exponential distribution. The Markov property only applies if the pdf of times between events is an exponential distri-bution. If the pdf has a coefficient of variation (CV) of less than 1, which is the CV for the exponential distribution, the point process will be qualitatively more regular than a Poisson process. In contrast, if the CV exceeds 1 the point process is even more irregular and clustered than a Poisson process. A gamma distribu-tion is a versatile form for the pdf of times between events (Exercise 3.3). The distribution of the time until the first event (T_1) can be different from that of the times between events, and this would be appropriate if the events were renewals of some component and T_1 was the residual lifetime of one already in use (Cox and Isham, 1980).

3.3.2   EAR1 process

It is possible to allow for correlations between intervals separating events by using autoregressive or moving average constructions. The simplest such model is the exponential autoregressive process of order 1 (EAR1). Let $\{T_i\}$ be the intervals between events, $\{M_i\}$ be a sequence of independent exponential random variables with mean $1/\lambda$, and $\{U_i\}$ be an independent sequence of Bernoulli random variables, i.e. U_i can take the value 0 with probability α or 1 with probability $1 - \alpha$. Then

$$T_0 = M_0$$

$$T_i = \alpha T_{i-1} + U_i M_i$$

The autocovariance of T_i at lag k, i.e. the covariance between T_i and T_{i+k}, is α^k/λ^2 (Cox and Isham, 1980).

3.3.3 Clustered point processes

A point process exhibits excess clustering if the variance of the number of events in disjointed intervals of fixed length exceeds the mean number of events in the intervals. A market trader thinks that requests to buy any particular share tend to be highly clustered. In a Bartlett–Lewis process clusters arrive according to a Poisson process with some rate λ_c. Within a cluster there will be N clients, and $(N-1)$ might have a Poisson distribution with mean μ. The times between clients within a cluster have an exponential distribution with some mean $1/\lambda_w$. Cox and Isham derive the quite complex expressions for the moments of the process in terms of the parameters. These can be used to estimate the parameters by substituting sample moments. Alternatively suitable values of the parameters for a simulation might be found by trial and error, subject to $\mu\lambda_c$ equalling the mean number of clients. A distribution of the number of shares bought or sold could be superimposed on the points of the Bartlett–Lewis process. The numbers of shares traded in consecutive deals might be correlated, and this could be allowed for in the model. Cox and Isham cite other applications of clustered point processes, including primary and secondary failures of computers.

A simple way of simulating a point process with high clustering, using a fixed time increment, is to let the rate depend on the time since the last event occurred. For example, if t_L is the time step at which the last event occurred,

$$\lambda_t = a + b\exp[-c(t-1-t_L)]$$

would drop from $a + b$ just after the event to a some time later. The relationship between the choice of the parameters a, b and c and the distribution of the number of events in fixed time periods could be investigated by simulation.

3.3.4 Superposition of point processes

If a large number of point processes are superimposed the resultant point process is approximately a Poisson process, under fairly general conditions. This result is analogous to the central limit theorem which implies an approximate normal distribution for the sum of a large number of random variables, superposition of point processes being analogous to adding random variables. In fact the correspondence is rather closer than an analogy because the Poisson distribution approaches a normal distribution as the mean increases. The superposition result explains why a Poisson process is a good model for the arrivals of calls to a busy telephone exchange.

3.4 Inventory control

3.4.1 Introduction

In modern manufacturing processes inventory is usually kept as low as is possible, and just-in-time strategies aim to remove the need for any inventory. For example, the Nissan plant in Sunderland orders car seats, from Ikeda-Hoover, to be delivered directly to the assembly line. No stock of seats is kept at Nissan. The system works well but the Ikeda-Hoover factory is on the Nissan site, which ensures there are no delays in delivery due to traffic jams, and is dedicated to manufacturing seats in various sizes and trims for Nissan vehicles, within a few hours of an order. However, just-in-time is not a sensible plan for low value and easily stored components such as nuts and bolts, and tyres.

Some car manufacturers now produce cars to a customer's exact specification within a few weeks of the order, rather than keep stocks of vehicles in a compound. With desktop publishing, a similar policy might be feasible for specialist technical books, but it would certainly not be appropriate for paperback novels. Lewis (1981) gave five principal reasons for holding stock, which are still relevant: to act as insurance against a higher than expected demand; to act as insurance against unexpected delays in supplier delivery times; to take advantage of quantity discounts; to take advantage of seasonal, or other, price fluctuations; and to avoid any delays in production caused by a lack of parts. The primary advantage of keeping a substantial level of stock is that customers' orders can be dispatched immediately; a failure to do so would probably lead to immediate loss of the order and possibly result in the loss of future business. This advantage must be balanced against the disadvantages of high stock levels: costs of storage; capital investment in stock with no return on the money; and stored material becoming obsolete or deteriorating. The last mentioned is particularly relevant for supermarkets.

There are two basic approaches for stock re-ordering policies although many variants are possible. The first is the re-order level policy. Supermarket staff at check-outs automatically record sales into computerised systems which can be programmed to re-order a product when stock has been reduced to a certain level. The re-order level for any particular product can depend on the rate at which it sells, its shelf life and storage costs. In manufacturing processes a re-order level policy is often implemented as a two-bin system. Whenever a bin is emptied, an order for a refill is made and the second, full, bin is begun. The second approach is the re-order cycle policy. Staff in a publishing company may meet at quarterly intervals to decide which books should be reprinted. An (s, S) policy is to order S if stock is below s at the time of the periodic review.

3.4.2   Re-order level and replenishment quantity

Suppose that the demand per time unit is normally distributed with mean μ and standard deviation σ, and that the time between placing an order and its delivery is L time units. If demands in consecutive time units are independent, a re-order level of

$$M = \mu L + z_\alpha \sigma \sqrt{L} \qquad (3.1)$$

corresponds to stock running out on $\alpha \times 100\%$ of the occasions on which re-orders are made.

Let the replenishment order size be q, the cost of keeping an item of stock over a time unit be c, and the cost of raising an order be d. The cost of raising an order might also include a machine set-up cost which is fixed regardless of the order size. Assume there is some fixed amount b of safety stock. Then the average cost per time unit is:

$$\left(\frac{q}{2} + b\right)c + \left(\frac{\mu}{q}\right)d$$

Differentiating with respect to q, and setting the derivative equal to zero, gives the optimum value of q, that is the value of q corresponding to minimum cost:

$$q_{\text{opt}} = \sqrt{(2\mu d/c)}$$

The customer service level will depend on the number of occasions on which re-orders are made, the probability that the stock runs out during the time before delivery of the order, and the average shortfall if stock does run out. The expected number of re-orders per time unit is μ/q_{opt}, and the probability that stock runs out is α in equation (3.1). It remains to find the average shortfall if stock does run out. Let D be the demand in time L, so $D \sim N(\mu L, \sigma^2 L)$. The average shortage, if stock runs out, is

$$E[D - M \mid D > M] = E[D \mid Z > (M - \mu L)/(\sigma\sqrt{L})] - M$$

$$= \mu L + (\sigma\sqrt{L})E[Z \mid Z > z_\alpha] - (\mu L + z_\alpha\sigma\sqrt{L})$$

where equation (3.1) has been used twice

$$= (\sigma\sqrt{L})\{E[Z \mid Z > z_\alpha] - z_\alpha\}$$

Now

$$E[Z \mid Z > z_\alpha] = \int_{z_\alpha}^{\infty} z\phi(z)\,\mathrm{d}z \Big/ \int_{z_\alpha}^{\infty} \phi(z)\,\mathrm{d}z$$

$$= \int_{z_\alpha}^{\infty} z\phi(z)\,\mathrm{d}z/\alpha$$

This has to be evaluated by numerical integration. To summarise, the expected shortfall per time period is:

$$(\mu/q_{\text{opt}}) \times \alpha \times (\sigma\sqrt{L})\{E[Z \mid Z > z_\alpha] - z_\alpha\}$$

If this is divided by μ we have the proportion of demand that is not met.

Example 3.5
A well-established textbook on the design and analysis of experiments sells an average of 200 copies per month. The unit cost of storage and loss of capital per month is 1.5 monetary units, and the set-up cost for a print run is 800 monetary units. The optimum print run is

$$q_{\text{opt}} = \sqrt{2 \times 200 \times 800/1.5} = 462$$

and the publisher rounds this up to 500. The standard deviation of monthly sales is 45. The publisher uses a re-order level policy, with α equal to 0.05, for any textbooks for which sales exceed 2000 per year. The time between ordering a print run and receiving the books is three months. If α is 0.05, then z_α equals

1.645 and

$$\{E[Z\,|\,Z > 1.645] - 1.645\} = 0.42$$

(from a table in Lewis, 1981). The percentage of the demand that is not met is

$$(0.05 \times 45 \times \sqrt{3} \times 0.42/500) \times 100\% = 0.33\%$$

3.4.3   Role of simulation in inventory control

The analysis in the preceding section is based on many simplifying assumptions. Sales levels will often vary throughout the year or even throughout the week. Sales may not be independent or normally distributed. Times between placing an order and the delivery may be better modelled as random variables. In practice there is likely to be some overshoot of the re-order level before it is noticed and an order is placed. The costings have been simplified. Although the analysis could be modified to allow for some of these points a simulation has greater versatility. Simulations also provide a clear visual display of typical stock balances and allow a manager to see the overall effects of varying parameters, e.g. STOKDEMO file in STATSMAN (Spooner and Lewis, 1995).

3.5   Summary

1. Uniformly distributed random numbers, over a range [0, 1], are the basis of any computer simulation which includes random components. They are available in most programming languages from pseudo-random number generators.
2. Uniformly distributed random numbers can be converted to random numbers from any other distribution, including multivariate distributions, by a variety of devices.
3. Computer simulations of random processes, such as queueing or inventory control, can be written with either fixed or variable time increments.
4. A point process is the sequence of time instants at which some event occurs. An event might be a customer arriving at a service point, for example. In a simple Poisson process the occurrences of events are random and independent and the underlying rate of occurrences is constant. The independence of occurrences implies that the time until the next event is independent of the entire past history of the process, including when the last event occurred. This is known as the Markov property and has the consequence that times between events have an exponential distribution.

    Independence of occurrences is a far stronger condition than times between events being independent. Times between metro trams might be independent and normally distributed with a mean of 10 minutes and a standard deviation of 1 minute. If I have just missed a tram I can expect to wait between 8 and 12 minutes for the next one, whereas if another passenger tells me he just missed one 5 minutes ago I expect to wait between 3 and 7 minutes. Now suppose a ticket inspector tells me that there was a 13-minute gap between the last trams. The consequence of the independence of times is that I have no reason to revise my estimates of waiting times, which are based on the mean of the distribution being 10 minutes. Bus services are rarely as regular as rail-based systems, because of the other traffic on the roads. If buses really do arrive randomly and independently at a mean rate of 6 per hour, the fact that I have been waiting an hour already tells me nothing about how much longer I must wait.

In a Poisson process, the number of events in fixed lengths of time has a Poisson distribution. If tanker collisions occur according to a Poisson process, the number of collisions in a year will have a Poisson distribution. A collision may involve only one tanker, e.g. if it collides with a wharf, or more tankers. The number of tankers involved in collisions in a year would not have a Poisson distribution.

The Poisson process is fundamental to the theory of point processes because the supposition of a large number of point processes is approximately a Poisson process. It is also the basis for a large number of more complicated models. In practice a Poisson process often gives a reasonable approximation to customer arrivals.

5. Suppose we have events in a Poisson process which occur at a rate λ per unit of time. In a simulation with fixed time increments, Δt, the probability of an event occurring at the next time step is $\lambda \Delta t$. If variable time increments are used the time until the next event is $-\ln(1 - R_i)/\lambda$ where $R_i \sim U[0, 1]$.

Exercises

3.1 The Deming bead experiment (Deming, 1986) is often used to teach people about variability. A bag contains 1000 beads of which 50 are red and 950 are white. Each participant is asked to draw out a sample of 20 beads and count the number of red ones, before replacing all the beads in the bag and passing it to the next person. Red beads represent defectives, and obtaining 3 or more is deemed unsatisfactory. If mixing is thorough the probability of drawing a red bead is almost constant at 0.05. Use the binomial distribution (Appendix A2.3) to find the probability of obtaining 3 or more red beads.

3.2 The Pascal variable X is the number of failures before the rth success in a sequence of independent trials in which the probabilities of success and failure are p and $(1 - p)$ respectively (Bernoulli trials, as in the binomial distribution). Explain why the probability mass function of the Pascal distribution is

$$P(x) = {}_{r+x-1}C_{r-1}\, p^r (1 - p)^x \quad \text{for } x = 0, \ldots$$

This generalises to the negative binomial distribution for non-integer values of r.

$$P(x) = \frac{\Gamma(r + x)}{\Gamma(r)x!}\, p^r (1 - p)^x \quad \text{for } x = 0, \ldots$$

By finding the moment-generating function or otherwise show that the mean and variance of X are $\mu = r(1 - p)/p$ and $\sigma^2 = \mu/p$ respectively.

The negative binomial distribution is sometimes defined, in an alternative form, as the distribution of the number of trials Y to obtain the rth success. Write down the values Y can take and its mean and variance.

The geometric distribution is the special case of the negative binomial distribution with $r = 1$.

3.3   The cth point in a Poisson process with rate λ occurs at time $T = T_1 + \cdots + T_c$ where the T_i are independent exponential random variables. Explain why the pdf of T satisfies

$$f(t)\delta t = \frac{(\lambda t)^{c-1}}{(c-1)!} e^{-\lambda t} \lambda \delta t$$

and hence deduce that

$$f(t) = \frac{\lambda(\lambda t)^{c-1}}{\Gamma(c)} e^{-\lambda t} \quad \text{for } 0 \le t;\ 0 < \lambda;\ 0 < c$$

This is the pdf of a gamma distribution. Although this derivation assumes c is an integer the pdf is well defined for any positive c. What is the special case of $c = 1$? Deduce that the moment-generating function of the gamma distribution is

$$M(\theta) = (1 - \theta/\lambda)^{-c} \quad \text{for } \theta < \lambda$$

Verify that the mean, variance, skewness and kurtosis are:

$$\mu = c/\lambda$$

$$\sigma^2 = c/\lambda^2$$

$$\gamma = 2/\sqrt{c}$$

and

$$\kappa = 3 + 6/c \quad \text{respectively}$$

3.4   The variable X has a Poisson distribution with parameter (λt). Follow a similar argument to that given in Appendix 2.3 to show that:

$$\mu = \mathrm{E}[X] = \sum_{x=0}^{\infty} x e^{-\lambda t} (\lambda t)^x / x! = \cdots = \lambda t$$

Show that $\sigma^2 = \mathrm{E}[(X - \mu)^2] = \lambda t$.

3.5   The variable T has an exponential distribution with parameter λ. Show that

$$\mu = \mathrm{E}[T] = \int_0^{\infty} t\lambda e^{-\lambda t}\, dt = \cdots = 1/\lambda$$

and $\sigma^2 = 1/\lambda^2$.

3.6   Conditional expectation (Ross, 1997)

(a)   Let θ be any random variable. Starting from the definition of variance, show that

$$\mathrm{var}(\theta) = \mathrm{E}[\theta^2] - (\mathrm{E}[\theta])^2$$

(b) Let X and Y be joint discrete random variables. The conditional expectation of X given that $Y = y$ is

$$E[X \mid Y = y] = \sum xP(x \mid y)$$

$$= \sum xP(x, y)/P(y)$$

Show that

$$E[E[X \mid Y = y]] = E[X]$$

where the outer expectation is with respect to the distribution of Y. It is customary, for example, to write $E[X \mid Y = y]$ as $E[X \mid Y]$.

(c) Conditional variance

$$\text{var}(X \mid Y) = E[(X - E[X \mid Y])^2 \mid Y]$$

$$\text{var}(X \mid Y) = E[X^2 \mid Y] - (E[X \mid Y])^2$$

$$E[\text{var}(X \mid Y)] = E[X^2] - E[(E[X \mid Y])^2] \tag{a}$$

also

$$\text{var}(E[X \mid Y]) = E[(E[X \mid Y])^2] - (E[X])^2 \tag{b}$$

By adding (a) to (b) obtain the conditional variance formula:

$$\text{var}(X) = E[\text{var}(X \mid Y)] + \text{var}(E[X \mid Y])$$

3.7 Antithetic variables.

Suppose h is some monotonic function. Let $U_1, \ldots, U_m$ be independent $U[0, 1]$ random numbers. Then

$$X_1 = h[U_1, \ldots, U_m] \quad \text{and} \quad X_2 = h[1 - U_1, \ldots, 1 - U_m]$$

will be negatively correlated and are described as antithetic. Compare the following sampling distributions.

(i) Generate 10 random numbers from $U[0, 1]$ and convert them to random numbers from an exponential distribution with mean 1. Calculate the mean. Repeat 1000 times and calculate the mean and standard deviation of the means.

(ii) Generate 5 random numbers $u_1, \ldots, u_5$ from $U[0, 1]$. Convert $u_1, \ldots, u_5, 1 - u_1, \ldots, 1 - u_5$ to random numbers from an exponential distribution with mean 1. Calculate the mean. Repeat 1000 times and calculate the mean and standard deviation of the means.

3.8 A model of market fluctuations (Mitrani, 1982).

Consider the behaviour of the market in a certain commodity. Assume that behaviour follows laws of supply and demand. These state, roughly, that

(i) the demand for the commodity is low when its price is high and vice versa (demand law);

(ii) if a high price is obtained for the commodity over a period of time, the supply during the next such period will be high and vice versa (supply law).

Now construct a model of the market based upon a formalisation of laws (i) and (ii). There is a single entity – the commodity which is produced and sold. That entity has two attributes – the quantity Q placed on the market and the price P at which it is sold. The pair (Q, P) defines the system state. Market performance figures are published at fixed intervals (say once a month), so the system state is considered only at those selected points. Thus an operation path for the market is a sequence of states $(Q_1, P_1), (Q_2, P_2), \ldots, (Q_i, P_i), \ldots$. The following is a simple model which includes random perturbations.

$$\text{demand} \quad Q_i = -aP_i + b + W_i; \quad a, b \geq 0$$
$$\text{supply} \quad Q_i = cP_{i-1} + d + X_i; \quad c, d \geq 0$$

where W_i and X_i are random variables. The values of a, b, c and d, and the distributions of W_i and X_i have to be determined empirically for different markets.

Write a program to simulate the system by implementing the following procedure:

1. Set $i = 0$. Fix a starting price P_0.
2. Increment i by one. Generate random X_i from the appropriate distribution. Using P_{i-1}, X_i and the supply law, calculate Q_i.
3. Generate random W_i from the appropriate distribution. Using Q_i, W_i and the demand law, calculate P_i. This determines the state (Q_i, P_i).
4. Repeat steps 2 and 3 for the required number of points in the operation path.
5. Plot Q_i against P_i.

Investigate the behaviour of the system, especially its stability, for different values of the parameters a, b, c and d and distributions of W_i and X_i.

(Reprinted from Mitrani (1982) with the permission of Cambridge University Press.)

3.9 A component has a lifetime T which has a cdf $F(t)$. The reliability function $R(t)$ is the complement of $F(t)$, and the hazard function is

$$h(t) = f(t)/R(t) \quad \text{for } 0 \leq t$$

which is proportional to the probability of imminent failure at time t.

(i)   The Markov property is that the future depends only on the present state, and in this context corresponds to a constant hazard rate, λ say. Prove that this implies T has an exponential distribution.
(ii) Show that the mean time before failure

$$E[T] = \int_0^\infty t f(t)\, dt = \int_0^\infty R(t)\, dt$$

(iii) A distribution has a hazard function t^a where a is constant. Find its cdf and pdf.

4
Obtaining market information

4.1 Introduction

CNN Quick Vote on 14th May 1999 asked, 'Do corporations, colleges etc. have an obligation to help the communities around them?' The results were summarised as follows:

Yes	76%	9575 votes
No	24%	3145 votes
	Total	12 720 votes

This poll is not scientific and reflects the opinion of only those internet users who have chosen to participate. The results cannot be assumed to represent the opinions of internet users in general, nor the public as a whole.

CNN's clear disclaimer applies despite the large sample size. The requirement for a scientific survey is that every person in the population has a known non-zero probability of being selected, and setting up such a random sampling scheme for a survey is expensive and time consuming. However, there are two substantial benefits from doing so. The first is that the results are unbiased, in so far as the average of results from a hypothetical very large number of surveys would equal the corresponding population value. The practical importance of this, when we have only one survey, is that all parties can agree that the results are fair, because no categories are given undue influence. Unbiased results from our particular survey are not necessarily close to the population values, but the second advantage of random sampling is that it is possible to give a measure of the precision of the result.

There are several compelling reasons for sampling a population rather than attempting to investigate every member. In a manufacturing context testing may be destructive. In market research the population may be hypothetical, e.g. all possible users of some product within the next year. Other reasons include cost, and a common finding that detailed study of a sample can provide more accurate information than superficial investigation of the whole population.

In some cases, such as whether to accept a delivery of 1000 windscreen wiper motors, it is straightforward to define the population, but when human populations are involved it is often less clear. Also, the fact that many people may choose not to respond somewhat undermines the advantages of an unbiased sampling scheme. For example, a national newspaper needs to decide whether its coverage of news, arts and sport is in the right proportion to satisfy its readers. A population could be defined as all adults in the country, all the people who read a paper daily, all the people who buy a paper daily, its current readers, or readers of it and similar papers. The editor's preferred population would probably be all potential readers, but identifying them and then drawing a random sample would be quite impractical. A more feasible option is to reduce the price for a week, enclose a questionnaire with the Saturday paper and offer people vouchers for free papers for another week if they return the completed questionnaire to their newsagent. Such ad hoc arrangements may provide valuable information for many marketing decisions, but a random sampling scheme is generally expected for research work or financial projections such as asset management plans. The requirement is that every member of the population has a known non-zero probability of selection and there are strategies, such as multi-stage sampling and cluster sampling, that circumvent enumeration of the entire population. The list on which the sampling scheme is based is known as the sampling frame.

When people refer to a random sample, without qualification, it is probably a simple random sample, or at least some approximation to it. The concept of the simple random sample is that the name of every member of the population is placed in a hat, using individual tickets. The tickets are thoroughly mixed, so that all are equally likely to be chosen, and the required number of tickets is drawn out. A hat is helpful for imagining the drawing of a random sample but it is not really a practical proposition, and thorough physical mixing is very difficult to achieve (Moore, 1979). The use of random numbers for drawing a simple random sample is described in Section 4.3.1.

Example 4.1

A research unit has been contracted to carry out a survey of people's exposure to newspapers, radio, TV and the internet in England. The target population is all families in England, but establishing this population requires some conventional definition of groupings that are to be considered as families and some means of identifying them. Barnett (1991) suggests that it would be far easier to sample addresses and study the population of households at these addresses. In such cases it may be useful to distinguish between the target population and the study population. We rely on the study population being similar to the original target population.

Rather than attempt to obtain lists of all households in the country, multi-stage sampling would be used. The first-stage sampling units might be districts within cities, towns and rural parishes. The first-stage sampling frame would be the list of all such units that cover the country. The second-stage sampling units might be polling districts. However, the sampling frame for the second stage would only need to cover the selected first-stage units. The final-stage units would be households, but the associated sampling frame would only need to cover selected polling districts.

The research unit might then send out a postal questionnaire. Non-response is a major practical problem with most surveys. It might be possible to speak to a representative of some of the households from which no reply had been received over the telephone. The researchers would attempt to elicit responses for some of the more important questions at least. For a final attempt, the research unit might send someone out to interview a random sample of the remaining households.

4.2 Non-response

Encouraging people to take part in surveys seems to be becoming increasingly difficult, so it is essential that the questionnaire is well designed. Trial versions should be tried out on colleagues and any sympathetic contacts in the target population. If the questionnaire is to be sent by post, the covering letter needs to convince possible respondents that the survey has some useful purpose. A small pilot survey should be undertaken before the main mailing.

In the analysis we usually either assume that responding to the survey is not associated with the answers to the questions, or we attempt to adjust for non-response on the basis of limited information from a random sample of those people who did not respond initially.

Example 4.2
A headline in the *Guardian* of 14th July 1999 was '62% against joining euro'. At the end of the article, by Ewen MacAskill, the source of the data was given: 'ICM interviewed a random sample of 1,200 adults aged 18-plus by telephone between July 9 and 11. Interviews were conducted across the UK and the results have been weighted to the profile of all adults'. The main finding of the poll was that if a referendum were held tomorrow, 25% would vote to join compared with 62% who would vote against and 13% who do not know. The study population is those people whose telephone numbers appear on a list provided by an agency, whereas the target population is all adults who are entitled to vote. The study population excludes people with no telephone, people whose telephone numbers are not listed, and people who have requested that their number be removed. The list may include some people who are not registered to vote in the UK. An ICM representative reported that nearly 12 000 numbers had to be tried in order to produce 1200 responses. Reasons for non-response included no answer to calls and discontinued numbers as well as refusals. This is a pragmatic approximation to a random sample from the target population, for this question. ICM have conducted similar surveys each month from January 1999, so the trends are likely to be more reliable than the percentages. The percentage in favour started at 29% and reached a high of 36% before decreasing to 25% for July. Opposition has increased steadily from 52% in January to 62% in July. An explanation

for the lack of enthusiasm is that the euro has lost nearly 13% of its value against the US dollar since it was launched in January.

Example 4.3
Jackson and De Cormier (1999) state that e-mail now reaches 15% of the US population and is predicted to reach 50% in the next five years (*Marketing News*, 1997). They compared the response rates of a targeted and a non-targeted group to a financial questionnaire which consisted of 11 questions designed to gather information on investments. The targeted group consisted of members of two financial Usenet newsgroups. The non-targeted group included members from two newsgroups, one with interests in journalism, the other with an interest in law enforcement. A total of 1000 questionnaires was sent by e-mail, 500 to each study group. The targeted group answered 73 (15% response rate) and returned 15 unmarked. The non-targeted group returned 53 answered and 62 unmarked. Many of the unmarked questionnaires were returned with notes, some of which threatened all kinds of legal action – presumably most of these were from the law enforcers. Despite some hostile responses, part of the conclusions drawn by Jackson and De Cormier was, 'The use of the Internet to collect data, and to conduct research shows a great potential for certain target markets. The group that responded positively to this survey would be a good target market for companies selling items connected to the stock market'.

Example 4.4
The main library at the University of Newcastle upon Tyne sent a postal survey to all academic staff and one-third of postgraduate research students (every third name on an alphabetical list) in February 1999. The questionnaire was designed to elicit opinions about current services and priorities for change. The response rate was 30% for academic staff and 18% for postgraduate research students. The corresponding survey for undergraduates and post-graduate students on taught courses is described in Example 4.13.

Example 4.5
The Sports Centre at the University of Newcastle upon Tyne sent a question-naire to a sample of 434 members, about 6% of the total. Every 20th name on an alphabetical list of students, and every 10th name on an alphabetical list of staff was taken. The questionnaire enquired about the use of facilities, the level of satisfaction with the service provided, and suggestions for improve-ments. The questionnaires to students were sent through the internal mail. Although questionnaires could be returned in the internal mail, 1 in 5 students on the alphabetical list of sampled names was sent a stamped addressed envelope (SAE) for return of the questionnaire. Some staff members gave a home address and others gave department addresses. The former staff group were sent an SAE for the reply. The response rates are shown below.

	Student	Staff
With SAE	24/63 (38%)	22/40 (55%)
Without SAE	65/255 (25%)	16/76 (21%)

The overall response rate was 29%. There is evidence that the inclusion of an SAE increased the response rate for both students ($\chi^2_{calc} = 3.98$, $P < 0.05$) and staff ($\chi^2_{calc} = 13.7$, $P < 0.00$) despite the availability of the internal mail (see Section 4.6 for details of the statistical test). This suggests that ease of return of the questionnaire is critical.

Example 4.6
The Office of National Statistics in the UK conducts an annual survey of family expenditure. It is not mandatory to respond, and the response rate for the 1996–97 survey was 61.9%. A few government questionnaires, however, do have to be answered. A census is carried out in the UK every ten years and 10% of the population is sent a more detailed questionnaire which also has to be completed. The Inland Revenue in the UK used to send out tax returns to a random sample of people who were taxed on the Pay As You Earn schedule which applied to most employees. Completing these was a legal requirement, but the practice stopped when the tax system changed in 1997.

Example 4.7
Ghosh and Taylor (1999) sent a questionnaire about advertising agency–client relationships to 54 New Zealand companies and 100 companies in Singapore. The response rates were 76% and 21% from the two countries respectively. An earlier, similar, study in the USA and the UK had reported a 44% response rate (Michell *et al.*, 1992). Raab and Donnelly (1999) discuss analytic methods for allowing for non-ignorable non-response in medical surveys, but these rely on the initial response being quite high, about 65% in their example. The underlying idea is to combine empirical estimates from people who respond with subjective assessments of those who do not.

4.3   Random sampling schemes

4.3.1   Simple random sampling

A number is associated with every member of the population of N members. A simple random sample of size n is obtained by reading down a random number table or generating numbers with a pseudo-random number generator. Every member of the population has the same probability (n/N) of being in the sample. However, other random sampling schemes can be set up with this property and it does not define a simple random sample. A formal definition of a simple random sampling scheme is that every possible sample of size n from the population of size N has the same probability of selection.

Example 4.8
A cruise ship has 148 lifeboats and the second officer decides to inspect six. The lifeboats are numbered on a plan of the ship, and this constitutes the sampling frame. One possibility would be for the officer to read down the first three columns of the fifth block of five columns of digits in Table A9.5 until six numbers between 001 and 148 are obtained, but it is quicker to associate five numbers with each lifeboat, i.e. 001, 201, 401, 601 and 801 all correspond

to lifeboat 1, and 148, 348, 548, 748 and 948 all correspond to lifeboat 148. The numbers are:

765   ignored as no corresponding lifeboat
930   lifeboat 130 selected
930   ignored as lifeboat 130 already in the sample
860   lifeboat 60 selected
175   ignored as no corresponding lifeboat
662   lifeboat 62 selected
456   lifeboat 56 selected
142   lifeboat 142 selected
202   lifeboat 2 selected

Thus the sample is lifeboats {130, 60, 62, 56, 142, 2}.

Example 4.9
Obtaining a random sample may be awkward, but careful planning can minimise the inconvenience. A business had commissioned the construction of a large container park from 400 000 concrete paving blocks (pavers). The cement content of pavers is important for strength and resistance to frost damage. About half the pavers had been laid, when a few spare were sent for cement content analysis. The results were lower than expected, but the supplier would only consider a claim if it was based on a random sample. The chemical analyses were very expensive, but the sample had to be large enough to have a high probability of detecting any substantial deficiency in cement content. The business manager decided on a sample of size 40 (reasons for this decision are given in Section 4.4), half to be selected at random from the paved area and half to be selected at random from the remaining pavers which were stacked on palettes. The paved area was identified on a plan of the site, and enclosed by a rectangle of dimensions 30 m × 50 m. Twenty pairs of independent uniform $U[0, 1]$ random numbers were multiplied by 30 and 50 respectively to give the coordinates of 20 pavers. The main practical difficulty was removing the selected pavers without damaging too many of the neighbouring ones. The second half of the sample was to be 20 pavers from 200 000, stacked in five 500s on 400 palettes. The palettes were numbered, and a numbering system within a palette was devised. Then 20 independent uniform $U[0, 1]$ random numbers were multiplied by 200 000 to obtain the numbers of 20 pavers. The pavers were found by first identifying the appropriate palette and then finding the current numbered block within the palette, e.g. 047 298 corresponds to paver 298 within palette 95.

4.3.2   Stratified sampling

Abdul-Aziz *et al.* (1998) sent a questionnaire on the use of statistical methods and other quality improvement techniques to a random sample of manufacturing companies in Malaysia. One of the main aims of the research was to establish the demand for government-funded advisory leaflets giving guidelines on the application of statistical methods in manufacturing. A second objective was to compare the proportions of companies in Malaysia which use these methods with the corresponding proportions from the same survey in the UK

(Abdul-Aziz, 2000). The UK government has a long history of promoting the assessment and registration of quality systems by third-party certification bodies to standards such as the BS 5750 and later ISO 9000. In contrast, the Malaysian government has placed more emphasis on Japanese quality practices through its 'Look East Policy'.

The sampling frame for the Malaysian survey was based on two lists of manufacturing companies maintained by the Ministry of International Trade and Industry (MITI), Malaysia. The target population was all Malaysian manufacturing companies. It is sometimes useful to distinguish between the target population and the study population. In this case the study population is listed manufacturing companies, and there are likely to be discrepancies due to companies being set up or going out of business since the list was last updated. The companies in the lists could be grouped into four categories. Those on the first list were manufacturing companies with paid-up capital of more than 2.5 million Malaysian dollars (RM) and will be referred to as larger companies. The second list was of manufacturing companies in Malaysia with paid-up capital of up to RM2.5 million (smaller companies). The latter includes small and medium-sized companies (SMEs) which constitute at least 80% of the total number of manufacturing companies in Malaysia. The amount of paid-up capital is commonly used by the Malaysian government to define SMEs. The growth of the latter is considered to be a vital aspect of the overall development of the manufacturing sector in Malaysia. The companies in the lists were also classified as resource-based or non-resource based (resource basis). Resource-based companies include companies from food, rubber, basic and fabricated metal, non-metallic minerals, leather, wood and paper manufacturing subsectors. Companies in the electrical, electronic, textile, chemical, machinery, transport equipment and plastic manufacturing subsectors are grouped under non-resource. This classification reflects the Malaysian government's current emphasis on the need to develop further the industries that are based on indigenous natural resources. The cross-classification of size by resource based, or not, gives the four categories shown below.

	Smaller	Larger
Resource based	1618	1954
Non-resource based	1055	1282

A sample of 11% of these companies was to be drawn, and this would require questionnaires to be sent to 650 companies. (Decisions on sample size are discussed in Section 4.4.) Some typical questions include whether a company uses just-in-time manufacturing strategies (JIT) and whether a company uses statistical methods when designing and analysing experiments (DOE), which was to be taken to include Taguchi methods. Before sending out the questionnaire, it seemed possible that the larger non-resource-based companies would be the most likely to have used JIT or DOE. A simple random sample of 650 companies might result in over-representation of this category, and hence over-estimation of the proportion using DOE. It is not reassuring to know that simple random samples will give unbiased results on average, if we can see that our particular sample is unrepresentative in respect of some criteria that are known for all members of the population. Replacing our simple random sample with another prejudices the principle of randomisation. The solution to the dilemma is to set up a stratified sampling scheme. The categories are known as strata, and simple random samples are taken from each stratum. The sample sizes within each stratum can be chosen independently, but it is often convenient to use the same sampling fraction throughout, as was done in this case. The same sizes for each stratum were therefore:

	Smaller	Larger
Resource based	178	215
Non-resource based	116	141

Abdul-Aziz *et al.* (1998) sent out 650 questionnaires in the main mailing, but only 38 responses were received. One hundred and ten blank questionnaires were returned as the addressee having 'gone away'. The questionnaire was sent out again, with a revised covering letter, to companies which had not responded, eliciting 95 more responses to give a 25% overall response. This low response is fairly typical for a postal questionnaire to manufacturing companies, but the consequential non-response bias could be large if agreeing to complete a questionnaire is associated with a tendency to apply statistical methods. A partial solution is to follow up a random sample of non-respondents with a telephone call, and to try to persuade them to answer the most important questions at least. This was done, and 11 out of 24 randomly selected non-respondents answered questions over the telephone. The overall response was 144 out of 540 (27%) and the corresponding response for the UK survey was slightly higher at 56 out of 183 (31%). However, Malaysian companies were much more likely to answer over the telephone (11 out of 24 against 6 out of 42). Some assessment of possible non-response bias can be made by comparing those companies which returned the postal questionnaire with those that responded over the telephone. However, this still leaves an estimated 41% of companies which would not respond. Reasons given by companies which were contacted by telephone, and still declined to take part, included lack of time, no one person in the company who could be expected to know what statistical techniques were being used, and commercial confidentiality. In this study the effects of non-response bias were mitigated for two reasons. The first is that those companies which are willing to respond to the questionnaire are perhaps more likely to use the guidelines, and it seems reasonable to produce these to meet the requirements of respondents. The second reason is that the comparison between the practices of Malaysian and UK companies is based on similar samples and it is more plausible to assume that the non-response bias is similar for the two countries than it is to assume that it is negligible.

Estimation

The intention had been to use the same sampling fraction (11%) in each of the four strata. In this case every company would have the same probability of selection, 0.11, but it would not be a simple random sample because of the constraints on the number of companies from each stratum. However, the different response rates in the different strata led to unequal sampling fractions. The proportions of respondents who say they use statistical design of experiments are summarised below. There were 10 missing responses which have been included as not using such methods.

	Smaller	Larger
Resource	4/36	3/26
Non-resource	1/30	5/32

We will assume that the non-response is not associated with the use of DOE and estimate the proportion of the population of manufacturing companies which use DOE. The estimate of the proportion of companies which use DOE is obtained by estimating the number of companies which use DOE as:

$$\frac{4}{36} \times 1618 + \frac{3}{26} \times 1954 + \frac{1}{30} \times 1055 + \frac{5}{32} \times 1282 = 640.7$$

and then dividing by the total number of companies:

$$1618 + 1954 + 1055 + 1282 = 5909$$

This gives an estimated proportion of 0.11. It is a weighted average of the four sample proportions.

To summarise, in stratified sampling the population is divided into groups, known as strata, on the basis of variables that are known for all members of the population. Simple random samples are taken from each stratum, and the results are combined, with appropriate weighting, to give unbiased estimates for the population. Comparisons can be made between strata. Optimal allocation of the overall sample size between strata is discussed in the Asset Management Plan case study (Section 4.5).

4.3.3   Cluster sampling and multi-stage sampling

The population is divided into groups called clusters. Then a random sample of clusters is drawn, and all the members of the selected clusters are taken to form the sample. If the clusters are drawn by a simple random sampling scheme, every person in the population has the same probability of selection. The clusters are usually based on administrative convenience. It is most efficient if the within-clusters variance exceeds the between-clusters variance.

If a random sample is taken from each cluster, we have a two-stage sampling scheme. The idea generalises to multi-stage sampling.

Example 4.10
A few years ago the Civil Aviation Authority (CAA) in the UK carried out a survey of the weights of packages stored in overhead bins as part of its safety regulation work. Flights were stratified by airline destination area, and time of departure. A random sample of flights was selected from each stratum. A simple random sample of six overhead bins was identified on each aeroplane, and the luggage in these bins was weighed. This is a two-stage sampling

scheme. The first-stage sampling units are flights and the second-stage sampling units are overhead bins. The bins are actually clusters of packages because all the packages in each bin were weighed.

Example 4.11
The Office of National Statistics in the UK is responsible for the annual Family Expenditure Survey. The Office uses a stratified multi-stage random sampling scheme in Britain. The area of Britain is divided into 672 postal sectors, and these were used as the first-stage sampling units for the 1996/97 survey. The postal sectors were stratified as metropolitan or non-metropolitan and by two 1991 census variables available for sectors, average socio-economic group and average car ownership. Post Office lists of addresses were used as the frame for the second-stage sampling within the selected postal sectors. Northern Ireland was treated separately and the Land Registry list of dwellings was used as a sampling frame.

Example 4.12
Reedy *et al.* (1980) compared the conditions of service, and the social, professional and occupational characteristics of nurses employed by area health authorities (AHA) and attached to general medical practices, and nurses employed by general practitioners. The AHA in England were grouped into clusters of three, in which there was at least one urban and one rural AHA, and stratified into three geographical areas: south-east, west and north. One cluster was randomly chosen from each stratum and four rural and five urban AHA were identified. The general medical practices within these nine AHA were classified by nursing resources: attached nurse only; employed nurse or nurses only; both attached and employed nurses; neither an attached nor an employed nurse. A stratified random sample of 41 practices with an attached nurse only, 32 practices with an employed nurse or nurses only, and 40 practices with both kinds of nurse was taken. The stratification was by AHA. From each of these practices, one nurse, of each kind for practices with both employed and attached nurses, was selected at random for interview with one of the research team. They interviewed 81 attached nurses and 72 employed nurses.

The probability of selection for a nurse was inversely proportional to the number of nurses of the same kind working in the practice. Since variables were being considered by nurse, rather than by practice, responses to the questions were weighted in inverse proportion to the nurses' probability of selection, i.e. the weights were the number of nurses of the same kind working in the practice. For example, one of the questions was: 'Have you attended any continuing professional education courses during the past twelve months?' Responses, and sampling weights, are given below.

Weight (number of nurses of same kind in practice)	Attached			Employed		
	Yes	No	Total	Yes	No	Total
1	55	20	75	19	37	56
2	3	3	6	5	8	13
3				2	1	3

The weighted percentage for the employed nurses is

$$\frac{19 \times 1 + 5 \times 2 + 2 \times 3}{56 \times 1 + 13 \times 2 + 3 \times 3} \times 100\% = 38\%$$

and the weighted percentage for the attached nurses is 70%.

4.3.4   Systematic sampling (list sampling)

A software company has an alphabetical list of all holders of licences for each of its products. There are 12 765 registered users of Quickstat and the company wishes to send a questionnaire to a sample of about 500 of them. An easy way to obtain the sample is to take a random starting point on the list between the 1st and 25th person, and then every 25th person to obtain a sample of 510 or 511. This is known as list sampling, or systematic sampling. It is a random sampling scheme and every person has the same probability of selection, 1/25. It can be thought of as a special case of clusters and we take one of them. It is usual to analyse the data as though they were drawn by a simple random sampling scheme. If the order of the listing has some geographical significance, list sampling will ensure the sample is spread over the area and will be advantageous. Systematic sampling should be avoided if there are periodicities in the ordering. For example, if there are 20 deliveries of concrete to a construction site each day do not sample every 20th delivery.

4.3.5   Quota sampling

Interviewers are instructed to question a given number of people in each of several categories, which are usually based on age group and sex. For example, a company which bakes bagels might ask a representative to set up a stand in a supermarket and identify at least 10 males and 10 females in the following age groupings, 6–10, 11–18, 19–25, 26–40, 40–60, 61 and above, and ask them to taste and comment on its bagels. It is not a random sample, but it dispenses with the need for a sampling frame and is quick, convenient and much cheaper than a random sampling plan. It also has the advantage of targeting shoppers. Quota sampling is often used to monitor changes, and the company might repeat the exercise after changing the recipe.

Example 4.13
Newcastle University library used a quota sampling scheme for the 10% campus sample of undergraduates and postgraduate students on taught courses. Quotas were set by sex and faculty. Students were asked to complete questionnaires at survey points which were set up around the campus (halls of residence, refectories, medical school, and computer clusters, as well as the library itself), and which operated at different times of day.

Example 4.14
Wicks and Bradshaw (1999) investigated differences between men's and women's perceptions of values currently shown and rewarded at work and of

desirable changes in the culture of business organisations. To begin with they asked 36 participants attending a seminar on organisational change to complete a questionnaire. These initial respondents were from 27 Canadian organisations representing a wide variety ranging from public service departments to private sector manufacturing companies. Each participant was asked to identify five male and five female managers in their organisation who would be willing to take part in the study and complete the questionnaire. The final sample size was 362. Wicks and Bradshaw discuss the limitations of this variant of quota sampling which they describe as a snowball sample, and give a breakdown of the sample by age, education, organisational level and function. The analysis of the questionnaire gives a score on each of three scales (see factor analysis in Chapter 5): dominance versus submissiveness, friendliness versus unfriendliness, and acceptance versus non-acceptance of authority; for four scenarios: values currently shown in the culture of the organisation, values that need to be shown in the future, values that women are presently rewarded for showing in their behaviour, and values that men are presently rewarded for showing in their behaviour. The widest range was in the mean friendliness scores, for men and women combined, which were 1.45, 6.34, 2.27 and 0.80 for the four scenarios respectively. We have to make a subjective assessment of how representative the results are of all Canadian organisations. There were no respondents from companies with fewer than 600 employees. A partial answer to the question of the effect of the particular organisation itself would be to analyse responses within and between the 27 organisations. It is possible that differences between male and female attitudes are less affected by bias in the selection procedure than are scores on the scales, although it may be that people who were willing to take part had more polarized views. However, even in a randomly selected sample, responding to the questionnaire is voluntary.

4.3.6   Size weighted means

A research worker asks people using an outdoor recreation centre, which is run by the local authority, how long their visit will last. Let t_i be the recorded times for a sample of n people. The estimate of the average visit length needs to take account of the fact that the probability of being in the sample is proportional to the length of visit. Therefore weights of $1/t_i$ should be applied to each observation, and the estimated mean time is:

$$\frac{\sum t_i \times (1/t_i)}{\sum (1/t_i)} = \left[\sum (1/t_i)/n \right]^{-1}$$

This is known as the harmonic mean. Dobbs (1993(a), (b)) discusses sample selection bias in this context in more detail.

Example 4.15
A sample of 10 people at a recreation centre were asked how long they expected to stay. Their responses in hours were: 1.5, 0.5, 2.0, 3.5, 0.5, 1.0, 2.5, 1.5, 4.0, 3.0. The estimated mean time per visit is 1.23 hours. If the arithmetic mean is used, an incorrect estimate of 2 hours would be obtained.

Example 4.16
One of the questions on the postal questionnaire sent to university staff who were members of the Sports Centre asked them to tick the category that most accurately described how often they used any of the facilities. The responses are summarised below.

4/week	9.8%
3/week	18.0%
2/week	21.4%
1/week	34.4%
1/month	16.4%

Average use is 1.74 visits/week. In a pilot study a student had asked the same question to members of staff using the main facility on a particular Thursday afternoon. The responses were:

4/week	2
3/week	4
2/week	8
1/week	2
1/month	0

This is not a random sample, but an estimate of the number of visits per week would be more plausible if responses are weighted in proportion to the product of the number responding with the reciprocal of the relative probabilities of being asked, i.e.

$$\frac{4 \times (2 \times 1/4) + 3 \times (4 \times 1/3) + 2 \times (8 \times 1/2) + 1 \times (2 \times 1/1)}{(2 \times 1/4) + (4 \times 1/3) + (8 \times 1/2) + (2 \times 1/1)}$$

$$= 2.04 \text{ visits/week}$$

Particular care needs to be taken whenever averaging times. Suppose, for example, that $\{t_i\}$ for $i = 1, \ldots, n$ are times between buses. The average wait per person is not $\frac{1}{2}\sum t_i/n$, but $\frac{1}{2}\sum t_i^2/\sum t_i$. This is because the number of people arriving at the bus stop is proportional to the length of the periods between buses, and these should therefore be weighted according to their lengths.

Example 4.17
The timetable states that the Number 234 bus runs at 20-minute intervals. The times (in minutes) between 10 buses were noted one morning:

$$4 \quad 25 \quad 10 \quad 32 \quad 18 \quad 27 \quad 13 \quad 35 \quad 16$$

The mean is 20 minutes but the average wait per person is estimated as 14.4 minutes. Hutchinson (1996) gives more examples.

4.4 Sample size

We start by considering the case of the manager of the business that commissioned the container park. The specification for the concrete paving blocks included a clause stating that the cement content of pavers should be between 16% and 20%. If the cement content drops below 16% the pavers will be susceptible to frost damage. Cement is also the most expensive ingredient in the concrete mix, so the main concern is that the cement content may be too low. The chemical analyses to determine cement content are expensive, but taking too small a sample, which is unlikely to provide evidence of a practically important deviation from the specification, would be a waste of money. Even if we do take a large enough sample to demonstrate evidence of a failure to comply with a specification, reliance on statistical methods is not well established in UK courts. A means of avoiding expensive litigation is to ensure that an acceptance sampling plan which is acceptable to all parties is incorporated in the original contract. The contract should specify the action to be taken if the supplied goods do not satisfy the requirements of the acceptance sampling procedure. This could include further testing at the supplier's expense, and concessions on price as well as outright rejection.

The formal hypothesis testing approach developed by Neyman and Pearson is useful for decision making in adversarial situations. To progress, we rephrase the specification in terms of a mean and standard deviation. If the cement contents are normally distributed, about 2 in 1000 will be more than three 3.09 standard deviations from the mean. Therefore, it seems reasonable to interpret the specification as a mean of 18% with a standard deviation of 2/3.09 which equals 0.65%. The specification could also be met by a mean of 17% and a standard deviation of 0.3%, for example, but tighter control of the process might cost the manufacturer more than the saving in cement. We will assume both parties agree a specified mean (μ) of 18% and standard deviation (σ) of 0.65% for the purpose of a hypothesis test about the mean cement content.

4.4.1 Estimating a population mean (standard deviation known)

Continuing with the pavers example, the null hypothesis (H_0) is that the specification is met, i.e. μ equals 18. The supplier is willing to agree to a probability of 0.1% of incorrectly rejecting H_0 when it is true (this is known as the Type I error and the probability of making it is conventionally denoted by α). We also have to specify whether the alternative hypothesis (H_1) is one-sided, here $\mu < 18$, or two-sided, here $\mu \neq 18$. A one-sided alternative hypothesis was agreed (given that the manufacturer is unlikely to use too much cement this is practically equivalent to the two-sided alternative with an α of 0.2%). The manager now calculates the consequences of the mean being 10% too low at 17.2%. Let X represent the cement content of a paver.

$$\Pr(X < 16 | X \sim N(1.72, (0.65)^2)) = \Pr(Z < -1.85) = 0.032$$

As many as 3% of pavers being below the specified cement content would not be satisfactory, and the manager decides that there should be a probability of 0.99 of rejecting H_0 in this case. The probability of rejecting H_0 $\mu = \mu_0$ when the mean is $\mu \neq \mu_0$ is a function of μ known as the power of the test, denoted by $1 - \beta$ (not rejecting H_0 when it is false is known as the Type II error).

We can now calculate the sample size (n). We will reject H_0 if the calculated value of the sample mean ($\bar{x}$) is less than some critical value c, which is itself below 18%. The sample mean $\bar{X}$ can be assumed to have a normal distribution. This would follow, to a very good approximation, from the central limit theorem even if the individual cement contents were not approximately normally distributed. If H_0 is true we require

$$\Pr(\bar{X} < c | \bar{X} \sim N(18, (0.65)^2/n)) = 0.001$$

This is equivalent to

$$\Pr(Z < (c - 18)\sqrt{n}/0.65) = 0.001$$

and since

$$\Pr(Z < -3.09) = 0.001$$

we have the requirement that

$$c = 18 - 3.09 \times 0.65/\sqrt{n}$$

The manager's requirement is

$$\Pr(\bar{X} < c | \bar{X} \sim N(17.2(0.65)^2/n)) = 0.99$$

and similar algebra leads to a second requirement that

$$c = 17.2 + 2.326 \times 0.65/\sqrt{n}$$

Consideration of the two requirements together leads to

$$18.0 - 17.2 = (3.09 + 2.326) \times 0.65/\sqrt{n}$$

and hence $n = 19.4$.

Confidence interval

The sample size calculation can be based on a less formal but somewhat quicker argument from a confidence interval. A 99.8% confidence interval for μ is given by

$$\bar{x} \pm 2.326\sigma/\sqrt{n}$$

We would reject H_0 that $\mu = \mu_0$ in favour of H_1 that $\mu \neq \mu_0$, at the 0.2% level (two-sided alternative), if this interval does not contain μ_0. Suppose we are willing to accept a 50:50 chance of rejecting H_0 if μ equals μ_1. Then there is a 50% chance that $\bar{X} + 2.326\sigma/\sqrt{n}$ is less than μ_0, and the sample size follows from the equation

$$\mu_0 - \mu_1 = 2.326\sigma/\sqrt{n}$$

In this case μ_0 and σ are 18 and 0.65 respectively, and if we choose μ_1 to be 17.7, for example, then n equals 25.4. The only limitation of this confidence interval argument is that we have to specify a mean which corresponds to a power of 0.5 rather than being able to set a mean corresponding to any power we choose.

In some cases sample size requirements may just be based on the width of a typical confidence interval.

4.4.2   Estimating a population proportion

A company produces a successful computer software package and is considering the development of a graphics supplement. It would market this at a special 'introductory price' and cover development costs if 4% of registered users bought it. The marketing manager therefore intends sending out questionnaires to a random sample of registered users. From past experience she thinks that about one-quarter of the users who express interest will buy the supplement, and the board of the company has decided to proceed with the project if at least 10% of users show enthusiasm.

Although this might not cover development costs the overall package would be improved and increased sales should justify the investment in the development. However, the board does need to be reasonably confident about the proportion (p) of interested users, and has stipulated that the lower end of the 90% confidence interval for the proportion should exceed 0.10. The decision to proceed would be borderline, if a confidence interval with a width of 0.06, for example, extended from 0.10 up to 0.16.

Let X be the number of people in a random sample of n who express interest in a graphics package. Then

$$X \sim \text{Bin}(n, p)$$

and to a good approximation, provided both np and $n(1 - p)$ exceed about 5,

$$X \sim N(np, np(1 - p))$$

Hence

$$X/n \sim N(p, p(1 - p)/n)$$

and an approximate $(1 - \alpha) \times 100\%$ confidence interval for p is given by

$$\hat{p} \pm z_{\alpha/2} \sqrt{\hat{p}(1 - \hat{p})/n}$$

where $\hat{p} = x/n$. The sample size is found by solving the equation

$$0.03 = 1.7\sqrt{(\hat{p}(1 - \hat{p})/n)}$$

with a value of 0.13 substituted for $\hat{p}$, and turns out to be 364. The manager had budgeted for 500 reply-paid questionnaires and decides to increase the sample size to 500, which is about 1 in 10 registered users. She takes a random number between 1 and 10, and sends out the questionnaire to the corresponding name and every tenth thereafter in an alphabetical list. This is not a simple random sample but it is an adequate approximation to one. There is no reason for any periodicity, which corresponds to registering enthusiasm about a graphics supplement, in the list. If the list happened to be geographically ordered, and this could possibly be linked to enthusiasm about buying a graphics supplement, the list sample would actually give a desirable spread. The marketing manager will interpret no response as no interest.

4.4.3   Comparison of means

Ferguson and Ketchen (1999) suggest that the failure of many studies to show a link between organisational configuration and performance has led prominent researchers to question the value of the concept of organisational configurations. They also point out that the lack of statistical evidence may be because sample sizes in many studies are too small to detect relationships between configurations and performance when they are present. After reviewing the recent literature and analysing 24 such studies they conclude that only half had an 80% or higher chance of detecting a medium effect, when testing the null hypothesis at the 10% level, only two out of 24 had a similar chance of detecting a small effect, and four out of 24 had less than an 80% chance of detecting a large effect. Bayesian hierarchical models (e.g. Section 2.5.5 and Ibrahim and Metcalfe, 1993) can be used for an overview study (meta-analysis), but this does not excuse individual studies having inadequate power. Small, medium and large effects are usually defined in standard units, and when this is done the power does not depend on the magnitude of the population standard deviations. To keep the algebra reasonably simple we assume random samples of the same size (n), from two populations with the same variance σ^2 but possibly different means μ_A and μ_B. We will ignore the nicety that we should use a t-distribution if we estimate σ from the samples. In our usual notation

$$\bar{X}_A - \bar{X}_B \sim N\left(\mu_A - \mu_B, \frac{\sigma^2}{n} + \frac{\sigma^2}{n}\right)$$

we will reject

$$H_0: \mu_A = \mu_B$$

in favour of

$$H_0: \mu_A < \mu_B$$

at the $\alpha \times 100\%$ level if

$$\bar{X}_A - \bar{X}_B < c = -z_\alpha \sqrt{\frac{\sigma^2}{n} + \frac{\sigma^2}{n}}$$

Define the effect as Δ, when the difference between μ_A and μ_B is $\Delta\sigma$. If we wish to have a $1 - \beta \times 100\%$ probability of rejecting H_0 when $\mu_A = \mu_B - \Delta\sigma$ we require that

$$c = -\Delta\sigma + z_\beta \sqrt{\frac{\sigma^2}{n} + \frac{\sigma^2}{n}}$$

Equating the two expressions for c, and making n the subject of the formula gives:

$$n = 2(z_\beta + z_\alpha)^2 / \Delta^2$$

So, for example, if we set α equal to 0.05, which corresponds to a two-sided alternative with $\alpha = 0.10$, and choose β equal to 0.20, we get the following table:

Effect	Δ	n	($z_{0.05} = 1.645, z_{0.20} = 0.8416$)
Small	0.2	309	
Medium	0.5	49	
Large	0.8	19	

Table 4.1   To what extent did the following represent important reasons for becoming BS5750 registered? (7-point scale: 7 = extremely important, 1 = not important at all). [Courtesy of Seddon *et al.*, 1993]

	Added significant value	Added some value	Neither added/ detracted	Detracted value
Number of respondents	176	278	158	31
Similar importance reported				
In order to gain market advantage	5.96	5.51	5.48	5.10
Customers demanding we achieve registration	4.88	5.01	4.98	4.87
More importance reported				
A drive to become more efficient	5.42	5.01	4.43	3.73
A drive to improve service to customers	5.69	5.20	4.76	4.10
A drive to improve the reliability of products	5.06	4.74	4.21	3.46
A drive to improve the reliability of services	5.37	4.80	4.46	3.76

It should be remembered that n is the sample size from each of the two populations, so the total sample size is $2n$.

Example 4.18
Seddon *et al.* (1993) sent a survey to a random sample of about 2000 organisations which had achieved registration to what was then BS 5750. The sampling frame was a list of BS 5750 registered companies maintained by BSI Consultants. Six hundred and forty-seven responses were received. At one point in the survey respondents were asked whether BS 5750 had added value to or detracted value from their business. They were also asked to answer several questions on a 7-point scale: 7 = extremely important, 1 = not important at all.

   Their results are quoted in Table 4.1. Standard deviations were not given but a standard deviation of about 1 for each question for each sub-group is probably a reasonable estimate (Exercise 4.6). The study is large enough to have a high probability of detecting any medium effects associated with the main sub-groupings, and any large effects associated with the 'detracted value' sub-group. The differences in the 'more importance reported' part of the table correspond roughly to medium or large effects. Seddon *et al.* (1993) concluded that people who felt they had success with the Standard took it on for broader reasons. But, a feeling of success with the Standard is not necessarily the same as a demonstration of tangible benefits. Seddon (1997) presents case studies based on visits to many of the companies in which people claimed success with the Standard, and argues that in most of these cases its implementation is not improving the quality of the services offered.

4.4.4 Matched pairs for comparing means

Muczyk (1978) describes a study to measure the impact of management by objectives (MBO) on organisational performance and staff motivation in banks. Ten economic indicators were used. Four were numbers of types of savings and loan accounts, and the remaining six were dollar values of accounts. The psychometric variables included role conflict, job involvement, perceived importance of skills, and job satisfaction. The study was conducted in a 41-branch bank in a suburb of Washington DC. Thirty-six of the branches were thought suitable to take part in the experiment. It was likely that variability between branches would swamp any changes brought about by adopting MBO. A valuable technique for providing a more precise measure of change is to pair branches so that they are as similar as possible. Within each pair, one branch is selected at random to implement MBO. Thirteen branches covering the range of branch sizes were identified. Comparable branches were then chosen on the basis of market area, volume of business, age and education of the manager, and the number of assistant managers. It is essential to randomise which branch in the pair is to use MBO, because the 13 branches first identified may have been selected for an impressive recent performance or a stated interest in applying MBO. It would also be advisable to randomise subject to a restriction that 6 or 7 of the original 13 branches implement MBO. If, for example, the first 13 branches had volunteered for MBO, and all 13 implemented it, any effect of the management technique would be indistinguishable from a volunteer effect. For somewhat similar reasons, the research team decided that each branch not implementing MBO would still be told that it was part of a study and be asked to complete the same forms and questionnaires as the other member of the pair. It is possible that being monitored, regardless of any accompanying management technique, affects performance and motivation. If both branches in the pairs were not treated the same in this respect, any apparent effect of MBO could equally be attributed to the effect of being monitored. Six branches were left outside the study, and in principle might provide a control group with which the monitored-only group could be compared. However, the sample sizes are too small for such an unpaired comparison to have much statistical power.

A decision about the sample size for a paired comparison experiment cannot be made without reference to the standard deviation of the differences between the matched pairs (σ_D). The aim of the matching is to make this standard deviation as small as possible. Suppose we have n pairs (X_{Ai}, X_{Bi}) and let D_i be the differences $X_{Ai} - X_{Bi}$. Then if we assume differences are normally distributed

$$\bar{D} \sim N(\mu_D, \sigma_D^2/n)$$

where μ_D is the mean of the population of all possible differences which equals the difference in the two population means $\mu_A - \mu_B$. A $(1 - \alpha) \times 100\%$ confidence interval for μ_D is given by

$$\bar{d} \pm t_{n-1,\alpha/2} s_d / \sqrt{n}$$

If we have some estimate of s_d, we can decide whether the width of this interval is acceptably narrow. In general an estimate of s_d can be based on results from any similar studies reported in the literature, a pilot study or perhaps a reasonably well-informed guess. If there is no substantial difference between the methods used for a pilot study and the main study it may be reasonable to combine them for the analysis. In the case of the bank, it is possible to make an estimate of s_d for the financial economic indicators, at least, from bank records. A typical measure of financial performance over the 12-month experimental period was to be the change in dollar value of savings accounts. Records of the changes over the past 12 months could be found for all branches. Differences in these changes between the paired branches can be calculated and the standard deviation of the differences gives an indication of a value for s_d. The differences will also give an indication of the success of the matching.

Suppose the mean and standard deviation of the 13 differences, in changes in savings accounts over the previous 12-month period, were 217 000 and 758 000 US dollars. Then a 95% confidence interval for the mean difference is

$$21\,700 \pm 2.179 \times 758\,000/\sqrt{13}$$

which gives

$$217\,000 \pm 458\,094$$

A similar width of confidence interval, about 1 million, should be expected for the difference attributable to MBO. This would be acceptable if a worthwhile business management technique is expected to make a difference of around a million dollars on this measure. Muczyk (1978) reported a 95% confidence interval of

$$185\,449 \pm 549\,109$$

This was based on 12 differences because the researcher was convinced that one branch selected for MBO was not adopting MBO. He therefore dropped it, and its paired branch, from the study. It could be argued that this branch is working within the MBO policy imposed by the organisation and should be left in the study. Perhaps the best solution is to present results with and without such cases.

4.4.5   Comparing proportions with independent samples

As part of a study of the effects of music in advertising Gorn (1982) divided 244 undergraduates on a first-year management course at McGill University into four groups. Group 1 listened to liked music during which a light blue pen was advertised, and Group 2 listened to disliked music during which a light blue pen was advertised. Groups 3 and 4 were treated in a similar way except that a beige pen was advertised. A pilot study had indicated that the students were indifferent about the pale blue or beige pen colour, liked the music from the film *Grease*, and disliked classical Indian music. At the end of the advertising presentation students were asked to take a free pen from either a box of light blue pens or a box of beige pens. They were also asked to express their reaction to the music on a scale from dislike very much (1) to like very much (5). The average score for the music from *Grease* in the pilot study was 4.3, whereas it was only 1.5 for the Indian music. Students were dropped from the analysis if they did not respond (1) or (2) to the Indian music and (4) or (5) to the music from *Grease*. Somewhat surprisingly, this left 195 students who were classified as below.

	Pen choice		Row total
	Advertised pen	Non-advertised pen	
Liked music	74	20	94
Disliked music	30	71	101
Column total	104	91	195

The standard analysis for the comparison of two proportions uses the binomial approximation. Let X_A out of n_A who listened to the liked music choose the advertised pen, and X_B out of n_B who listened to the disliked music choose the advertised pen. Let the proportions in the corresponding populations be p_A and p_B. Then, provided $n_A p_A$ and $n_A(1 - p_A)$ exceed about five, we have to a good approximation

$$X_A/n_A \sim N(p_A, p_A(1 - p_A)/n_A)$$

Similarly for X_B/n_B and since the two estimators are independent:

$$\frac{X_A}{n_A} - \frac{X_B}{n_B} \sim N\left(p_A - p_B, \frac{p_A(1 - p_A)}{n_A} + \frac{p_B(1 - p_B)}{n_B}\right)$$

If we write $\hat{p}_A$ and $\hat{p}_B$ for X_A/n_A and X_B/n_B respectively, we have the following approximate $(1 - \alpha) \times 100\%$ confidence interval for $p_A - p_B$:

$$(\hat{p}_A - \hat{p}_B) \pm z_{\alpha/2}\sqrt{(\hat{p}_A(1 - \hat{p}_A)/n_A + \hat{p}_B(1 - \hat{p}_B)/n_B}$$

More formally, if we wish to detect a difference in proportions Δp with a probability $1 - \beta$ when testing a null hypothesis

$$H_0: p_A = p_B$$

against a one-sided alternative

$$H_1: p_A < p_B$$

at the $\alpha \times 100\%$ level

$$\Delta p = (z_\beta + z_\alpha)\sqrt{\frac{p_A(1 - p_A)}{n_A} + \frac{p_B(1 - p_B)}{n_B}}$$

Use of this formula requires some plausible values for p_A and p_B. It is usual to choose $n_A = n_B$, in which case we can just write n for both. Although the maximum value of n for a given Δp corresponds to $p_A = p_B = 0.5$, a given Δp will usually have least practical significance when p_A and p_B are near 0.5. In the advertising case Gorn could have taken $n_A = n_B$, $p_A = p_B = 0.5$ and required an 80% chance of detecting a difference of 0.2. Then when testing H_0 at the 2.5% level with a one-sided alternative hypothesis,

$$n = (0.842 + 1.96)^2 \times (2 \times 0.5 \times 0.5)/(0.2)^2 = 98$$

The results of his experiment lead to a 95% confidence interval for $p_A - p_B$ as [0.365, 0.609] and there is strong evidence that the liked music has had an effect. The main limitation of the experiment is that the corresponding population is somewhat nebulous. Even if McGill first-year management undergraduates were fairly typical of students in the early eighties any inference to the general adult population is rather subjective. Pitt and Abratt (1988) carried out a similar experiment with an Australian undergraduate marketing class of 172 students. The liked music was now Stevie Wonder's 'I just called to say I love you', and the disliked music was an aria from Wagner's *Lohengrin*. A substantial difference in the studies was that the product, condoms in a red or blue pack, was an example of what advertisers then referred to as unmentionable products. Whether this accounts for the fact that their results were markedly different from Gorn's is debatable but it is a distinct possibility.

	Product package choice		
	Advertised package	Non-advertised package	Row total
Liked music	40	34	74
Disliked music	30	42	72
Column total	70	76	146

A 95% confidence interval for the difference in proportions preferring the advertised package in the corresponding population is [−0.04, 0.28].

4.4.6   McNemar's test

McNemar's test is a paired comparison procedure for proportions. Water companies in England and Wales are expected to demonstrate that any schemes undertaken to improve a discoloured water supply have a noticeable effect on customers. There were several complaints about discoloured water in a particular area and a water company undertook remedial work. Before starting the work the company sent a brief questionnaire to a random sample of 80 addresses in the area that was affected. The main question was whether the customer had noticed discoloured water within the past six months. Six months after the work was completed the same questionnaire was sent to the same addresses. Sixty-two questionnaires were returned on both occasions, and the results are summarised below.

		After	
		Discolour	No discolour
Before	Discolour	2	12
	No discolour	4	44

McNemar's test focuses on the respondents who change their minds, 16 in this case. Let X be the number who change from claiming discoloured water to reporting no discoloured water. If the work has had no effect, then

$X \sim \text{Bin}(16, \frac{1}{2})$ and

$$\Pr(X \geq 12 | X \sim \text{Bin}(16, \tfrac{1}{2})) = 0.038$$

There is evidence that the work has been successful ($P = 0.04$ with a one-sided alternative). The sample size must be large enough to give at least 5 people who claim discoloured water before the work starts. If all of these reported no discolouration after the work was completed, and if none of the people who reported no discolouration before the work changed his or her assessment, the company would just have evidence of an improvement ($P = 0.03$ with a one-sided alternative). Aiming for over 10 people claiming discoloured water before the work starts would be much safer.

4.4.7 Unknown standard deviations

Some decisions about sample sizes do rely on estimates of variability in populations. Although standard deviations may not be known precisely, it is usually possible to make some reasonable assessment from general knowledge of the variable, published results from similar studies, or a pilot study. If there is no substantial change in a questionnaire it may be reasonable to include the results from a pilot study with the main study. If a t-distribution will be used with fairly small sample sizes, it is straightforward to iterate for the estimated sample size. However, the precise width of the confidence interval also depends on the sample value of s which is unknown.

4.4.8 Finite population correction

If the population size is large compared with the sample size, the precision of the sample mean depends almost only on the sample size. That is, if the population has a mean μ and variation σ^2, then the sample mean $\bar{X}$ has mean μ and variance σ^2/n. The precision of estimates of mean household expenditure made in the Family Expenditure Survey depends on the sample size, but hardly at all on the proportion of the population sampled which is less than 1 in 1000 households. Sometimes the population is hypothetical and infinite, as for the CAA survey of weights and packages stored in overhead bins on aeroplanes. The population is all possible packages if present passenger behaviour continues. In contrast to these cases, there is sometimes a clearly defined finite population $x_1, \ldots, x_N$, and the main purpose of the survey may be to estimate the population total

$$T = \sum_{j=1}^{N} x_j$$

rather than the mean, although they are directly related. The population mean μ and variance σ^2 are defined by

$$\mu = \sum_{j=1}^{N} x_j / N$$

$$\sigma^2 = \sum_{j=1}^{N} (x_j - \mu)^2 / N$$

Suppose we take a random sample of size n without replacement $\{X_i\}$. Define the sample mean and variance by

$$\bar{X} = \sum_{i=1}^{n} X_i/n$$

$$S^2 = \sum_{i=1}^{n} (X_i - \bar{X})^2/(n-1)$$

Then $\bar{X}$ is an unbiased estimator of μ

$$E[\bar{X}] = \mu$$

and to a very close approximation (Barnett, 1991)

$$\mathrm{var}[\bar{X}] = \frac{\sigma^2}{n}\left(1 - \frac{n}{N}\right)$$

Notice that if n/N is negligible we revert to σ^2/n, and if n/N equals 1 we have sampled the whole population and the variance of the estimator is 0. An unbiased estimator of the population total T is

$$\hat{T} = N\bar{X}$$

and its variance is given by

$$\mathrm{var}(\hat{T}) = N^2\sigma^2(1 - n/N)/n$$

The factor of $(1 - n/N)$ is known as the finite population correction (FPC). It can be applied when estimating the variance of other statistics if an appreciable proportion of the population is sampled. If n/N is 0.1, then the standard deviation of estimators will be reduced by about 5%. When calculating confidence intervals σ^2 can be replaced by s^2 and a t-distribution used.

Example 4.19
The following case arose during the development of computer software for a quarrying company. Rocks are transported from the quarry to a wharf, by lorries. At the wharf, the rocks are stacked in distinct piles within bunkers according to their masses. The bunker mass ranges are from 1–3, 3–5, ..., up to 11–13 tonnes. The cranes on the lorries are fitted with load cells, and the quarrying company knows the masses of all the rocks in each pile although it does not identify individual rocks.

Rocks are taken off by small ore-carrying ships. The boatswains try to load complete piles but often finish with only part of a pile, either because the ship is fully loaded or because they wish to catch the tide. The cranes on the ships are not equipped with load cells. It is reasonable to assume that rocks are taken in a random order from a pile.

The quarrying company wanted an assessment of likely errors if it charged on a pro-rata basis for the last pile. In general a pile consists of N rocks which have a mean mass μ and a standard deviation of mass σ, and n of these rocks are loaded onto a ship. Let $\bar{X}$ be the mean mass of the n rocks and let W be their total mass. If rocks are taken at random

$$\bar{X} \sim N(\mu, \sigma^2(1 - n/N)/n)$$

where the normal approximation will be reasonable unless n is 1 or $N - 1$. Since

$$W = n\bar{X}$$

$$W \sim N(n\mu, n\sigma^2(1 - n/N))$$

and there is a 95% probability that W is within:

$$n\mu \pm 1.96\sigma(n - n^2/N)^{1/2}$$

For example, the mean and standard deviation of a pile of 40 rocks had been recorded as 6.12 and 0.56 tonnes respectively. A boatswain has loaded 30 of these. There is a 95% chance that the total mass is within

$$30 \times 6.12 \pm 1.96 \times 0.56 \times (30 - 30 \times 30/40)^{1/2}$$

$$183.6 \pm 3.0$$

In percentage terms the error in total mass for the last pile is unlikely to be more than 2% and will be negligible in terms of the total mass of the cargo.

4.5   Asset management plan

4.5.1   Introduction

In 1989 the water authorities in England and Wales were taken out of public ownership and sold on the London stock market under the terms of the Water Act. A Director-General of Water Services (DG) was appointed to regulate the industry. Before the flotation each water authority was required to prepare an asset management plan (AMP) which provided an estimate of the expenditure required over a 20-year period to improve and maintain assets so that specified standards of performance and levels of service could be achieved. The proposed new water companies would be allowed to use their AMPs to justify increasing charges by more than the retail price index. AMPs were also a move away from current cost accounting, which relies on notional lives for assets, towards renewals accounting which assumes assets will be maintained indefinitely. Renewals accounting is far more realistic for a water supply network. Relatively short sections of pipe will be cleaned, relined or replaced as necessary and reservoirs and treatment works are refurbished rather than rebuilt. The AMPs would also be used as business plans. If the AMPs were to be effective it was essential that they would be perceived as fair and that they would give an indication of the precision of the cost estimate. Therefore, the water authorities were required to involve independent engineering consultants during the preparation of AMPs, so that the consultants could cross-certify the final submission.

The assets of a typical water company include reservoirs, water treatment works, pumping stations, thousands of kilometres of water mains, and office buildings. A complete survey of all these would be impractical and unnecessary. Instead samples were investigated. A Bayesian approach was described in Chapter 2. However, most water authorities took a more formal approach before the 1989 flotation. The considerable knowledge about the water supply system was used for stratification. The following is loosely based on work I did for the former Northumbrian Water Authority and was a result of helpful discussions with staff there and the independent consultants. It also incorporates some ideas that arose from more recent meetings with engineers at Northumbrian Water Ltd and Yorkshire Water Ltd.

The water supply network was split into the raw water system, dams, water treatment works, the strategic supply network and the local distribution network. I shall describe the method for estimating the total cost for the local distribution network (LDN) together with an approximate 90% prediction interval. The local distribution network was divided into 146 zones which were defined so that they were approximately self-contained, inasmuch as work in one would not significantly affect its neighbours. Zones are typically spurs off the trunk main which is part of the strategic supply network. A sample of zones was surveyed in detail and schemes that would need to be carried out within the following 20-year period were identified and costed. There are many potential sources of error. Apart from the uncertainty over the specification of individual schemes, which will at least tend to average out, there is the possibility of a tendency to systematically overestimate or underestimate the work needed to complete schemes. This would be a source of bias and would persist, however many schemes are assessed. Similar considerations apply to the costings of the schemes. Unit costs will vary for different schemes but such errors will tend to average out. In contrast, errors in the average unit costs used in the exercise will bias the result. Sampling error is relatively unimportant and is straight-forward to estimate if a random sampling scheme is used.

4.5.2   Stratified sampling

Two statistical devices were used to increase the precision of the estimate of total cost for a given sample size. Stratification was carried out by dividing the population of zones into sub-populations called strata, so that zones within a stratum would be relatively similar. The other device was to allow for the fact that within strata the costs for zones would be likely to be in proportion to their sizes. After some discussion we decided that the most practical measure of zone size was the number of properties and that within-strata zone costs would be assumed proportional to zone size. The cost per property is likely to be higher in rural areas, because there is more mains pipe per property, so some stratification on an urbanisation scale was needed. Five categories were defined: rural, mixed but mainly rural, mixed but mainly urban, suburban, and inner city. The urbanisation categorisation was crossed with a prior assessment of zones as having a high or low level of problems per property. This was an effective way of taking account of prior knowledge. The use of subjective judgement for constructing strata cannot lead to bias, because every zone has a known probability of selection and is weighted accordingly. The cross-classification led to 10 strata but two were combined to give the eight strata shown in Table 4.2. There was not enough time for a formal pilot study, but several surveys of small areas of the system were available, and a sample size of 33 zones was thought adequate. If zones were of equal size, the optimum allocation of the sample zones over strata would be to make the number taken from each

Table 4.2 The eight strata and allocation of sample over stratified zones

	Fewer prior identified problems/property	More prior identified problems/property
Rural		2 zones of sizes 1184 and 1400 properties. Both in sample
Mixed (mainly rural)	13 zones in stratum. Smallest 179 properties, largest 1886 properties. Sample of 3 drawn	12 zones in stratum. Smallest 212 properties, largest 3539 properties. Sample of 3 drawn
Mixed (mainly urban)	28 zones in stratum. Smallest 370 properties, largest 14 627 properties. Sample of 5 drawn	
Suburban	51 zones in stratum. Smallest 104 properties, largest 14 404 properties. Sample of 9 drawn	19 zones in stratum. Smallest 908 properties, largest 12 900 properties. Sample of 5 drawn
Inner city	12 zones in stratum. Smallest 727 properties, largest 3075 properties. Sample of 3 drawn	9 zones in stratum. Smallest 1740 properties, largest 10 000 properties. Sample of 3 drawn

stratum: proportional to the number of zones within the stratum, proportional to the standard deviation within the stratum, and inversely proportional to the square root of the expense of surveying a zone within the stratum.

Optimum allocation

A population consists of N equal sized zones. There is a cost y_i associated with each zone and the objective is to estimate the total y_{TOT} cost for all zones. The zones are divided into k strata of sizes $N_1, \ldots, N_k$. A sample of size n is to be distributed over the strata: n_1 from stratum $1, \ldots, n_k$ from stratum k. Suppose the standard deviation of zones within stratum i is σ_i and the expense of surveying the zone is e_i. If the sample means from each stratum are $\bar{Y}_i$

$$y_{TOT} = \sum_{i=1}^{k} \bar{Y}_i N_i$$

$$\text{var}(y_{TOT}) = \sum_{i=1}^{k} (\sigma_i^2 / n_i) N_i^2$$

Optimum allocation for a given survey expenditure (E) will minimise $\text{var}(y_{TOT})$ subject to the constraint

$$\sum_{i=1}^{k} e_i n_i = E$$

The method of Lagrange multipliers is the most convenient way of doing this. Let

$$\psi = \sum_{i=1}^{k} (\sigma_i^2 / n_i) N_i^2 + \lambda \left(\sum_{i=1}^{k} e_i n_i - E \right)$$

where λ is the Lagrange multiplier. Now differentiate partially with respect to the n_i and set the derivatives equal to 0. Then

$$\frac{-\sigma_i^2 N_i^2}{n_i^2} + \lambda e_i = 0$$

which can be rearranged to give

$$n_i \sqrt{\lambda} = \sigma_i N_i / \sqrt{e_i}$$

This is the essential result, but it can be made neater by substituting for $\sqrt{\lambda}$. Multiply both sides by e_i and add over strata to obtain

$$\sum e_i n_i \sqrt{\lambda} = \sum \sigma_i N_i \sqrt{e_i}$$
$$\sqrt{\lambda} = \sum \sigma_i N_i \sqrt{e} / E$$

Hence

$$n_i = \frac{E \sigma_i N_i / \sqrt{e_i}}{\sum_{i=1}^{k} \sigma_i N_i \sqrt{e_i}}$$

This could only be used as a very general guide because standard zones were not of equal size and standard deviations and costs could only be estimated roughly. The allocation is shown in Table 4.2. At least three zones were taken from each stratum, except for that with only two zones, so that a reasonable estimate of the sample variance could be made. The allocation for the larger strata was based on a proportional allocation with more emphasis placed on the suburban stratum with the higher assessment of problems per property, because it was expected to have a larger variance. The 33 sample zones were surveyed, and engineering schemes were identified and costed using a unit cost database which was developed as a separate investigation. The sum of the scheme costs within a zone gives a zone cost. For the moment we will ignore the uncertainty about the zone cost ascertained from the survey.

4.5.3   Ratio estimator

Suppose we have a population, or stratum, of N zones with known sizes (x_j) and unknown costs (y_j). The objective is to estimate the total cost for the population

$$y_{TOT} = \sum_{j=1}^{N} y_j$$

The ratio of cost to size in the population (r) will be defined as

$$r = \sum_{j=1}^{N} y_j \bigg/ \sum_{j=1}^{N} x_j$$

We now draw a random sample of n zones and ascertain the costs (Y_i). Denote the sample as

$$(X_i, Y_i) \quad \text{for } i = 1, \dots, n$$

where the X_i are the sizes of the sample zones. If costs are even roughly proportional to size it will be much more efficient to estimate y_{TOT} by estimating r and multiplying this estimate by $\sum_{j=1}^{N} x_j$ than by multiplying $\bar{Y}$ by N (Exercise 4.7). The sample estimate of r is:

$$\hat{r} = \frac{\sum Y_i}{\sum X_i} \tag{4.1}$$

where the summations are over the sample, i.e. for $i = 1, \ldots, n$. The error is

$$(\hat{r} - r) = \frac{\sum Y_i - r \sum X_i}{\sum X_i} = \frac{\sum (Y_i - rX_i)}{n\bar{X}}$$

We now make some simplifying assumptions. The $(Y_i - rX_i)$ are independent and the variance of $\bar{X}$ is small compared with the variance of $(Y_i - rX_i)$ for large samples. So, treating $\bar{X}$ as a constant gives

$$\text{var}(\hat{r} - r) = \frac{\sum \text{var}(Y_i - rX_i)}{n^2 \bar{X}^2}$$

Now since r is a constant the variance of $(\hat{r} - r)$ is the same as the variance of r. Also, since the expected value of $(Y_i - rX_i)$ is 0:

$$\text{var}(Y_i - rX_i) = E[(Y_i - rX_i)^2]$$

The sample estimate of

$$\sum E[(Y_i - rX_i)^2] \quad \text{is} \quad \sum (Y_i - \hat{r}X_i)^2$$

and an approximate estimator of the variance of $\hat{r}$ is

$$\widehat{\text{var}}(\hat{r}) = \frac{\sum (Y_i - \hat{r}X_i)^2}{n^2 \bar{X}^2} \tag{4.2}$$

and the estimated standard deviation of $\hat{r}$ is the square root of this.

Example 4.20
A random sample of three zones from a stratum of 12 has been costed. The sizes (i.e. number of properties) for all the zones, and the costs for the surveyed zones, monetary units (mu), are given in Table 4.3. The formulae give

$$\hat{r} = 643 \, \text{mu/prop}$$

with an estimated standard deviation of 104 mu/prop. There are 21 320 properties in the zone so the estimate of the total cost for the stratum is 13 708 760 mu and the estimated standard deviation of this estimate is 2 217 280 before any finite population correction is applied. The FPC would reduce this by a factor of $\sqrt{(1 - 3/12)}$ to give 1 920 221, though this is rather approximate as it takes no account of the varying zone sizes.

Other strata are costed in a similar way and the results are summarised in Table 4.4. The total cost for the LDN is 204 842 153 mu. The sampling variances are added to give a variance for this total cost, and hence the standard deviation

Table 4.3 Zone sizes (number of properties) for a stratum and costs (monetary units) for a random sample of 3

Zone size	Cost of surveyed zones
212	
629	675 524
712	
858	351 308
1068	
1541	
2011	
2316	
2582	1 588 297
2714	
3138	
3539	

Table 4.4 Estimated average costs per property, strata costs, sampling standard deviations and variances of strata costs. R is rural, ..., IC is inner city, and FP, MP are fewer or more prior identified problems/property

Stratum	No. properties	Cost/property	Cost	Stdev	Variance
$R + MR/FP$	10 738	581	6 238 778	0	0.00000E + 00
R/MP	2584	2105	5 439 320	1 345 912	1.81148E + 12
MR/MP	21 320	643	13 708 760	1 920 221	3.68725E + 12
MU	189 958	537	102 007 446	1 539 775	2.37091E + 12
S/FP	402 109	97	39 004 573	689 431	4.75315E + 11
S/MP	131 176	204	26 759 904	1 379 031	1.90173E + 12
IC/FP	22 124	113	2 500 012	1 766 410	3.12020E + 12
IC/MP	47 830	192	9 183 360	2 375 340	5.64224E + 12
Total	827 839		204 842 153		1.90091E + 13

of the total cost is calculated to be 4 359 945 mu, 2.13% of the total cost. However, we have not yet allowed for possible engineering bias when estimating the work needed to complete schemes, or for any bias in the average unit costs. Neither have we allowed for variability of work for individual schemes and unit costs about their expected values, but this is of less practical importance because it will reduce in percentage terms as more schemes are added. We can combine the work needed and unit cost sources of error if we apply a calibration factor, based on out-turn to predicted costs for past schemes, to our initial estimates of unit costs. This will not enable us to separate engineering bias from unit cost errors, but there is no need to do so for the AMP.

4.5.4 Calibration and its uncertainty

Calibration factor

We calculated a calibration factor by taking a random sample of recently completed schemes and comparing out-turn costs with costs estimated using

AMP methods and the interim unit cost database. We defined ratios (r_i) as

$$r_i = \frac{\text{out-turn cost}}{\text{retrospective predicted cost using interim database and AMP methods}}$$

$$= \frac{Y_i}{X_i}$$

estimated the ratio in the population (r) by the ratio estimator $\hat{r}$ (equation (4.1)), and calculated the estimated standard deviation of $\hat{r}$ (equation (4.2)). We also calculated the standard deviation of ratios for individual schemes about the average, as

$$\text{sd}(r_i) = \sqrt{\frac{\sum(r_i - \hat{r})^2}{n - 1}}$$

The interim unit cost database was constructed from known costs of materials and typical hourly rates for people working in the industry. Costs were brought to a common time point using prices indices for the industry. Engineers costed the schemes using only the information that would be available if they had been identified during an AMP survey and without knowing the actual out-turn cost.

Example 4.21
An interim unit cost database has been set up for work in water quality zones, and will be calibrated from data for eight recently completed schemes given in Table 4.5. Calculations using equations (4.1) and (4.2) give

$$\hat{r} = 1.29$$

$$\widehat{\text{sd}}(\hat{r}) = 0.12$$

$$\widehat{\text{sd}}(r_i) = 0.34$$

As a rough check, $\widehat{\text{sd}}(\hat{r})$ should be approximately equal to $\widehat{\text{sd}}(r_i)/\sqrt{n}$ because it is a weighted average of n ratios. The interim unit cost database was calibrated

Table 4.5   Predicted and out-turn costs (monetary units) for eight historic water distribution schemes

Predicted cost	Out-turn cost	Ratio
81	94	1.16
20	18	0.90
98	107	1.09
79	75	0.95
69	52	0.75
144	216	1.50
119	176	1.48
124	207	1.67

by multiplying all unit costs by 1.29. It is more convenient to express $\widehat{sd}(\hat{r})$ and $\widehat{sd}(r_i)$ in percentage terms:

$$\widehat{sd}_\%(\hat{r}) = \widehat{sd}(\hat{r}) \times 100/\hat{r} = 9.3$$

$$\widehat{sd}_\%(r_i) = \widehat{sd}(r_i) \times 100/\hat{r} = 26$$

Allowing for uncertainty in calibration

The total cost for the LDN is 204 842 153. The sampling variance is 1.9×10^{13}, and taking square root gives the sampling standard deviation as 4 359 945. If this is expressed as a percentage of the total cost, the sampling standard deviation is 2.13%. Since all the work has been costed using the same unit cost database any errors in it will persist, and the estimated standard deviation of the calibration factor is 9.3%. The standard deviation of individual schemes is 26% but this will reduce as more schemes are added. If all the schemes had the same estimated cost this component of uncertainty would reduce in proportion to the reciprocal of the square root of the number of schemes added. The variation in scheme costs will make the reduction less dramatic (Exercise 4.8). We assume variation of individual schemes about their expected values is independent, and calculate

$$\frac{\left[\sum_{\text{all schemes}} (\text{scheme cost} \times 26/100)^2 \right]^{1/2}}{\sum_{\text{all schemes}} \text{scheme cost}} \times 100\%$$

Suppose this equals 2.7%. It is reasonable to suppose that sampling errors are independent of error in the calibration factor, and that variation of individual schemes about the average is independent of both other sources of uncertainty. Then the overall uncertainty in the total is obtained as the square root of the sum of the variances.

Component of uncertainty	Standard deviation (%)	Variance (%^{2})
Sampling	2.13	4.54
Calibration factor	9.30	86.49
Individual schemes about their expected values	2.70	7.29
		98.32

The overall uncertainty, expressed as a standard deviation (%) is 9.9%. It follows that an approximate 90% limit of prediction for the total cost is

$$205 \pm 34 \text{ million mu}$$

where the 34 is $1.645 \times 9.9 \times 205/100$. Increasing the sample size would reduce the sampling error but have a negligible effect on the overall error.

4.6 Analysis of contingency tables

Surveys usually include questions for which respondents are asked to select one from two or more categories; e.g. 'Does the company use statistically designed experiments?' with 'Yes' or 'No' as possible responses; 'How often do you visit a supermarket?' with a choice of response from 'More than once a week', 'About once a week', 'About once a month', and 'Rarely or never'. The scale does not have to be ordinal; e.g. 'What is your preferred colour for a metallic paint finish on a car?' with 'Silver', 'Bronze', 'Green' and 'Blue' as possible responses. Respondents can themselves be classified, by age and by sex for example. An m-way contingency table is a classification of a sample of n objects by m categorical variables. The entries in the cells are counts and they add to give the sample size n. A two-way table is described as $r \times c$ if the row and column variables have r and c categories respectively. The usual null hypothesis to be tested is that the variables of classification are independent.

4.6.1 Analysis of $r \times c$ contingency tables

Prabhu (1999) classified a sample of 298 manufacturing companies in the north-east of England by 'world-class status'. He defined four main categories: 'potential winners' (w) if they demonstrated high levels of performance and practice; 'promising' (p) if they have adopted good practice but have yet to realise the improvements in performance; 'vulnerable' (v) if they are achieving high performance levels without the support of good practices; and 'could do better' (c) if they show relatively low practice and performance levels. In his paper Prabhu includes a contingency table (Table 4.6) of companies classified by world-class status and number of employees in categories: 0–49, 50–149, 150–249, 250 and more. We will test a null hypothesis

H_0: world-class status independent of size

against an alternative H_1 that the two classifications are related. Let p_{ij} be the probability that a randomly selected company is in size category i and status category j, $p_{i.}$ be the probability that a randomly selected company is in size category i, and $p_{.j}$ be the probability that a randomly selected company is in

Table 4.6 Company size by world-class status cross-tabulation

Company size		WC and PW	Vuln	Prom	Rfl and CdB	Total
				World-class status		
	0 to 49	35	41	3	26	105
		31.50%	55.40%	17.60%	27.40%	35.20%
	50 to 149	23	15	6	47	91
		20.70%	20.30%	35.30%	49.40%	30.50%
No. of	150 to 249	31	14	6	15	66
employees		27.10%	18.90%	35.30%	15.80%	22.20%
	250 plus	23	4	2	7	36
		20.70%	5.40%	11.80%	7.40%	12.10%
	Total	112	74	17	95	298
		100%	100%	100%	100%	100%

status category j. Then H_0 can be expressed as

$$H_0: p_{ij} = p_{i \cdot} p_{\cdot j}$$

The observed count in the ijth cell, n_{ij}, has a binomial distribution

$$n_{ij} \sim \text{Bin}(n, p_{ij})$$

and if np_{ij} exceeds about 5 we have to a close approximation

$$\frac{n_{ij} - np_{ij}}{\sqrt{np_{ij}(1 - p_{ij})}} \sim N(0, 1)$$

If H_0 is true and p_{ij} is relatively small when compared with 1,

$$\frac{n_{ij} - np_{i \cdot} p_{\cdot j}}{\sqrt{np_{i \cdot} p_{\cdot j}}} \sim N(0, 1)$$

Let the sum of the counts in row i and column j of the table be R_i and C_j respectively. Estimates of $p_{i \cdot}$ and $p_{\cdot j}$ are

$$\hat{p}_{i \cdot} = R_i/n \quad \text{and} \quad \hat{p}_{\cdot j} = C_j/n$$

and the expected value of n_{ij}, when H_0 is true, can be estimated by

$$e_{ij} = n\hat{p}_{i \cdot} \hat{p}_{\cdot j} = R_i C_j/n$$

We also have the approximate distributional results:

$$\frac{n_{ij} - e_{ij}}{\sqrt{e_{ij}}} \sim N(0, 1) \quad \text{and therefore} \quad \frac{(n_{ij} - e_{ij})^2}{e_{ij}} \sim \chi_1^2$$

A chi-square goodness-of-fit statistic, W, is defined as

$$W = \sum_{j=1}^{c} \sum_{i=1}^{r} \frac{(n_{ij} - e_{ij})^2}{e_{ij}}$$

Since the sum of independent chi-square variables has a chi-square distribution with degrees of freedom equal to the sum of the degrees of freedom of the summands, it is plausible that W might have an approximate χ_n^2 distribution when H_0 is true. However, the n_{ij} are not exactly independent because they are constrained to add to n. We lose one degree of freedom for each of the parameters estimated from the data, i.e. $r - 1$ of the p_i, since they must sum to 1, and $c - 1$ of the p_j. In terms of constraints, if we specify $r + c - 1$ row and column totals then n and the remaining row or column total are determined. Kendall and Stuart (1980), for example, give a proof that if H_0 is true then

$$W \sim \chi_{n-r-c+1}^2$$

is a good approximation to the multinomial distribution of W, provided the expected values exceed about 5. If H_0 is not true we anticipate large differences between observed and expected values, and hence large values of W. Therefore, we reject H_0 at the $\alpha \times 100\%$ level if W exceeds $\chi_{n-r-c+1,\alpha}^2$. If there is evidence against H_0, the magnitudes and signs of contributions to W will indicate the significant differences. A Minitab analysis of Table 4.6 is shown in Table 4.7. There is strong evidence that there is an association between company size and world-class status. Two cells have expected counts less than 5.0, but as both of these

Table 4.7 Chi-square test for company size by world-class status cross-tabulation

	W	V	P	C	Total
1	35	41	3	26	105
	39.46	26.07	5.99	33.47	
2	23	15	6	47	91
	34.20	22.60	5.19	29.01	
3	31	14	6	15	66
	24.81	16.39	3.77	21.04	
4	23	4	2	7	36
	13.53	8.94	2.05	11.48	
Total	112	74	17	95	298

Chi-Sq $= 0.505 + 8.545 + 1.492 + 1.668 + 3.669 + 2.554 + 0.126 + 11.156 + 1.547 + 0.348$
$+ 1.327 + 1.734 + 6.628 + 2.729 + 0.001 + 1.746 = 45.776$

DF $= 9$, p-value $= 0.000$.
Two cells with expected counts of less than 5.0.
Expected counts are printed below observed counts.

expected values exceed 2.0 and the corresponding contributions to W are small the approximation is unlikely to be misleading. If the upper two categories for company size are combined we obtain a calculated value of 42.45 for W. The conclusion is unaltered if we compare this with a χ_6^2 distribution. The main findings are that the smallest companies are more likely to be 'vulnerable', the largest companies are more likely to be 'potential winners', and that a higher than expected number of companies with between 50 and 149 employees are in the 'could do better' category.

4.6.2 Comparison of proportions

In a 2×2 contingency table the hypothesis that the two variables of classification are independent is identical to the hypothesis that two proportions are equal, although the data are usually collected differently. For a contingency table a random sample of size n, from the bivariate population, is classified by the two variables, and the marginal totals are not known in advance. In contrast to this, if the original aim is to compare two proportions we would probably take random samples from the two populations. The statistical test based on the chi-squared goodness-of-fit statistic is identical to the test for the equality of two proportions which uses normal approximations without continuity corrections (Exercise 4.4). Little (1989) considers other possible analyses.

Example 4.22
One of the many possible two-way tables that can be drawn up from the results of the survey of Malaysian manufacturing companies is:

		Carry out market research	
		No	Yes
Carry out market research	No	55	19
	Yes	26	21

There were 12 missing values. The calculated value of W is 4.692, and this is significant at the 3% level with a two-sided alternative. There is evidence that companies with ISO 9000 are more likely to carry out market research. However, an association does not necessarily imply that having ISO 9000 registration makes companies more inclined to carry out market research. Another possible explanation is that Malaysian companies which are keen to increase their market share carry out market research and aim to become registered to ISO 9000. Another suggestion is that non-resource-based companies are more likely to carry out market research and more likely to become registered to ISO 9000. This can be investigated by looking at the table of resource or non-resource by ISO registration:

	Registered to ISO 9000	
	No	Yes
Resource based	46	19
Non-resource based	44	24

and the table of resource or non-resource by market research:

	Carry out market research	
	No	Yes
Resource based	35	25
Non-resource based	39	22

with 12 missing values. There is no support for the last suggestion. If there had been, the methods of Section 4.6.3 could be used.

Example 4.23
Ghosh and Taylor (1999) interviewed samples of managers from New Zealand and Singapore who had switched advertising agencies. They compared their reasons for the account switches. The proportions responding that the switch was due to a deteriorating standard of creative work are given below.

		NZ	Singapore
Deteriorating standard of creative work given as an	Yes	28	12
important reason for switch	No	13	9

This is naturally considered as a comparison of two proportions. Let p_{NZ} and p_{SI} be the proportions who would consider a deteriorating standard of creative work as an important reason for switching agencies. The null hypothesis is

$$H_0: p_{NZ} = p_{SI}$$

If H_0 is true then, to a reasonable approximation,

$$\frac{\hat{p}_{NZ} - \hat{p}_{SI}}{\sqrt{\hat{p}(1-\hat{p})\left(\dfrac{1}{n_{NZ}} + \dfrac{1}{n_{SI}}\right)}} \sim N(0, 1)$$

where

$$\hat{p} = (n_{NZ}\,\hat{p}_{NZ} + n_{SI}\,\hat{p}_{SI})/(n_{NZ} + n_{SI})$$

In this case the calculated value of the statistic is 0.87 and there is no evidence against the hypothesis. With such small samples the between-country comparison is not very powerful. A 95% confidence interval for the difference in proportions is more informative because it emphasises the lack of precision in the estimates:

$$(28/41 - 12/21) \pm 1.96 \times \sqrt{(28/41)(13/41)/41 + (12/21)(9/21)/21}$$

$$0.11 \pm 0.26$$

The next example illustrates how the analysis should be modified if proportions are calculated from a sample in which different members have had different probabilities of selection.

Example 4.24
In order to elicit nurses' opinions about their professional relationships, Reedy *et al.* (1980) sought their response to the statement 'General practitioners always treat nurses as colleagues'. The weighted responses expressed as percentages were:

	Attached nurses (81)	Employed nurses (72)
Agree	51%	74%
Neither	18%	9%
Disagree	31%	17%

We will test whether the difference in percentages agreeing is statistically significant by constructing an approximate 95% confidence interval for the differences in proportions in the corresponding populations. Suppose there are n_1, n_2 and n_3 nurses employed in sample practices with 1, 2 and 3 employed nurses respectively. Let X_i be the response of the ith nurse where 1 is agree and 0 is either disagree or neither agree nor disagree, and let p be the proportion of all employed nurses who agree with the statement. We will assume that p is the same for employed nurses regardless of the number of other nurses in the practice. Let $\hat{p}$ be the sample estimate of p.

$$\hat{p} = \frac{\sum_{i=1}^{n_1} X_i + 2\sum_{j=1}^{n_2} X_j + 3\sum_{k=1}^{n_3} X_k}{n_1 + 2n_2 + 3n_3}$$

Since

$$\mathrm{var}(X_i) = \mathrm{var}(X_j) = \mathrm{var}(X_k) = p(1-p)$$

$$\mathrm{var}(\hat{p}) = \frac{n_1\,p(1-p) + 4n_2\,p(1-p) + 9n_3\,p(1-p)}{(n_1 + 2n_2 + 3n_3)^2}$$

In this study the values of n_1, n_2 and n_3 were 56, 13 and 3 respectively. Hence $\mathrm{var}(\hat{p})$ can be estimated from the above formula if p is replaced by $\hat{p}$ equal to 0.74, which gives 0.003 137. If we now write p_E for p to distinguish the employed nurses from the attached nurses we can summarise the results so far as

$$\hat{p}_E = 0.74 \quad \widehat{\text{var}}(\hat{p}_E) = 0.003\,137 \quad \widehat{\text{sd}}(\hat{p}_E) = 0.056$$

Similar calculations for the attached nurses give

$$\hat{p}_A = 0.51 \quad \widehat{\text{var}}(\hat{p}_A) = 0.002\,773 \quad \widehat{\text{sd}}(\hat{p}_A) = 0.053$$

An approximate 95% confidence interval for the difference in proportions $p_E - p_A$ is

$$0.74 - 0.51 \pm 1.96 \times \sqrt{0.003\,137 + 0.002\,773}$$

$$0.23 \pm 0.15$$

Since this does not include zero there is evidence that a higher proportion of the employed nurses agree with the statement. Allowing for the weighting does not increase the estimated variances from $p(1-p)/n$ in this example because the weights are not very extreme. We have not taken any account of the fact that 40 practices had both kinds of nurse in this analysis. We could apply McNemar's test to these practices. The responses of nurses in these practices can be summarised as:

		Employed nurse	
		Agree	$\overline{\text{Agree}}$
Attached nurse	Agree	21	2
	$\overline{\text{Agree}}$	7	10

McNemar's test looks at the nine practices in which nurses' opinions differed. Let X be the number of practices in which the employed nurse does not agree with the statement. If there is no difference in opinions on this matter between the two kinds of nurse then $X \sim \text{Bin}(9, 0.5)$.

$$\Pr(X \leq 2 | X \sim \text{Bin}(9, 0.5)) = 0.09$$

There is only rather weak evidence of a difference from this sub-sample of practices.

Paradoxical results from combining contingency tables

Combining contingency tables can lead to misleading inferences. Wardrop (1995) gives an example concerning occurrences of streak shooting in basketball, that is, in pairs of free throws the second is more likely to be a hit if the first is a hit. Wardrop quotes statistics from the 1980–81 and 1981–82 seasons.

Larry Bird				Rick Robey		
	Second shot				Second shot	
First shot	Hit	Miss		First shot	Hit	Miss
Hit	251	34		Hit	54	37
Miss	48	5		Miss	49	31

Both players shot slightly better after a miss, though not statistically significantly so, and there is no suggestion of streak shooting. The substantial difference is that Bird is a far better shooter. If the tables are combined we have:

		Second shot	
		Hit	Miss
First shot	Hit	305	71
	Miss	97	36

Now 81% of first shot hits are followed by a second shot hit, compared with 73% if the first shot is a miss. This misleading comparison arises from combining the results from two players. A similar situation can arise from survey data if key variables are not allowed for in the analysis. Logistic regression is a convenient means of allowing for more variables, but, in practice, the sample size limits the number that can reasonably be considered.

4.6.3 Binary logistic regression

One of the questions on the questionnaire to Malaysian companies was whether they had employee suggestion schemes. The proportions of companies responding that they do are given below.

	Smaller companies	Larger companies
Resource based	6/36	9/26
Non-resource based	7/30	11/32

This is a three-way contingency table with variables: size, whether or not resource based, and whether or not use suggestion schemes. Everitt (1992) describes log-linear models for contingency tables, but he points out that if one of the variables is considered as a response then an equivalent analysis in terms of a linear-logistic model can be used.

Minitab Release 12 provides convenient logistic regression routines. For the binary logistic regression there are just two responses, e.g. 'yes' and 'no'. Here, there are four sub-populations: smaller resource-based companies, ..., larger non-resource-based companies. Let p_i be the proportion of companies in sub-population i which use suggestion schemes. Define two predictor variables, x_{1i} which is -1 for smaller companies and $+1$ for larger companies, and x_{2i} which is -1 for resource-based companies and $+1$ for non-resource-based companies. The logit of p is defined as $\ln(p/(1-p))$ and an example of a binary logistic regression model is:

$$\ln(p_i/(1-p_i)) = \beta_0 + \beta_1 x_{1i} + \beta_2 x_{2i}$$

and the number of companies responding 'yes' (y_i) in a sub-sample of size n_i has a binomial distribution with parameters n_i and p_i. The parameters are estimated by the values of β_j which maximise the likelihood

$$\mathcal{L} = \prod_{i=1}^{4} {}_{n_i}C_{y_i}\, p_i^{y_i}(1-p_i)^{n_i-y_i}$$

where the p_i are related to the β_j and x_i by

$$p_i = \exp(\beta_0 + \beta_1 x_{1i} + \beta_2 x_{2i})/(1 + \exp(\beta_0 + \beta_1 x_{1i} + \beta_2 x_{2i}))$$

and the $_{n_i}C_{y_i}$ factors are irrelevant to the maximisation. These maximum likelihood estimates have to be found by some numerical optimisation procedure. The likelihood principle also provides approximate standard deviations for the estimates (Appendix 4). Minitab gives the following results:

Predictor (β_j)	Coef. ($\hat{\beta}_j$)	St. dev.	Z	P	Odds ratio	95% CI Lower	Upper
Constant	−1.0260	0.2078	−4.94	0.000			
Small/large	0.3735	0.2083	1.79	0.073	1.45	0.97	2.19
Resource/non-resource	0.0907	0.2073	0.44	0.662	1.09	0.73	1.64

The Z column is the ratio of the coefficient to its standard deviation and the P column is the probability that the modulus of a standard normal variable will exceed the Z-value. The ratio of p to $1 - p$ is known as odds. The odds ratio column is the increase in odds for a unit change in the predictor variable. If x_{1i} changes from -1 to 1, $\ln(p_i/(1 - p_i))$ increases by 2×0.3735 and $p_i/(1 - p_i)$ increases by a factor of 2.11. The factor of 1.45 corresponds to a change of 1 unit in x_{1i} and is the square root of 2.11. The 95% CI is a 95% confidence interval for the odds ratio. Other statistics given by Minitab include the log-likelihood (-70.014) and 'Test that all slopes are zero: $G = 3.655$, $DF = 2$, P-value $=$ 0.161'. G is the difference in -2 log-likelihood between a model that has only a constant term and the fitted model. It provides a test of a hypothesis that β_1 and $\beta_2 = 0$. If the hypothesis is true, G has a chi-square distribution with degrees of freedom equal to the number of predictor variables (excluding the constant). In this case G does not provide any convincing evidence against this null hypothesis. However, the best estimate of β_1 corresponds to large companies being twice as likely to use employee suggestion schemes, which is a substantial increase, and considered on its own it is statistically significant beyond the 10% level. The sub-sample sizes are rather small and the coefficient has not been estimated with high precision. We can include an interaction term $x_3 = x_1 \times x_2$ in the model. Notice that the coding of x_1 and x_2 results in x_1, x_2 and x_3 being uncorrelated. As a consequence the estimates of β_1 and β_2 do not change very much. They change slightly because the model is not linear in the unknown parameters. Minitab gives the following results:

Predictor (β_j)	Coef. ($\hat{\beta}_j$)	St. dev.	Z	P
Constant	−1.0204	0.2084	−4.90	0.000
Small/large	0.3971	0.2084	1.82	0.069
Resource/non-resource	0.1023	0.2084	0.49	0.623
Interaction	−0.1076	0.2084	−0.52	0.606

Log-likelihood $= -69.881$, $G = 3.922$, $DF = 3$, P-value $= 0.270$.

There is no evidence of an interaction. See Exercise 4.9 for an explanation of how estimates of standard deviations are available despite no remaining

Table 4.8 Classification of Malaysian companies by sector and production technology

	Small batch	Large batch	Continuous
Electrical	13	13	0
Mechanical	24	12	3
Chemical	20	9	13
Timber	12	6	1

degrees of freedom. Logistic regression can still be used when there is no clear response variable. For example, companies can be classified into a three-way contingency table by size, whether or not the company is resource based, and the production technology in operation, e.g. small batch, large batch or continuous. A logistic regression can be used if any one of the variables is chosen to be treated as the response. For example, the logit of the proportion of small companies could be regressed on whether or not the company is resource based and indicator variables for the production technology. That is:

$$x_1 = -1 \text{ for resource based, } +1 \text{ for non-resource}$$
$$x_2 = 1 \text{ for large batch and 0 otherwise}$$
$$x_3 = 1 \text{ for continuous and 0 otherwise}$$

where both x_2 and x_3 equal to 0 correspond to small batch.

4.6.4 Correspondence analysis and multiple correspondence analysis

Correspondence analysis is a method of analysing two-way contingency tables which leads to a graphical display of relationships between the variables. The companies which responded to the Malaysian survey can be classed by manufacturing sub-sector (1. Electrical; 2. Mechanical; 3. Chemical; 4. Timber) against production technology (1. small batch; 2. large batch; 3. continuous) as in Table 4.8. In general, suppose we have m row and p column classifications and the observed count in the ijth category is n_{ij}. Row and column totals R_i and C_j can be calculated but the a_i and b_j are unknown values which are yet to be determined.

	Production technology				Row sum	Sub-sector value
	1	2	...	c		
Sub-sector						
1	n_{11}	n_{12}	...	n_{1c}	R_1	a_1
2	n_{21}		...		R_2	a_2
⋮	⋮			⋮	⋮	⋮
r	n_{r1}	n_{r2}	...	n_{rc}	R_r	a_r
Column sum	C_1	C_2	...	C_c		
Production technology value	b_1	b_2	...	b_c		

The following succinct description of the principle is based on Manly (1994). Cox and Cox (1994) give a clear description of the mathematical details. The objective is to find numerical values for the a_i and b_j which maximise the correlation between sub-sector and production technology with respect to the empirical bivariate distribution given by the n_{ij}. The solution is obtained from the equations:

$$a_1 = [(n_{11}/R_1)b_1 + \cdots + (n_{1c}/R_1)b_c]/\theta$$
$$\vdots$$
$$a_r = [(n_{r1}/R_r)b_1 + \cdots + (n_{rc}/R_r)b_c]/\theta$$
$$b_1 = [(n_{11}/C_1)a_1 + \cdots + (n_{r1}/C_1)a_r]/\theta$$
$$\vdots$$
$$b_c = [(n_{1c}/C_c)a_1 + \cdots + (n_{rc}/C_c)a_r]/\theta$$

where θ is the correlation coefficient. These can be written in matrix form as

$$a = R^{-1}Nb/\theta \qquad (4.3)$$

$$b = C^{-1}N^T a/\theta \qquad (4.4)$$

where R is an $r \times r$ diagonal matrix with R_i in the ith row and ith column and C is a $C \times C$ matrix defined in a similar way. The equations are intuitively reasonable, inasmuch as a relatively high count, n_{ij}, will tend to make a_i and b_j similar. In an extreme case, if r equals c and the only counts are along the diagonal then θ is 1 for any choices of the a_i provided b_i is set equal to a_i. The equations are called reciprocal averaging. If the expression for b in the second matrix equation (4.4) is substituted into the first (4.3), the result is an eigenvalue problem:

$$[(R^{-1/2}NC^{-1/2})(R^{-1/2}NC^{-1/2})^T](R^{1/2}a) = \theta^2(R^{1/2}a)$$

The highest correlation between sub-sector and production technology corresponds to the square root of the largest eigenvalue. The corresponding eigenvector is $R^{1/2}a$. Hence the a_i can be calculated and the b_j follow from equation (4.4). Notice that there is an arbitrary scaling factor for a and b, but it is convenient to use a normalised eigenvector with a positive first element. We can now plot the a_i and b_j along a number line, attaching the category labels. Production technology and sub-sector categories that are close together along the line tend to be associated in the population of companies. We can make a more interesting plot if we also consider the next largest eigenvalue. This is another solution which has a correlation equal to the square root of the second highest eigenvalue. The corresponding eigenvector is uncorrelated with the first, usually described as 'orthogonal' in matrix algebra, and we can calculate another set of a_i and b_j which give position along a line orthogonal to the first.

Example 4.25
For the data in Table 4.8

$$N = \begin{pmatrix} 13 & 13 & 0 \\ 24 & 12 & 3 \\ 20 & 9 & 13 \\ 12 & 6 & 1 \end{pmatrix}$$

$$R = \mathrm{diag}(26 \quad 39 \quad 42 \quad 19)$$

$$C = \mathrm{diag}(69 \quad 40 \quad 17)$$

where diag( ) is a square matrix with the given numbers along the leading (descending) diagonal, and 0s elsewhere. The eigenvalue problem is

$$\begin{pmatrix} 0.2567 & 0.2645 & 0.2025 & 0.1895 \\ 0.2645 & 0.3199 & 0.2953 & 0.2259 \\ 0.2025 & 0.2953 & 0.4229 & 0.1980 \\ 0.1895 & 0.2259 & 0.1980 & 0.1603 \end{pmatrix} (R^{1/2}a) = \theta^2(R^{1/2}a)$$

The eigenvalues are 1, 0.1457, 0.0142 and 0. The normalised eigenvector corresponding to the eigenvalue 0.1457, and hence a correlation of 0.382, is

$$[0.5453 \quad 0.2130 \quad -0.7809 \quad 0.2180]^T$$

and hence

$$a = [0.1069 \quad 0.0341 \quad -0.1205 \quad 0.0500]^T$$
$$b = [0.0151 \quad 0.0664 \quad -0.2179]^T$$

If the sub-sector and production technology labels are plotted on a line, 'chemical' (at -0.1205) will be close to 'continuous' (at -0.2179). The chi-square goodness-of-fit analysis (Exercise 4.10) shows that the observed count in the chemical and continuous cell (13) is well above its mathematical expected value of 5.67. The solution corresponding to the second eigenvalue of 0.0142 is

$$a = [0.1382 \quad -0.0829 \quad 0.0357 \quad -0.0979]^T$$
$$b = [-0.0795 \quad 0.1124 \quad 0.0582]^T$$

but the correlation of 0.12 is too low to really justify a second axis for the plot.

Multiple correspondence analysis

The mathematical principle is unchanged because the multi-way contingency table is first rewritten as an $n \times c$ two-way table. The columns are a coding of all the variables and there is a row for every member of the sample.

Example 4.26
The Malaysian companies to which the questionnaire was sent can be classified by size, whether or not they are resource based, and whether or not they responded.

	Smaller companies	
	Responded	No response
Resource based	36	150
Non-resource based	30	78

	Larger companies	
	Responded	No response
Resource based	26	181
Non-resource based	32	117

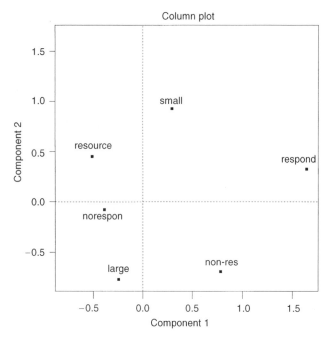

Figure 4.1 Multiple correspondence analysis for Malaysian manufacturing companies e.g. non-resource companies are more likely to respond.

This is a three-way table for 650 companies. Each of the three variables has two categories, so the matrix for the correspondence analysis is 650 × 6. All entries are 0 or 1. There will be a 1 in column 1 if a company responded, and a 1 in column 2 if it did not. If a company is resource based there will be a 1 in column 3, and 0 if it is not, otherwise there will be a 1 in column 4. Smaller companies have a 1 in column 5 and larger companies have a 1 in column 6. Thus the first 36 rows will be identical and equal (1 0 1 0 1 0), these will be followed by 30 identical rows equal to (0 1 1 0 1 0), and so on, finishing with 117 rows of the form (0 1 0 1 0 1). The graph from a Minitab correspondence analysis is shown in Figure 4.1. The inertias in the Minitab text output correspond to the squared correlations. In this example they are 0.3696 and 0.3497 for the two axes.

4.7 Summary

1. Random sampling appears to be fair and leads to unbiased estimators of population parameters. It also provides a basis for gauging the precision of estimates.
2. In a random sampling scheme every member of the population has a known non-zero probability of selection. If these probabilities are not equal appropriate weightings are used.
3. Non-response to surveys is a common problem. If possible, some information should be obtained from a random sample of people who did not respond initially.

4. The efficiency of a survey can be improved if values of some variables are known for all members of the population. These variables can be used to divide the population into relatively homogeneous sub-populations, known as strata. It may also be helpful to relate the variable of primary interest to these other variables by analysing ratios or by fitting regression models (see Chapter 5).
5. In two-stage sampling a large population is divided into sub-groups. A list of these sub-groups is a sampling frame for the first stage, and a random sample is drawn. The members of the selected sub-groups are then listed to form a sampling frame for the second stage, and a random sample is drawn. The principle can be extended to multi-stage sampling. In cluster sampling all the members of the selected sub-groups form the sample.
6. Sample size calculations should be made before the survey begins. It will be a waste of resources if the study is too small to have any chance of providing the information that is required. The calculations rely on estimates of population standard deviations. These can be based on past experience of similar studies or a pilot sample.
7. Survey responses are often categorical and may be analysed by categories of respondents. Tables of frequencies of responses corresponding to cross-classifications are known as contingency tables. They are usually analysed in a similar manner to the comparison of proportions. The difference in the sampling procedures is that the sample sizes for each category of respondent are selected in advance when comparing proportions. Logistic regression is a convenient and versatile means of analysing contingency tables. Minitab Release 12 provides binary, ordinal and nominal logistic regression routines, e.g.

Stat ▷ Regression ▷ Binary Logistic Regression

Correspondence analysis provides a graphical display of associations between variables from contingency tables. Minitab Release 12 provides a routine:

Stat ▷ Tables ▷ Multiple Correspondence Analysis

Exercises

4.1 A tax inspector wished to estimate the proportion of people who withhold information on their tax returns. The inspector suspected that, despite guarantees of anonymity, people would not answer a direct question honestly. An alternative question was tried. The question read:

'Flip a coin. Now either *A* or *B*.

    A. You get a head.

Flip the coin a second time. Put a tick in the box if you get a tail, and put a cross in it if you get a head.

    B. You get a tail.

Put a tick in the box if you withhold information on your tax return, otherwise put a cross in the box.'

Answer ☐

What would you deduce if the response is 65% ticks?

4.2   A population consists of N zones with associated costs c_j. A random sample of n zones is selected without replacement, in order to estimate the total population cost (T). That is

$$T = c_1 + \cdots + c_N$$

The probability of selecting zone i at the first draw is p_i. The total will be estimated by

$$\hat{T} = N \sum_{i=1}^{n} \frac{c_i}{p_i} \bigg/ \sum_{i=1}^{n} \frac{1}{p_i}$$

Consider a special case of $N = 3$, $n = 2$ and p_1, p_2, p_3 equal to 0.4, 0.4 and 0.2 respectively.

(i)   Write down the three possible samples and their probabilities of occurrence.

(ii) Show that $\hat{T}$ becomes $3 \times (c_1 + c_2)/2$ for the sample consisting of zone 1 and zone 2, and $(c_1 + 2c_3)$ for the sample consisting of zone 1 and zone 3. What is $\hat{T}$ for the other possible sample?

(iii) Show that $\hat{T}$ is an unbiased estimator of T.

4.3   A population of size N is divided into two strata of equal size $N/2$. A sample of size n is divided equally over the two strata. The proportions of people in stratum 1 and stratum 2 who are satisfied with the service provided is p_1 and p_2 respectively. The overall proportion of people who are satisfied is $p = (p_1 + p_2)/2$. Compare the variances of a stratified estimator of p and an estimator based on a simple random sample from the entire population. What is the numerical ratio of these variances if:

(i) $p_1 = 0.1$ and $p_2 = 0.9$
(ii) $p_1 = 0.3$ and $p_2 = 0.7$
(iii) $p_1 = p_2 = 0.5$.

4.4   Show that the chi-square goodness-of-fit statistic applied to a 2×2 contingency table is algebraically identical to a test for the equality of two proportions based on independent random samples n_1 and n_2, i.e.

$$\frac{\dfrac{X_1}{n_1} + \dfrac{X_2}{n_2}}{\sqrt{\hat{p}(1 - \hat{p})\left(\dfrac{1}{n_1} + \dfrac{1}{n_2}\right)}} \sim N(0, 1)$$

where

$$\hat{p} = (X_1 + X_2)/(n_1 + n_2)$$

4.5   Singular value decomposition.
      If A is an $n \times p$ matrix of rank r, then A can be written as

$$A = U\Lambda V^T$$

where $\Lambda = \operatorname{diag}(\lambda_1, \ldots, \lambda_r)$ and $U^T U = VV^T = I$. The λ_i are called the singular values of A, and λ_i^2 are the non-zero eigenvalues of the symmetric matrices AA^T and $A^T A$.

(i)   Verify these results for $A = [1 \quad 2]$.

(ii) Find singular value decompositions for the matrices

$$\begin{pmatrix} 4 & 1 \\ 1 & 2 \end{pmatrix} \quad \text{and} \quad \begin{pmatrix} 4 & 1 \\ 0 & 2 \end{pmatrix}$$

4.6   A question is answered on a scale from 1 to 7 by a sample of n people. What is the variance, and the standard deviation, of the score if there are equal numbers of 1s, 2s, ..., 7s? What is the maximum possible variance of the score? What would the standard deviation be if the score is approximately normally distributed?

4.7   Use equation (4.2) to show that,

$$\text{var}\left(\hat{r} \sum_{j=1}^{N} x_j \right) \simeq (N/n)^2 \sum_{i=1}^{n} (Y_i - \hat{r} X_i)^2$$

Compare this with $\text{var}(N\bar{Y})$.

4.8   Individual schemes have a standard deviation of 40%. What is the standard deviation, in percentage terms, of:

(i)   the sum of 16 independent schemes, each of which has an expected value of 100 mu;

(ii) the sum of 16 independent schemes, 14 of which have expected values of 50 mu and 2 of which have expected values of 450 mu;

(iii) the sum of 16 independent schemes, 15 of which have expected values of 20 mu and 1 of which has an expected value of 1300 mu?

4.9   Sixteen out of a random sample of 115 Network North commuter trains arrived more than 5 minutes late, compared with 8 out of 120 for Service South.

(i)   Calculate an approximate 90% confidence interval for the difference in the proportions of late trains.

(ii) Test the hypothesis that the proportions are equal against an alternative that they differ at the 10% level.

(iii) The binary logistic regression model

$$\ln(p_i/(1 - p)_i) = \beta_0 + \beta_1 x_i$$

was fitted where p_i is the proportion late, and x_i is 0 for Network North and 1 for Service South. The fitted model was

$$\text{logit}(p_i) = \begin{array}{cc} -1.8225 - 0.8165 x_i \\ (0.2694) \quad (0.4545) \end{array}$$

Test that all slopes are zero:

$$G = 3.413, \quad DF = 1, \quad P\text{–value} = 0.065$$

How many degrees of freedom are left?

4.10 Refer to Table 4.8 and use a chi-square test, pooling as appropriate, of the hypothesis that production technology is independent of sector.

5

Explanation and prediction

5.1 Introduction

Suppose you are the managing director of a company that owns ten small shops which sell wooden artefacts. You wish to expand the business and can afford to buy one of two new shops that are offered for sale. The estate agent provides the floor area and estimates of how many pedestrians pass by during a day, which is an indicator of how busy the shopping area is, for both shops. You have similar information for your ten existing shops, together with the yearly sales. A method for predicting sales from the floor area and the number of pedestrians passing by, based on the data from these ten shops, would help you decide which of the new shops is the better prospect.

A business development manager in a company that manufactures marine paints needs to estimate sales of paint for the next two years. Various publications list new building contracts for most of the world's shipyards by type of vessel and dead-weight. As part of this project, the manager needs to make estimates of the areas that will require painting on these new ships, from the detailed records of the painting jobs the company has supervised over the past few years. The objective is to relate the area painted to the dead-weight.

A research worker has asked employees at a company that manufactures industrial machinery to complete a questionnaire which has been designed to find out about communications within the company. There are 70 statements, and respondents are asked to give their level of agreement with each one. An industrial psychologist has suggested that there are two main factors that influence respondents' replies to all these questions, access to clear information and social relations within the company. The research worker will investigate this hypothesis for different categories of questions. For example, five of the questions relate to teamwork, five relate to efficiency, and so on.

The common feature of all these cases is that we have measured several variables for each shop, ship or employee. In the first two cases we wish to predict one variable which is crucial for the business decision, be it the number of sales or the area to be painted, from other variables which are known for the prospective shop purchases or for the ships under construction. A technique for making such predictions, called regression, is covered in the first half of this chapter. In the third case we are more concerned with explaining relationships between variables. In the second half of the chapter we look at some of the multivariate techniques that can be used to answer the questions that prompt such surveys.

5.2   Regression with one predictor variable

5.2.1   The standard regression model

The relationship between the dead-weight and area to be painted is likely to depend on the class of ship, so the first step was to stratify the population into tankers, large bulk carriers, small bulk carriers, ferries, and so on. The manager also wished to analyse different sections of the ship separately, because the coatings would differ. For example, the freshwater tanks, decks and wetted area of the hull all receive distinct paint finishes. The dead-weights and boot-top areas (hull above water line), in coded units, for 103 bulk carriers above a certain dead-weight are listed in TableWSdwt, and plotted in Figure 5.1. The data are from painting contracts that the company has recently supervised, and these jobs are considered to be representative of current ship building design. It would be quite impractical to look at a random sample of plans for new building contracts.

Random variation about an underlying linear relationship is a plausible model for the data in Figure 5.1. That is

$$Y_i = \beta_0 + \beta_1 x_i + E_i \quad \text{for } i = 1, \ldots, n \qquad (5.1)$$

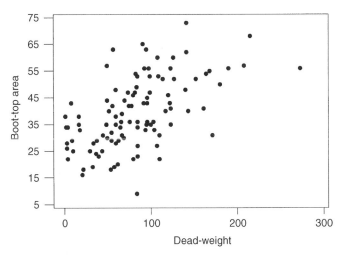

Figure 5.1   Boot-top area versus dead-weight; both variables have been scaled for commercial confidentiality

where x_i is the coded dead-weight, Y_i is the coded area and n is the number of ships in the sample, 103 in this case. The E_i represent random variation about the hypothetical underlying linear relationship. In general, the random variations can include measurement error as well as inherent variation in the population and it is convenient to refer to them as 'errors'. The errors are assumed to have a mean of zero and to be independent of the predictor variable (x). These two assumptions, which we will refer to as $A1$ and $A2$, are crucial and cannot be checked from the data. In particular it is essential to ensure that measuring equipment is properly calibrated. If the mean of the errors is not zero, it will be incorporated into the estimate of the constant (β_0). If the errors are linearly associated with x, this will be incorporated into the estimate of the slope (β_1). For example, if errors tend to be positive for small x and negative for large x, then the slope is likely to be underestimated. A subtle corollary of the second assumption is that errors made when measuring the values of x_i should be small in comparison with the standard deviation of the E_i, and if this is not realistic the methods described in Section 5.4 should be used. There are three other assumptions about the errors which are made in the standard model: $A3$ – that they are independent of each other; $A4$ – that they have a constant variance (σ^2) regardless of the value of x; and $A5$ – that they are normally distributed. These assumptions can be tested, at least to some extent, from the data, and the analysis can be modified if alternative assumptions would be more realistic. A representation of the standard model is shown in Figure 5.2.

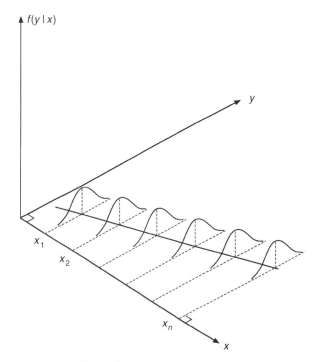

Figure 5.2   3D representation of a linear regression

The model is fitted by applying the 'principle of least squares', a method that was first used by A.M. Legendre and Karl Gauss, working independently at the end of the eighteenth century, to estimate orbits of comets. Referring to the model of equation (5.1), the sum of squared errors is

$$\Psi = \sum (Y_i - (\beta_0 + \beta_1 x_i))^2$$

and the least squares estimators of β_0 and β_1 are the functions of the x_i and Y_i which make Ψ a minimum. The least squares estimators of β_0 and β_1 will be denoted by $\hat{\beta}_0$ and $\hat{\beta}_1$ respectively. The algebraic detail is postponed until the section on multiple regression, because this includes a single predictor variable as a special case. When we have the data pairs (x_i, y_i) the least squares estimates of β_0 and β_1 are the numbers which minimise the sum of squared errors

$$\Psi = \sum (y_i - (\beta_0 + \beta_1 x_i))^2 \tag{5.2}$$

The least squares estimates will also be denoted by $\hat{\beta}_0$ and $\hat{\beta}_1$. The distinction between the estimator, a random variable, and the estimate, a number, should be clear from the context. The formulae are

$$\hat{\beta}_1 = \sum (x_i - \bar{x})(y_i - \bar{y}) \Big/ \sum (x_i - \bar{x})^2 \tag{5.3a}$$

and

$$\hat{\beta}_0 = \bar{y} - \hat{\beta}_1 \bar{x} \tag{5.3b}$$

The line,

$$y = \hat{\beta}_0 + \hat{\beta}_1 x$$

which is often more conveniently expressed as,

$$y = \bar{y} + \hat{\beta}_1 (x - \bar{x})$$

is the fitted regression line. The parallel to the y-axis distance from a point (x_i, y_i) to the line is,

$$|y_i - (\hat{\beta}_0 + \hat{\beta}_1 x_i)|$$

and it follows from equation (5.2) that the fitted regression line is the line for which the sum of squared parallel to the y-axis distances from the points to the line is a minimum (see Figure 5.3). We define the fitted values as

$$\hat{y}_i = \hat{\beta}_0 + \hat{\beta}_1 x_i \quad \text{for } i = 1, \ldots, n$$

and the residuals by

$$r_i = y_i - \hat{y}_i$$

The residuals are estimates of the errors,

$$e_i = y_i - (\beta_0 + \beta_1 x_i)$$

The errors are unknown, because β_0 and β_1 are unknown, but we do know that

$$\sum r_i^2 \leq \sum e_i^2$$

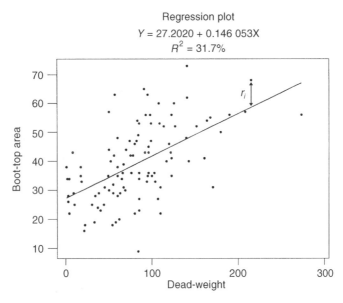

Figure 5.3　Regression of boot-top area on dead-weight

is a direct consequence of applying the principle of least squares. An unbiased estimate (s^2) of the variance of the errors (σ^2) is given by

$$s^2 = \sum r_i^2 / (n - 2)$$

where we have lost two degrees of freedom because two parameters were estimated from the data. The fitted regression line for area (y) on dead-weight (x) is

$$y = 27.202 + 0.146\,05x$$

The estimated standard deviation of the errors (s) is 10.94.

5.2.2　Logarithmic regression

In 1989 the water authorities in England and Wales were taken out of public ownership and sold on the London stock market. Ten companies were formed. These companies were allowed to fund improvements to the system by increasing charges to customers beyond the increase in retail price index, provided they identified a need for the improvements in their asset management plans (AMPs). Asset management plans were, and still are, subject to scrutiny by independent assessors and the government agency that oversees the industry (OFWAT). The companies were expected to give estimates of uncertainty associated with the predicted costs of improvements.

One aspect of the AMP, in at least one water company, was the cost of completing schemes to prevent flooding during extreme storms. Suppose 41 schemes have been identified, on the basis of recent customer complaints, and classified by the number of properties at risk, as a measure of the size of the

Table 5.1 Predicted mean costs and variances of individual schemes by number of properties

Number of properties	Number of schemes	Mean log (cost)	Mean cost	Individual variance (10^9)
1	28	9.0562	20 702	2.07
2	5	9.3397	27 487	3.65
3	4	9.6232	36 496	6.43
5	2	10.1901	64 340	19.99
6	1	10.4736	85 427	35.25
8	1	11.0406	150 601	109.55

scheme, as shown in the two left-hand columns of Table 5.1. A manager has been asked to estimate the cost, at current prices, of completing this work, together with 90% limits of prediction, from the data in TableWSpropfld. The data in Table 5.1 are fictitious, but those in TableWSpropfld are genuine.

A first step in any statistical analysis should be to plot the data. The cost of the scheme is plotted against the number of properties affected, and the fitted regression line is shown in Figure 5.4(a). We are 95% confident that the hypothetical line lies within the dotted 95% confidence interval lines. The outer chained lines represent 95% prediction intervals for the costs of individual schemes. A tendency for costs to increase with the number of properties is just discernible, but the assumption of the errors having a constant variance is not realistic. In Figure 5.4(b) the natural logarithms of the costs are plotted against the number of properties. The assumptions underlying the regression appear more plausible, and this model has the advantage of excluding negative costs. The assumption that the errors of the logarithms of the variable have a constant variance is equivalent to assuming constant variance multiplicative errors for the original variable. However, there is a complication when estimating the mean cost for a scheme. Let Y be the logarithm of cost (C). The regression gives us an estimate of the mean value of Y for a given number of properties (x), $\mu_{Y|x}$, and an estimate (s) of the standard deviations of the errors (σ) about this mean. If we assume Y is normally distributed the mean is identical to the median. Taking exponential of the mean of Y gives the median of the distribution of C. But C has a log-normal distribution, which is positively skewed and has a mean (Exercise 5.1):

$$\mu_{C|x} = \exp\left(\mu_{Y|x} + \frac{\sigma^2}{2}\right) \tag{5.4}$$

The sample estimate is obtained by replacing $\mu_{Y|x}$ with the prediction from the fitted regression line and substituting s for σ. A 95% confidence interval for $\mu_{Y|x}$ can be transformed to an approximate 95% confidence interval for $\mu_{C|x}$ by taking exponential and multiplying by $\exp(s^2/2)$. In contrast to the construction of the confidence interval, a 95% prediction interval for C is simply exponential of the 95% prediction interval for Y. No factor is applied because:

$$Pr(L < Y < U) = Pr(\exp(L) < \exp(Y) < \exp(U))$$
$$= Pr(\exp(L) < C < \exp(U))$$

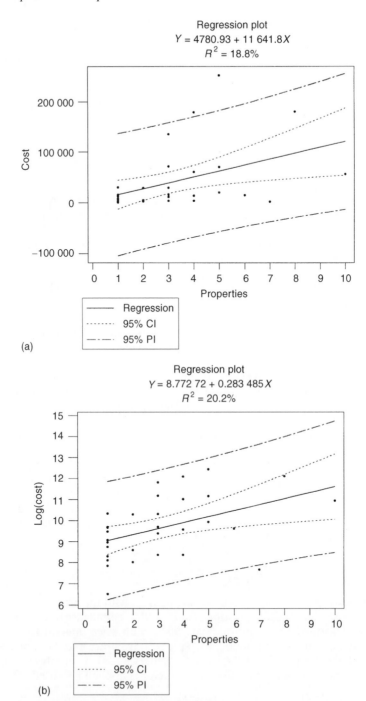

Figure 5.4   Regression of (a) cost of scheme to alleviate flooding and (b) natural logarithm of cost against the number of properties affected

The formula for a $(1 - \alpha) \times 100\%$ confidence interval (CI) for the mean value of Y when x equals x_p is:

$$\hat{y} \pm t_{n-2,\alpha/2} s \sqrt{(1/n) + (x_p - \bar{x})^2 / \sum (x_i - \bar{x})^2} \qquad (5.5)$$

where

$$\hat{y} = \bar{y} + \hat{\beta}_1 (x_p - \bar{x})$$

The first term under the square root, $(1/n)$, allows for the sampling variability of $\bar{Y}$, and the second, $(x_p - \bar{x})^2 / \sum (x_i - \bar{x})^2$, allows for uncertainty in the estimation of β_1. The expression,

$$s \sqrt{(1/n) + (x_p - \bar{x})^2 / \sum (x_i - \bar{x})^2}$$

is the standard deviation of the predicted mean. The formula for a $(1 - \alpha) \times 100\%$ prediction interval (PI) for a single value of Y when x equals x_p includes the variance of Y about its mean:

$$\hat{y} \pm t_{n-2,\alpha/2} s \sqrt{1 + (1/n) + (x_p - \bar{x})^2 / \sum (x_i - \bar{x})^2} \qquad (5.6)$$

If the sample size is large this is roughly:

$$\hat{y} \pm z_{\alpha/2} s$$

The PI is sensitive to the assumption that the errors are normally distributed.

Example 5.1
Fit a regression line for predicting the natural logarithm of the cost (y) from the number of properties (x), using the data in TableWSpropfld. Construct 90% confidence intervals for the mean logarithm of cost and the mean cost, when there are five properties in a scheme. Calculate 95% prediction intervals for the logarithm of cost and cost of a single scheme of 5 properties. The summary statistics are:

$$n = 31 \quad \bar{x} = 3.129\,03 \quad \bar{y} = 9.659\,75$$

$$\sum (x_i - \bar{x})^2 = 161.484 \quad \sum (x_i - \bar{x})(y_i - \bar{y}) = 45.7782$$

$$\sum (y_i - \bar{y})^2 = 64.1367$$

The formulae lead to the regression line

$$y = 8.7727 + 0.2835x$$

with $s = 1.328$. The estimate of $E[Y | x = 5]$ is 10.190, and the 90% confidence interval is

$$[9.666, 10.714]$$

The 90% prediction interval for a single value of Y when x equals 5 is

$$[7.873, 12.507]$$

The estimate of mean cost when there are five properties is

$$\exp(10.190 + (1.328)^2/2) = 64\,331$$

An approximate 90% confidence interval for the mean cost is

$$[\exp(9.666 + (1.328)^2/2), \exp(10.714 + (1.328)^2/2)]$$

which equals $[38\,093, 108\,640]$. The 90% prediction interval for the cost of a single scheme of five properties is

$$[\exp(7.873), \exp(12.507)]$$

which equals $[6341, 652\,647]$.

The objective of the analysis was to estimate the total cost for the schemes in Table 5.1, and to provide 90% prediction limits for the total cost. The steps are summarised in Table 5.1. The mean log (cost) is given by the regression equation and the mean cost is calculated by taking exponential and multiplying by $\exp(s^2/2)$, which equals 2.415. The individual variance is estimated from the relationship between the variances of Y and C if Y is normally distributed,

$$\sigma^2_{C|x} = (\mu_{C|x})^2(\exp(\sigma^2) - 1)$$

with σ^2 replaced by s^2, i.e. the variance is the product of the square of the mean with 4.83. The estimate of the total cost is the sum of the estimated mean costs for the 41 proposed schemes and equals 1 227 790 monetary units. There are two sources of uncertainty in this estimate. The first is that the regression relationship is calculated from a sample of 31 schemes, and sampling error in the position of the regression line will be common to all the estimates. The mean number of properties for the proposed schemes is 1.805. A 90% approximate confidence interval for the mean cost when the number of properties is 1.805 is $[16\,282, 41\,556]$. The point estimate of the mean is 25 998 so the 90% confidence interval is roughly 63% of the point estimate to 160% of the point estimate. The same percentages apply to our estimate of the total cost, and a rather approximate standard deviation is $(37 + 60)\%/(2 \times 1.7)$, which equals 29%. The second source of uncertainty is the variation of individual schemes about their mean values. If such variation is assumed to be independent, the sum of the 41 variances is 2.867×10^{11}. The standard deviation of the predicted total about its mean is therefore 5.355×10^5, which is about 44% of the point estimate of the total cost. If the two components of uncertainty are independent, the standard deviation of the predicted total in percentage terms is

$$((44\%)^2 + (29\%)^2)^{\frac{1}{2}} = 53\%$$

The variation of the total about its mean will be approximately normally distributed because it is the sum of 41 independent components. The uncertainty in the mean is not normally distributed but it is being added to an approximate normal variable with a larger variance. So a very approximate 66% prediction interval for the total is plus or minus 53%. This is very wide and warns the company that the estimate is extremely uncertain. Ninety per cent limits can be formally calculated as approximately plus or minus $1.7 \times 53\%$ which equals plus or minus 90%, but the symmetry that follows from assuming a normal distribution gives a rather unrealistic lower limit.

If no account is taken of the fact that the proposed schemes involve fewer properties, on average, than completed schemes, the estimate of the total AMP cost would be 41 times the average cost of the 31 completed schemes. This equals 1 689 548 which is substantially more than the estimate of 1 227 790 obtained from the regression analysis.

5.2.3   Intrinsically linear models

Cohen *et al.* (1996) set up a model for investigating the trade-off between performance and time-to-market. As part of the model, they assume that the rate of improvement in performance (dQ/dt) is proportional to some power of the size of the development team (L). That is:

$$\dot{Q} = kL^a \tag{5.7}$$

where $\dot{Q}$ is the time derivative of Q, i.e. dQ/dt. A company has recently acquired a chain of 77 hotels. The new management intends improving the quality of service and decides to take the percentage increase in room occupancy over the next three months as a measure of the rate of improvement. An analyst defines L as the ratio of the average number of staff during the three-month period to a full complement under the old management. The analyst will compare the fit of equation (5.7), to the data from the 77 hotels, with an alternative model

$$\dot{Q} = \frac{bL}{L+c} \tag{5.8}$$

In neither case is $\dot{Q}$ linearly related to L but it is possible to rewrite the equations so that they are linear in the unknown parameters. If we take logarithms of both sides of equation (5.7) we obtain

$$\ln(\dot{Q}) = \ln(k) + a\ln(L)$$

and we can estimate a and $\ln(k)$, and hence k, by the standard regression procedure if we assign $\ln(L)$ and $\ln(\dot{Q})$ to x and y respectively. Equation (5.8) can be rewritten by taking reciprocal of both sides,

$$\frac{1}{\dot{Q}} = \frac{1}{b} + \frac{c}{b}\frac{1}{L}$$

and we can take x and y as $1/L$ and $1/\dot{Q}$ respectively. If we linearise the equations we should remember that it is not consistent to assume $N(0, \sigma^2)$ errors about both the original and the linearised relationships. Also, as explained in Section 5.2.2, the linearised equations provide an estimate of $E[f(\dot{Q})]$ and this is not the same as $f(E[\dot{Q}])$. One way around this is to scale the fitted relationship so that the average of differences between observed and fitted $\dot{Q}$ is zero. Other options are to assume $\dot{Q}$ has a log-normal distribution (Section 5.2.2) or to use a non-linear regression procedure described in Section 5.3.3.

No such complications arise if the dependent variable is linearly related to some function of the predictor variable. Raj (1985) related percentage market share for brand i (Y_i) to the square of the percentage of customers who are loyal to brand i (W_i) for five different product categories: food, pet food, personal care, non-durable household goods and tobacco. The assumed relationship is

$$Y_i = \beta_0 + \beta_1 w_i^2 + E_i$$

where E_i are independent errors with mean 0. Data were obtained from the Target Group Index Report for 1976 (TGI, 1976). The typical sample size for TGI surveys of people's purchasing habits was about 20 000 households. The simple model was a surprisingly good fit for all the product categories, the correlations between y_i and w_i^2 being: 0.71, 0.58, 0.53, 0.78 and 0.37 for food, ..., tobacco. The household goods product category included 15 products such as paper napkins, aluminium foil and disposable cups. There were 138 brands and the parameter estimates $\hat{\beta}_0$ and $\hat{\beta}_1$ were 0.60 and 0.008 respectively. Although the model is unsophisticated, the study did provide evidence of the importance of encouraging brand loyalty. Raj (1985) then looked at all 86 products across the five product categories. There were 1091 brands altogether. For each of the 86 products he calculated the average percentage of loyal users (y_i). He then fitted a regression of y_i on the reciprocal of the number of brands (u_i) of the product. The model is

$$Y_i = \beta_0 + \beta_1 \frac{1}{u_i} + E_i$$

where E_i are independent errors with mean 0. The correlation between y_i and $1/u_i$ was 0.69, and the estimates $\hat{\beta}_0$ and $\hat{\beta}_1$ were 19.63 and 149.87 respectively.

Example 5.2
The data in Table 5.2 are US logistics costs between 1981 and 1994 broken down by inventory carrying, transportation and administrative costs (Delaney, 1995) reproduced in Bowersox and Closs (1996). Over this period there has been a concerted effort to reduce inventory carrying costs. The ratio of inventory cost to transport cost (r_t) has certainly decreased over the 14 years. We will predict the ratio in three years' time using three different models.

Table 5.2 US logistics costs between 1981 and 1994

Year	Inventory	Transport cost	Ratio	Time	In (ratio)	In (time)
1981	283	236	1.199 15	1	0.181 615	0.000 00
1982	255	240	1.062 50	2	0.060 625	0.693 15
1983	228	244	0.934 43	3	−0.067 823	1.098 61
1984	257	250	1.028 00	4	0.027 615	1.386 29
1985	240	265	0.905 66	5	−0.099 091	1.609 44
1986	233	271	0.859 78	6	−0.151 080	1.791 76
1987	243	288	0.843 75	7	−0.169 899	1.945 91
1988	266	313	0.849 84	8	−0.162 707	2.079 44
1989	311	331	0.939 58	9	−0.062 325	2.197 22
1990	298	352	0.846 59	10	−0.166 538	2.302 59
1991	270	360	0.750 00	11	−0.287 682	2.397 90
1992	243	379	0.641 16	12	−0.444 475	2.484 91
1993	250	394	0.634 52	13	−0.454 890	2.564 95
1994	277	425	0.651 76	14	−0.428 072	2.639 06

(i) $r_t = ae^{-kt}F_t$

where t is time in years from 1980 and F_t is random variation with a log-normal distribution with median 1. The linearised form of the model is:

$$\ln(r_t) = \ln(a) - kt + E_t$$

where $E_t \sim N(0, \sigma^2)$. The correlation between $\ln(r_t)$ and t is -0.923. The fitted model is

$$\ln(r_t) = 0.158\,52 - 0.0423t$$

with $s = 0.076\,95$. The predicted ratio when t equals 17 is

$$\exp[0.158\,52 - 0.0423 \times 17 + (0.076\,95)^2/2]$$

which equals 0.57. A 95% PI for the ratio is (0.47, 0.70).

(ii) $r_t = at^b F_t$

which linearises to

$$\ln(r_t) = \ln a + b\ln(t) + E_t$$

The correlation between $\ln(r_t)$ and $\ln(t)$ is -0.881. The fitted model is

$$\ln(r_t) = 0.234\,09 - 0.218\,41\ln(t)$$

with $s = 0.094\,42$. The prediction when t equals 17 is

$$\exp(0.234\,09 - 0.218\,41 \times \ln(17) + (0.094\,42)^2/2)$$

which equals 0.68.

(iii) $r_t = a + bt + E_t$

The correlation between r_t and t is -0.919 and the fitted model is

$$r_t = 1.138\,93 - 0.0362t$$

The prediction when t equals 17 is 0.52.

To summarise, models (i) and (iii) give slightly better fits to the existing data, but assuming a linear trend will continue, even over the next three years, using model (iii) for the prediction is not realistic. Model (ii) is perhaps the most realistic for predicting several years ahead. It would make more physical sense to include some lower limit (L). Then the model would be:

$$r_t = L + at^b$$

This is now non-linear in the unknown parameters (see Section 5.4.1). However, it is unlikely that we can obtain any reliable estimate of L from these data. We could assume some value such as 0.1. If this is done a regression of $\ln(r_t - 0.1)$ on $\ln(t)$ gives a prediction of 0.69 when t equals 17.

5.3 Regression with several predictor variables

Rosewood is a company which sells wooden artefacts. It now owns ten shops, situated in shopping centres in the provinces. The average number of pedestrians passing per hour, and average weekly sales, over the past two years, are given in Table 5.3, together with the floor areas. The data are from a marketing consultancy, and were developed as an exercise by the Department of Management Studies at the University of Glasgow. The company can afford

Table 5.3　Pedestrians passing, floor area and sales for Rosewood shops

Pedestrians/hr	Floor area	Sales
564	650	980
1072	700	1160
326	450	800
1172	500	1130
798	550	1040
584	650	1000
280	675	740
970	750	1250
802	625	1080
650	500	876

to buy one of two shops, *A* and *B*, that are for sale in different shopping centres. Relevant sales figures are not available for these shops, but the floor areas and numbers of pedestrians passing by are available from the estate agent:

Shop	Pedestrians/hour	Floor area (ft^2)
A	475	1000
B	880	550

Fixed and operating costs would be similar, properties in the busier centre being more expensive per square foot. We would expect sales to depend on the number of pedestrians passing, but a larger floor area should be an advantage as well because a wider range of merchandise can be displayed. Shop *A* has the larger floor area, but fewer people pass by. The objective of the analysis is to fit a model for predicting sales from one or both of pedestrians/hour and floor area, using the data from the ten shops, and to use the model to decide which of the shops *A* and *B* is the better investment. As a first step, the correlations between the variables are:

	Pedestrians/h	Area	Sales
Pedestrians/h	1		
Area	0.17	1	
Sales	0.90	0.43	1

As expected, there is a high correlation between sales and pedestrians per hour. There is a more modest positive correlation between sales and area, and even this can be accounted for, to some extent, by the positive correlation between area and pedestrians/hour. There are several ways of displaying data triples graphically, but I find the Excel bubble plot one of the clearest. Figure 5.5 is a plot of sales against pedestrians per hour, with the point markers having areas proportional to the floor areas of the shop with 400 subtracted, so that the differences show up. The data in the Excel worksheet need to be in adjacent columns in the order: pedestrians/hour, sales and (area −400). There is a slight

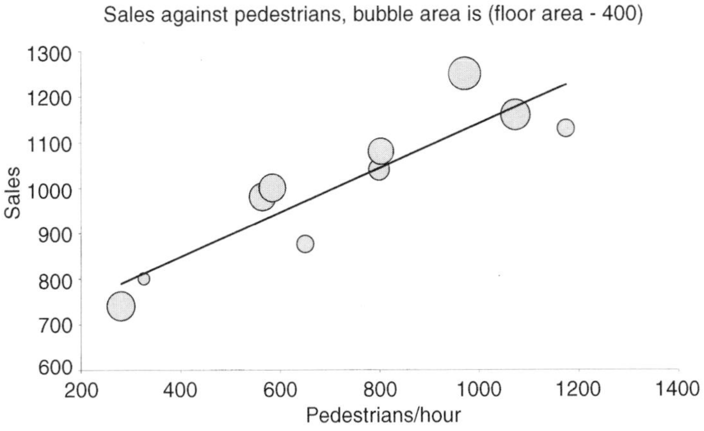

Sales against pedestrians, bubble area is (floor area - 400)

Figure 5.5   Excel bubble plot of sales against number of pedestrians with bubble size indicating floor area

tendency for the larger circles to lie above the regression line of sales on pedestrians/hour, but it is not convincing. The algebraic analysis which follows enables us to be more precise about the effect of floor area.

5.3.1   The multiple regression model

The multiple regression model is

$$Y_i = \beta_0 + \beta_1 x_{1i} + \beta_2 x_{2i} + \ldots + \beta_k x_{ki} + E_i \tag{5.9}$$

where Y_i are the values of the variable we will, in future, wish to predict, known as the response variable in Minitab, $x_{1i}, \ldots, x_{ki}$ are the predictor variables and E_i are the errors. It is referred to as a linear model because it is linear in the unknown parameters β_i, which are referred to as the regression coefficients. The E_i will be assumed to have a zero mean and to be independent of the explanatory variables. In the standard form of the model the errors are also assumed to have a constant variance, σ^2, to be independent of each other, and to have a normal distribution. A sample of size n can be modelled by

$$\begin{bmatrix} Y_1 \\ \vdots \\ Y_n \end{bmatrix} = \begin{bmatrix} 1 & x_{11} & \cdots & x_{k1} \\ \vdots & & & \vdots \\ 1 & x_{1n} & \cdots & x_{kn} \end{bmatrix} \begin{bmatrix} \beta_0 \\ \vdots \\ \beta_k \end{bmatrix} + \begin{bmatrix} E_1 \\ \vdots \\ E_n \end{bmatrix}$$

or, more concisely,

$$Y = XB + E$$

where Y, X, B and E are $n \times 1, n \times (k+1), (k+1) \times 1$ and $n \times 1$ matrices, respectively. The sum of squared errors is

$$\psi = \sum E_i^2 = (Y - XB)^T (Y - XB)$$

The following calculus results are used to find the formula for the least squares estimators. Let $\phi(\zeta)$ be a scalar function of an $m \times 1$ array ζ, where

$$\zeta^{\mathrm{T}} = [\zeta_1, \ldots, \zeta_m]$$

That is, ϕ is a function of m variables $\zeta_1, \ldots, \zeta_m$. Define the array of partial derivatives

$$\frac{\partial \phi}{\partial \zeta} = \left[\frac{\partial \phi}{\partial \zeta_1}, \ldots, \frac{\partial \phi}{\partial \zeta_m} \right]^T$$

The results given below are consequences of this definition and the usual rules of calculus. Let c be an $m \times 1$ array of constants and M be an $m \times m$ matrix of constants. Then:

R1 $\dfrac{\partial}{\partial \zeta}(\zeta^T c) = c$

R2 $\dfrac{\partial}{\partial \zeta}(\zeta^T M \zeta) = M\zeta + M^T \zeta$

R3   A necessary requirement for ϕ to have a maximum or minimum is that $\partial \phi / \partial \zeta = 0$

It is straightforward to verify these results for the case of m equal to 2, and the general results follow the same principles.

The normal equations

We will apply the principle of least squares and find expressions for the β_j which will minimise ψ. To begin with

$$\psi = (Y - XB)^T (Y - XB)$$

$$= Y^T(Y - XB) - B^T X^T (Y - XB)$$

$$= Y^T Y - Y^T XB - B^T X^T Y + B^T X^T XB$$

All the terms on the right-hand side are 1×1 matrices, that is to say, scalars, which therefore equal their transpose. In particular,

$$Y^T XB = (Y^T XB)^T = B^T X^T Y$$

Therefore

$$\psi = Y^T Y - 2B^T X^T Y + B^T X^T XB$$

We can now differentiate with respect to B. Applying R1 and R2 and noting that $(X^T X)^T = X^T X$, gives

$$\frac{\partial \psi}{\partial B} = -2X^T Y + 2(X^T X)B$$

Provided $X^T X$ has an inverse, R3 leads to the *normal equations*

$$\hat{B} = (X^T X)^{-1} X^T Y \tag{5.10}$$

The addition of more explanatory variables can only decrease the residual sum of squares. The practical question is whether this decrease is sufficient to give more accurate predictions. The properties of the estimators are needed to answer this. The least squares estimators are unbiased, as can be seen by taking expectation and using the assumption that the E_i are independent of the predictor variables:

$$E[\hat{B}] = (X^T X)^{-1} X^T E[Y] = (X^T X)^{-1} X^T X B = B$$

The variance–covariance matrix of the estimators, C, is defined as

$$C = \begin{bmatrix} \text{var}(\hat{\beta}_0) & \text{cov}(\hat{\beta}_0, \hat{\beta}_1) & \text{cov}(\hat{\beta}_0, \hat{\beta}_2) & \cdots \\ \text{cov}(\hat{\beta}_0, \hat{\beta}_1) & \text{var}(\hat{\beta}_1) & \text{cov}(\hat{\beta}_1, \hat{\beta}_2) & \\ \vdots & & \text{var}(\hat{\beta}_2) & \\ & & & \ddots \\ & & & & \text{var}(\hat{\beta}_k) \end{bmatrix}$$

in which

$$\text{cov}(\hat{\beta}_i, \hat{\beta}_j) = E[(\hat{\beta}_i - E[\hat{\beta}_i])(\hat{\beta}_j - E[\hat{\beta}_j])]$$

Since $E[\hat{B}] = B$, C can be written concisely as,

$$C = E[(\hat{B} - B)(\hat{B} - B)^T]$$

which equals $(X^T X)^{-1} \sigma^2$. The proof follows.

$$C = E[(\hat{B} - B)(\hat{B} - B)^T]$$
$$= E[(X^T X)^{-1} X^T (Y - E[Y])(Y - E[Y])^T X (X^T X)^{-1}]$$

Now note that $Y - E[Y]$ is a column of E_i. $E[E_i^2]$ is σ^2 and $E[E_i E_j]$ is 0 if i is different from j. So, we have

$$C = (X^T X)^{-1} X^T \sigma^2 I X (X^T X)^{-1}$$

As σ^2 is a scalar it can take any position in the matrix product. Hence

$$C = (X^T X)^{-1} \sigma^2$$

In most applications σ^2 will not be known, so it will have to be estimated from the data. The residuals, r_i, which are estimates of the errors, are calculated by subtracting the fitted values ($\hat{y}_i$) from the observed y_i
The fitted values are given by,

$$\hat{y}_i = \hat{\beta}_0 + \hat{\beta}_1 x_{1i} + \ldots + \hat{\beta}_k x_{ki}$$

and the residuals are defined by

$$r_i = y_i - \hat{y}_i \tag{5.11}$$

An unbiased estimate of σ^2 is

$$s^2 = \sum r_i^2 / (n - k - 1)$$

where $k + 1$ degrees of freedom are lost because $k + 1$ parameters are estimated from the data (Appendix 7). This theory will now be applied to the shop sales case.

The regression of sales (y) on pedestrians per hour (x_1) is

$$y = 651.8 + 0.490x_1$$

with $s = 74.0$

The regression of sales on both pedestrians per hour and floor area (x_2) is

$$y = 385.7 + 0.464x_1 + 0.471x_2$$

with $s = 59.5$. The estimated standard deviations of the coefficients are given below.

Predictor variable	Coefficient (β_i)	Estimated coefficient ($\hat{\beta}_i$)	Standard deviation of $\hat{\beta}_i$
1	β_0	385.7	125.6
Pedestrians/hour	β_1	0.4642	0.0674
Floor area	β_2	0.4708	0.2027

The inclusion of the floor area has resulted in a substantial decrease in the estimated standard deviation of the errors, and the positive coefficient is consistent with our commonsense explanation that sales will tend to increase if floor areas increase. I consider the model with both predictor variables an improvement, although with a rather small sample size, even a 90% confidence interval for β_2 is quite wide, i.e.

$$0.4708 \pm t_{10-3,0.05}0.2027$$

which gives $[0.09, 0.82]$

5.3.2   Making predictions

The next step is to make predictions for the two shops. It is shown in Appendix 7 that under the standard assumptions a $(1 - \alpha) \times 100\%$ confidence interval for $E[Y \mid x = x_p]$ is

$$x_p^T \hat{B} + t_{n-k-1,\alpha/2} s \sqrt{x_p^T (X^T X)^{-1} x_p}$$

and $(1 - \alpha) \times 100\%$ limits of prediction for a single value of Y when $x = x_p$ are

$$x_p^T \hat{B} \pm t_{n-k-1,\alpha/2} s \sqrt{1 + x_p^T (X^T X)^{-1} x_p}$$

Minitab implements these formulae and gives the following point predictions and limits of prediction for the two shops.

Shop	Predicted sales	90% limits of prediction
A	1077	(878, 1276)XX
B	1053	(931, 1176)

XX denotes a row with very extreme x values, and the warning should be heeded. The range of floor areas used when fitting the model was between 450 and 750. The floor area for Shop A is 1000 square feet and well outside this range. The predicted sale involves extrapolation rather than interpolation. The 90% limits of prediction are wide, but only because of uncertainty about the estimates of the coefficients, and depend on the assumption that the model, in this case a regression plane in three dimensions, remains valid when extrapolated. There is no way of checking from the data that this assumption is reasonable. It is possible that such a large floor area would be unhelpful, because a full range of goods can easily be shown in smaller premises. The predicted sales for the two shops are close in comparison with the width of the 90% limits of prediction. The prediction for Shop A is particularly unreliable and highly dependent on its large floor area. I would advise the management to buy Shop B.

5.3.3   Checking assumptions

If a regression of y is on a single predictor x, it is possible to see from the scatter plot whether the assumption of an underlying linear relationship with constant variance errors is reasonable. Outlying points can be seen clearly and should be investigated. However, if we use more predictor variables we have to rely on the residuals for checking assumptions. In matrix terms the column of fitted values is given by

$$\hat{Y} = X\hat{B} = X(X^T X)^{-1} X^T Y$$

which can be written as

$$\hat{Y} = HY$$

where

$$H = X(X^T X)^{-1} X^T$$

H is known as the hat matrix. The ith diagonal element, h_{ii}, of the hat matrix is known as the leverage of the ith observation. The leverage depends only on the predictor variables, and a high leverage indicates that the vector of predictor values $(1, x_{1i}, \ldots, x_{ki})$ is relatively far away from the mean vector (Exercise 5.2). It follows that the corresponding observed response, y_i, has a large influence on the estimates of the regression coefficients. The column of residuals R can be calculated from

$$R = Y - \hat{Y} = (I - H)Y$$

Substituting

$$Y = XB + E \quad \text{leads to} \quad R = (I - H)E$$

Hence the covariance matrix of the residuals is (Exercise 5.3):

$$(I - H)I\sigma^2(I - H)^T = (I - H)\sigma^2$$

The variance of the ith residual is $(1 - h_{ii})\sigma^2$, which is always slightly less than σ^2. The explanation for this is that the parameters are fitted so as to minimise the sum of squared errors, with the consequence that points with high leverage will tend to attract the fitted surface. Minitab will provide residual plots based on the residuals, the standardised residuals, or the Studentised deleted residuals. The standardised residuals are the residuals divided by their estimated standard deviations, i.e. $r_i/(s\sqrt{1-h_{ii}})$. The Studentised deleted residuals are calculated from the prediction error sum of squares (PRESS) residuals. The ith PRESS residual, $r_{(i)}$, is the difference between the y_i and the predicted value of y from the regression fitted with the datum $(1, x_{1i}, \ldots, x_{ki}, y_i)$ excluded. It is not necessary to calculate the n regressions because

$$r_{(i)} = r_i/(1-h_{ii})$$

A Studentised deleted residual is a PRESS residual divided by its estimated standard deviation and, given the assumptions of the regression model, it has a t-distribution with $n-k-1$ degrees of freedom. I prefer to use the simple residuals, although the Studentised deleted residuals are good for identifying unusual values. The residual plots usually display similar patterns for any choice of type of residuals. Another variation on a residual is the Cook statistic, D_i, which is a composite measure of size of residual and its leverage:

$$D_i = r_i^2 h_{ii}/((k+1)s^2(1-h_{ii})^2)$$

Cook (1977) suggested checking observations if D_i exceeds the upper 5% point of an F-distribution with $1+k$ and $n-k-1$ degrees of freedom.

It is not possible to detect any linear association between the errors and a predictor variable because this will just be absorbed into the estimate of the co-efficient. In algebraic terms, it is a consequence of the least squares algorithm that

$$\sum r_i(x_{ji} - \bar{x}_j) = 0$$

Nevertheless, it can be useful to plot residuals against predictor variables because obvious curvature indicates that quadratic and cross-product terms should be considered. A plot of residuals against the fitted values is a useful check on whether an assumption of constant variance is reasonable. If it is not, a generalised least squares analysis would be preferable although the most important modification is to adjust the width of prediction intervals. This could be done by grouping residuals according to the magnitude of the fitted value, calculating the standard deviation within groups and then regress-ing these standard deviations on the fitted values at the mid-points of the grouping intervals. The assumption that errors are normally distributed is also mainly important for prediction intervals. It can be checked by looking at a histogram and a normal plot of the residuals. If the observations can be put into time order, or any other ordering which has some physical significance, the assumption that the errors are independent can be checked by calculating their serial correlations, which are also known as autocorrelations. The serial correlation at lag l, where l is an integer, is the correlation between a column of residuals and the same column moved by l positions. This is usually calcu-lated for the first few l, e.g. $1, \ldots, 5$, from the formula

$$\sum_{i=1}^{n-l} r_i r_{i+l} \Big/ \sum_{i=1}^{n} r_i^2$$

There is no need to subtract $\bar{r}$ because the identity $\sum_{i=1}^{n} r_i = 0$ follows from the least squares criterion, and $\sum_{i=1}^{n-l} r_i$ will be very close to 0. If there is appreciable positive serial correlation, the estimated standard deviations of the coefficients will be too low and statistical significance will be overestimated. Intuitively, the positive correlation reduces the effective sample size. If there are substantial serial correlations a generalised linear regression should be used. Other aspects of serial correlation are discussed in sections 6.2.3. and 8.3.

5.3.4   Interpretation of regressions when predictor variables are correlated

Sandpiper Software Services

A software engineer left a large software company, Skipper Software, which specialises in computer packages for small boat design, to start her own business, Sandpiper Software Services. Skipper have retained her as a consultant to install one of the company's best-selling products, Waveskimmer, on clients' computer networks. Until now she has been paid by the hour for this work, but Skipper have recently asked her to submit a quotation for completing an installation before she starts work.

She will base quotations on the times taken for the last 40 jobs. Her records include her initial assessment of difficulty, on a scale from 1 to 5, the number of snags actually encountered, and the time taken, in hours. The data are given in TableWSsyspr, and some summary statistics are:

	Min	Max	Mean	Standard deviations
Difficulty	1	5	2.350	1.189
Snags	1	33	15.75	7.81
Time	10	46	20.62	8.68

The correlations between the variables are shown below.

	Time	Snags
Snags	0.306	
Difficulty	0.028	0.747

The low correlation between 'difficulty' and 'time' suggests that she cannot expect her initial estimate of difficulty to improve, noticeably, her estimate of time taken. The regression is

$$\text{time} = 20.1 + 0.20 \times \text{difficulty}$$

The standard deviation of the errors is 8.79, and the coefficient of 'difficulty' has an estimated standard error of 1.18 which is considerably larger than the coefficient itself. So, she decides to give a fixed-price quotation, based on an average of 20.6 hours per installation.

Although it will not help with her quotation, she notices the higher correlation between 'time' and 'snags' and the highest correlation between 'difficulty' and 'snags'. A regression of 'time' on 'snags' and 'difficulty' is rather surprising:

$$\text{time} = 17.1 + 0.72 \times \text{snags} - 3.32 \times \text{difficulty}$$

The coefficients of 'snags' and 'difficulty' have estimated standard errors of 0.25 and 1.63 respectively, and both estimated coefficients are statistically significant (the *P*-values are 0.006 and 0.049 respectively). The coefficient of difficulty is now negative, but this can be explained by the presence of 'snags' in the model. A large number of snags predicts a long time, but this prediction decreases as 'difficulty' increases. That is, a large number of snags occurring when the initial estimate of difficulty was low is expected to take longer to sort out than the same number of snags when a high level of difficulty was anticipated.

Substantial correlations between predictor variables can lead to difficulties of interpretation. In survey work, for example, such correlations are inherent to the population, and are often of considerable interest to the investigator. However, when designing experiments it is usually possible to arrange for the predictor variables to be uncorrelated, or approximately so. It is also often worthwhile scaling variables to avoid correlations between linear and squared, and cross-product, terms when including the latter in a regression. The advantages are that rounding errors are reduced, and linear and quadratic effects are clearly separated. The only disadvantage is that cross-product terms may have a physical significance, e.g. the product of area and depth of a soldered joint is proportional to its volume. High correlations between predictor variables lead to a near singular $X^T X$ matrix, and hence numerical instability of the algorithm. This is known as multicollinearity, and if it occurs at least one variable will have to be dropped from the model.

5.3.5   Quadratic terms

The multiple regression model given in equation (5.9) is often referred to as a linear model because it is linear in the unknown parameters. It can be used to fit polynomial curves or surfaces by defining extra predictor variables as powers or products of the original variables.

Rosewood

The original predictor variables are pedestrians/hour (x_1) and floor area (x_2) and the objective is to predict sales (y). It is possible that the size of the increase in sales due to some fixed increment in pedestrians/hour will tend to reduce as the number of pedestrians/hour increases. This can be allowed for by including a quadratic term in pedestrians/hour, i.e. x_1^2, in the model. If this is done, the fitted relationship is

$$y = 266.960 + 0.9247x_1 + 0.4344x_2 - 0.000\,318\,9x_1^2$$

with an estimated standard deviation of the errors (s) equal to 56.0, which is a slight reduction from 59.5 with only x_1 and x_2. The sample values of x_1 range from 280 to 1172, and the plot of x_1^2 against x_1 is remarkably close to a straight line. The sample correlation between x_1 and x_1^2 is 0.982. Such high correlations can make the matrix inversion an ill-conditioned numerical problem. It is better practice, and it makes interpretation easier, if x_1 is mean corrected before squaring. The mean of x_1 is 721.8, and the correlation between x_1 and $(x_1 - 721.8)^2$ is -0.022, so x_1 and $(x_1 - \bar{x}_1)^2$ are almost uncorrelated. The fitted regression is

$$y = 433.1 + 0.4644x_1 + 0.4344x_2 - 0.000\,318\,9(x_1 - 721.8)^2$$

which is algebraically identical to the previous equation.

Adding a quadratic term for area and the cross-product of pedestrians/hour with area does not improve the fit since the estimated standard deviation of the errors increases to 57.8 (Example 5.4).

Example 5.3
Danaher (1997) asked hotel staff to distribute questionnaire packs to 270 randomly selected guests at check-in. A prepaid addressed envelope was included and participants were told that their names would be entered in a draw for a bottle of champagne. Forty-seven completed questionnaires were returned. Respondents were asked to give their overall level of satisfaction on a seven-point scale, ranging from very dissatisfied to very satisfied, and their opinion of check-in, room and check-out on a three-point scale: -1 for worse than expected, 0 for as expected, and $+1$ for better than expected. The objective of the study was to determine the relative importance of the quality of these three services. Danaher fitted a regression model with quadratic terms and obtained the following equation.

$$\text{Overall-satisfaction} = 5.3 + 0.18\text{check-in} + 0.97\text{room} + 0.52\text{check-out}$$
$$+ 0.18(\text{check-in})^2 - 0.3(\text{room})^2 - 0.29(\text{check-out})^2$$

The coefficients of the squared terms were not statistically significant at the 10% level. The model with linear terms only was:

$$\text{Overall-satisfaction} = 5.32 + 0.22\text{check-in} + 0.93\text{room} + 0.45\text{check-out}$$
$$(0.30)\ (0.22) \qquad\qquad (0.23) \qquad (0.21)$$

where the estimated standard deviations of the coefficients are bracketed. The room is considered most important. Danaher also carried out a similar study on airline passengers and obtained the equation:

$$\text{Overall-satisfaction} = 5.25 + 1.06\text{crew} + 0.57\text{meals} + 0.88\text{comfort}$$
$$- 0.39(\text{crew})^2 - 0.3(\text{meals})^2 - 0.19(\text{comfort})^2$$

The results of the questionnaire data were compared with a conjoint analysis, and this work will be referred to in Chapter 7.

5.3.6   Indicator variables and the Fulmar ferry

Categorical variables with c categories can be included in regression models by setting up $(c - 1)$ indicator variables. A ferry, the *Fulmar*, sails from port A to port B and from there to port C before returning to port A. Each voyage therefore consists of three 'legs' AB, BC and CA. The *Fulmar* always takes on fuel at port A. A fuel additive, which the manufacturers claim will improve fuel consumption through a catalytic effect on the oxidation of the fuel, was included in the fuel line to the engine on a random selection of 80 from 160 voyages. For each leg of each voyage the passage time, fuel consumed, work done, draught and weather were noted. The work done was measured with an integrating power meter which was a standard fitting on the *Fulmar*. The weather was classified as strong following wind, slight following wind, calm, slight head wind and strong head wind. The main purpose of fitting regression models was to investigate whether the additive had any effect.

The leg of the voyage was coded with two indicator variables, x_2 and x_3. They were set as:

	x_2	x_3
Leg AB	0	0
Leg BC	1	0
Leg CA	0	1

It follows that the coefficients β_2 and β_3 represent the differences in fuel consumption, relative to leg AB, for legs BC and CA, respectively. The indicator variable x_4 was used to code legs for which the additive was in use, when it was set at 1. It was set at 0 if no additive was used on a voyage. Because the additive was included in the fuel line, rather than mixed in the fuel tank, there was no reason to allow for a 'carry-over' effect when investigating any catalytic action. The fuel consumption should be proportional to the work done, to a close approximation, and it was clear from a plot that any other effects were relatively small. The following regression model was eventually fitted:

$$\text{fuel used on leg} = \beta_0 + \beta_1 \text{ (work done on leg)}$$

$$+ \beta_2 x_2 + \beta_3 x_3 + \beta_4 x_4 + \text{error}$$

with the result, in coded units for commercial confidentiality:

$$\text{fuel used on leg} = 3.24 + 1.07 \text{ work} + \left\{ \begin{array}{l} 0 \quad \text{leg } AB \\ 0.57 \text{ leg } CA \\ -0.28 \text{ leg } BC \end{array} \right. + \left\{ \begin{array}{l} 0 \quad \text{no additive} \\ -0.37 \text{ additive} \end{array} \right.$$

Addition of draught or weather variables did not improve the fit.

A 90% confidence interval for β_4 was $[-0.70, -0.04]$. Although this was some weak evidence that the additive had some effect, it was expensive and a 2% reduction in fuel consumption was needed to save its cost. The average fuel consumption was about 20 units, and the operations manager had decided that evidence of at least a 1% reduction was required to justify continuing trials with the additive. Since the 90% confidence interval for its effect extended above -0.20, trials were brought to a halt. There was some statistical evidence that the leg had an effect on fuel consumed. A possible explanation for this finding is that the speed might be held more constant on one leg, thereby giving a slight improvement in engine efficiency.

At an anecdotal level, the chief engineer reported that use of the additive left the engine much cleaner. If the additive does have a cleaning effect, this might also be a mechanism for saving fuel during the period between engine overhauls. The analysis described would not detect any such effect. One approach would be to compare trends in fuel consumption for the *Fulmar*, using the additive over the period, with another ferry which was not using the additive. Alternatively, we might compare trends in fuel consumption between overhauls for the *Fulmar* when the additive was used, with trends between overhauls when it was not used.

5.3.7 How many predictor variables?

A regression model is just an empirical approximation to some underlying physical process, and there is no definitive answer to this question. If the predictor variables are significantly correlated among themselves, several models, which may look quite different, can give similar results when making predictions within the domain of values used when fitting them. Extrapolation beyond this domain is inadvisable unless there are good physical reasons for doing so, and even then should not be relied on far beyond it. The following guidelines may be useful.

1. Include variables which have an obvious physical relationship with the dependent variable in the model.
2. Only add further predictor variables if the estimated standard deviation of the errors, s, decreases sufficiently to compensate for the additional complexity. Note that a small reduction in s can be accompanied by wider intervals of prediction, away from the centroid of the predictor variables, because of the increased uncertainty in the estimates of the parameters.
3. A $(1 - \alpha) \times 100\%$ confidence interval for an individual coefficient is given by:

$$\hat{\beta}_j \pm t_{n-k-1,\alpha/2} \hat{\text{stdev}}(\hat{\beta}_j)$$

If the confidence interval does not include zero, we have evidence that the inclusion of x_j leads to more precise predictions than a regression based only on the other $n - k - 2$ predictor variables. However, we may not wish to consider predictor variables in isolation. For example, we may wish to compare a quadratic surface with a plane. The following approach can be used to decide whether the addition of a set of variables gives a significantly improved fit. Suppose model 1 is based on p predictor variables and model 2 is based on k predictor variables, where $p < k$, and consider a null hypothesis (H_0) that $\beta_j = 0$, for $p + 1 \leq j \leq k$. If H_0 is true

$$\frac{(\text{residual sum of squares for model 1} - \text{residual sum of squares for model 2})/(k - p)}{s_{\text{model 2}}^2} \sim F_{k-p,n-k-1}$$

If the calculated ratio exceeds $F_{k-p,n-k-1,\alpha}$ there is evidence to reject H_0 at the $\alpha \times 100\%$ level.

The numerator is often written in an alternative, but identical form, as:

$$\frac{(\text{regression sum of squares for model 2} - \text{regression sum of squares for model 1})}{(k - p)}.$$

Example 5.4

In the Rosewood example the regression of sales (y) on pedestrians per hour (x_1) and floor area (x_2) was given as

$$y = 385.7 + 0.464x_1 + 0.471x_2$$

This can be represented by a plane in three dimensions (Figure 5.6(a)). The residual sum of squares can be found from s by squaring it and multiplying

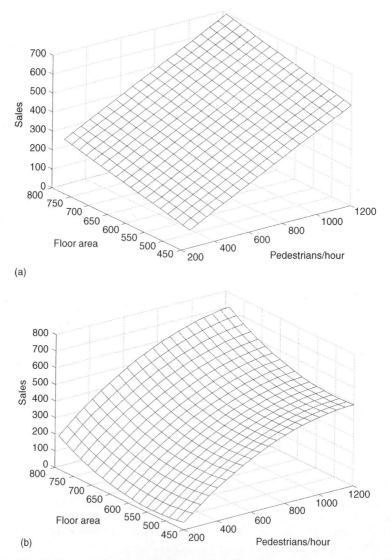

(a)

(b)

Figure 5.6 (a) Plane representing sales on floor area and number of pedestrians; (b) quadratic surface when squares and cross-product of floor area and number of pedestrians included

by $(n - k - 1)$ which is 7 in this case. However, it is more convenient to look at the analysis of variance (ANOVA) table that is given by both the Excel and Minitab regression routines. This is based on the following breakdown of the total (corrected) sum of squares, $\sum (y_i - \bar{y})^2$.

$$\sum (y_i - \bar{y})^2 = \sum ((y_i - \hat{y}_i) + (\hat{y}_i - \bar{y}))^2$$
$$= \sum (y_i - \hat{y}_i)^2 + \sum (\hat{y}_i - \bar{y})^2 + 2 \sum (y_i - \hat{y}_i)(\hat{y}_i - \bar{y})$$

The first term on the right is known as the sum of squared residuals, referred to by Excel as 'Residual' and by Minitab as 'Residual error'. The second term is the sum of squares attributed to the regression, and the sum of products which makes up the third term is 0 (see Exercise 5.6). The Minitab table for the Rosewood shops regression on x_1 and x_2 is

Source	DF	SS	MS	FF	P
Regression	2	211 302	105 651	29.87	0.000
Residual error	7	24 761	3537		
Total	9	236 062			

Here DF is the degrees of freedom, SS is the breakdown of the total sum of squares, and MS is the SS divided by the DF. The residual error mean square is s^2 and the expected value of the residual error mean square is the variance of the errors in the regression model (σ^2). In general, the expected value of the regression mean square is σ^2 plus some non-negative function of the unknown coefficients $\beta_1, \ldots, \beta_k$ and the values of the predictor variables $x_1, \ldots, x_k$. If all the $\beta_1, \ldots, \beta_k$ are zero then this expected value is just zero σ^2 and the ratio of the regression MS/residual MS has an F distribution with an $F_{k,n-k-1}$ distribution. High values of F are evidence that not all of $\beta_1, \ldots, \beta_k$ are zero. The P-value is the

$$\Pr(F > F_{\text{calc}} \,|\, F \sim F_{k,n-k-1})$$

where F_{calc} is the calculated value of the ratio. The regression of sales which included quadratic terms was given as

$$y = 392 + 0.462x_1 + 0.477x_2 - 0.000\,410(x_1 - 721.8)^2 + 0.002\,67(x_2 - 605)^2$$
$$+ 0.000\,19(x_1 - 721.8)(x_2 - 605)$$

This can be represented by a quadratic surface in three dimensions (Figure 5.6(b)). The ANOVA table is

Source	DF	SS	MS	FF	P
Regression	5	222 709	44 542	13.34	0.013
Residual error	4	13 354	3338		
Total	9	236 062			

The second model is of the general form

$$Y_i = \beta_0 + \beta_1 x_{1i} + \beta_2 x_{2i} + \beta_3 x_{3i} + \beta_4 x_{4i} + \beta_5 x_{5i} + E_i$$

where x_{3i} equals $(x_{1i} - 721.8)^2$ etc., and testing the null hypothesis

$$H_0: \beta_3 = \beta_4 = \beta_5 = 0$$

against the alternative hypothesis

$$H_1: \text{At least one of } \beta_3, \beta_4, \beta_5 \text{ not equal to zero}$$

is the test of the hypothesis that the relationship between the expected value of sales and the predictor variables, x_1 and x_2, is a plane. The calculated value of the test statistic is

$$\frac{(222\,709 - 211\,302)/(5 - 2)}{3338} = 1.14$$

and we refer this to $F_{3,4,0.10}$, which equals 4.19, to test at the 10% level. There is no evidence to reject the null hypothesis. This does not imply we have proved the null hypothesis true, but it is a useful means for deciding whether additional terms are justified. If there is no evidence against H_0, the empirical estimates of the coefficients β_3, β_4 and β_5 are probably not precise enough to be relied on.

4. A commonly quoted statistic in regression analyses is the proportion of the variability accounted for by the regression (R^2), which is known as the coefficient of determination. It is calculated from the formula:

$$R^2 = 1 - \sum r_i^2 \Big/ \sum (y_i - \bar{y})^2 = \left(\sum (y_i - \bar{y})^2 - \sum r_i^2 \right) \Big/ \sum (y_i - \bar{y})^2$$

Example 5.4 (continued)
The regression of y on x_1 and x_2 has an R^2 of $211\,302/236\,062$ which equals 0.895, or 89.5%. The regression of y on the five predictor variables has an R^2 of $222\,709/236\,062$ which equals 94.3%. R^2 will always be between 0 and 1, and 1 corresponds to an exact fit. This is its limitation. R^2 can only increase if an extra predictor variable is added to the regression because the coefficient of that variable is chosen to minimise $\sum r_i^2$. If we have $n - 1$ predictor variables we are bound to have an exact fit. A modification which allows for this is the adjusted R^2 defined by

$$R_{adj}^2 = 1 - \left(\sum r_i^2 \Big/ (n - k - 1) \right) \Big/ \left(\sum (y_i - \bar{y})^2/(n - 1) \right) = 1 - s^2/s_y^2$$

where s_y^2 is the variance of the y_i, which corresponds to fitting a model with only β_0 and no predictor variables. However, R_{adj}^2 is just a linear transform of s^2 and will increase or decrease as s^2, and hence s, decreases or increases. There is therefore no substantial advantage to using R_{adj}^2 rather than s itself, which has a straightforward interpretation as the estimated standard deviation of the errors, when considering the effect of adding more explanatory variables. Another possible use for R_{adj}^2 is comparing regressions using Y and, for example, $\ln Y$ as dependent variables. However, the R_{adj}^2 for a model with $\ln Y$ as the dependent variable would change slightly if it were to be based on differences between the y_i and predictions of the $E[y_i]$ made from the logarithmic model (Exercise 5.7).

Although R^2 is a useful statistic it does give a rather optimistic impression of the predictive capability of regression models. The ratio of the estimated standard deviation of the errors about the regression (s) to the standard deviation of the original y_i (s_y) is

$$s/s_y = \sqrt{(n-1)(1-R^2)/(n-k-1)}$$

Example 5.5
A regression with five predictor variables is fitted to a set of 100 data. The R^2 is 95%. Then the ratio

$$s/s_y = \sqrt{99 \times (1-0.95)/94} = 0.23$$

If we tried to predict Y without fitting a regression, approximate 95% limits of predictions, assuming Y is normally distributed, would be

$$\bar{y} \pm 2s_y$$

Using the regression would reduce the width of limits of prediction by a factor of approximately 0.23.

   If there is only one predictor variable R^2 is the square of the correlation coefficient, i.e. $|r| = \sqrt{R^2}$.

Example 5.6
Barrett and Weinstein (1998) wrote that corporate entrepreneurship (*CE*), flexibility (*F*), and market orientation (*MO*) are being recognised as key success factors in an increasingly competitive global economy. They investigated the relationships between these variables and business performance (*BP*). The findings are useful because all three factors can be controlled by management. The Tennessee Association of Business provided a mailing list of over 1800 diverse businesses. Barrett and Weinstein set up a sampling frame of 750 members in manufacturing companies with more than 25 employees. Questionnaires were sent to senior managers and 142 responses were received. Nine questions were designed to measure corporate entrepreneurship. There were seven questions to measure flexibility and 20 for market orientation. Respondents were also asked for their opinion of the previous year's overall performance of their business and their overall performance relative to leading competitors. Most questions were answered on a seven-point Likert scale. The correlation matrix is:

	F	MO	BP
CE	0.533	0.477	0.344
F		0.438	0.335
MO			0.477

The highest correlation between *BP* and any factor is 0.477 between *MO* and *BP*. The fitted regression is

$$BP = 1.35 + 0.72 \times MO$$

with an R^2 of 23%. If all three factors are included the regression equation is

$$BP = 0.83 + 0.17 \times CE + 0.16 \times F + 0.55 \times MO$$

with an R^2 of 25%. It is not a statistically significant improvement but it is interesting to see the result of fitting the three-factor model. In particular, the effect of MO given that CE and F are in the model is less than the effect of MO if no other predictor variables are included. This is because MO is positively correlated with both CE and F. In the model with MO, as the only predictor variable, increasing MO by one unit will tend to be associated with increases in CE and F, and BP increases by 0.72. In the model with all three predictor variables, increasing MO by one unit subject to the constraints that CE and F do not change only leads to an increase of 0.55 in BP. Barrett and Weinstein tried various models which include interaction terms, such as $F \times MO$, e.g.

$$BP = \beta_0 + \beta_1 \times CE + \beta_2 \times F + \beta_3 \times MO + \beta_4 \times F \times MO$$

which had an R^2 of 27%. The interaction allows for the effects of changes in F to depend on the value of MO, but it was not a statistically significant improvement and no numerical values of the coefficients were given. They also tried fitting the models to small, medium and large companies only. An alternative would have been to introduce indicator variables, e.g.

	IV1	IV2
Small	1	0
Medium	0	1
Large	0	0

Including $IV1$ and $IV2$ would allow for different intercepts (β_0) for different sizes. Inclusion of interactions such as $IV1 \times MO$ and $IV2 \times MO$ would allow for different coefficients of MO within the different size categories.

5.3.8   Generalised least squares

The assumptions that the E_i are independent of each other and have a constant variance can be replaced by any other assumed covariance matrix. Suppose that

$$Y = XB + E$$

where

$$E[E] = 0 \quad \text{cov}(E) = V\sigma^2$$

and the elements of E are independent of the covariates. This is a generalised linear regression model. Any covariance matrix V can be written in the form:

$$V = Q^2$$

It is straightforward to find Q, because any real symmetric matrix is diagonalisable, i.e.

$$V = MDM^{-1} \quad \text{and} \quad Q = MD^{1/2}M^{-1}$$

where D is a diagonal matrix of the eigenvalues, M is a matrix of corresponding eigenvectors, and $D^{1/2}$ is the matrix with square roots of the eigenvalues along the leading diagonal. If

$$F = Q^{-1}E$$

then

$$\mathrm{cov}(F) = \mathrm{E}[FF^T] = \mathrm{E}[Q^{-1}EE^T(Q^{-1})^T]$$

Since Q is a symmetric matrix of constants

$$\mathrm{cov}(F) = Q^{-1}\mathrm{E}[EE^T]Q^{-1}$$
$$= Q^{-1}Q^2Q^{-1}\sigma^2 = I\sigma^2$$

The original model premultiplied by Q^{-1}

$$Q^{-1}Y = Q^{-1}XB + Q^{-1}E$$

is of the form

$$W = UB + F$$

where W contains the dependent variables, U contains the explanatory variables, and the elements of F satisfy the usual assumptions. The practical problem is that V is unlikely to be known, and it is usual to make some assumptions about the form of V, possibly from initially fitting the model of Section 5.3.1 and looking at the residuals.

5.4   Non-linear regression and robust regression

5.4.1   Non-linear regression

An oil company wishes to establish a relationship between its market share of petrol sales at garage forecourts (y) and its price relative to a weighted average of other brands (x). The company reviews its pricing policy each week and an analyst will fit a model to weekly records (x_t, y_t) over the past two years ($t = 1, \ldots, 104$). One commonly used model for market share (e.g. Cohen *et al.*, 1996) is

$$y = \frac{a}{1 + bx^c}$$

where a, b and c are the parameters to be estimated. It is not linearisable.

A model for the relationship between sales y in period t and advertising expenditure x in period t (e.g. Lambin, 1997) is

$$y_t = a + b\sum_{i=0}^{m} c^i x_{t-i}$$

The expression $\sum_{i=0}^{m} c^i x_{t-i}$ is a finite exponentially weighted moving average.

A general non-linear model is

$$Y_i = f(x_i, \theta) + E_i \quad \text{for } i = 1, \ldots, n$$

where f is an arbitrary function, x_i is a vector of predictor variables, θ is a vector of unknown parameters and the E_i are errors which are defined in the same way as for the multiple regression model of Section 5.3.1. If the principle of least squares is applied, the problem is to find the θ which minimises

$$\psi = \sum_{i=1}^{n} (y_i - f(x_i, \theta))^2$$

The parameter values can be found numerically by an optimisation algorithm (Appendix 5) and SPSS has a convenient routine which implements an optimisation algorithm automatically once the function is defined. However, there may be several local minima, and convergence can be sensitive to the choice of initial values. Also, in many non-linear models there are constraints on the parameter values, e.g. $0 < c < 1$ in the model for sales on advertising expenditure. The simplest approach is to hope the algorithm naturally converges to values which satisfy the constraints. If it does not, the parameter c could be replaced by $(e^d/(1 + e^d))$, so the constraint will be satisfied for any value of the new parameter d. Problems arise if the model is over-parameterised, i.e. two parameters have a very similar effect so that it is hard to distinguish between them. The estimated variance of the errors is given by

$$s^2 = \sum (y_i - f(x_i, \hat{\theta}))^2/(n - p)$$

where p is the number of parameters estimated from the data. There are two strategies for obtaining approximate standard deviations of the estimators of the parameters. One is to linearise the model about the estimated values of the parameters using Taylor series. The other is to use bootstrap methods (Appendix 6).

Example 5.7
The data in TableWSPinkham are yearly sales and advertising expenditure for the Lydia E. Pinkham Medicine Company 1907–60 in thousand US dollars (Pankratz, 1991). There is no obvious overall trend, but the decrease in the early thirties is presumably due to the Depression. I used the non-linear regression routine in SPSS to fit the model

$$\text{sales} = a + bx\_0 + bcx\_1 + bc^2x\_2 + bc^3x\_3 + bc^4x\_4$$

where x_k is advertising expenditure k years ago. I guessed initial values for a, b and c of 700, 1 and 0.1 respectively. The results are given below.

Non-linear Regression Summary Statistics   Dependent Variable SALES

Source	DF	Sum of squares	Mean square
Regression	3	192 535 571.970	64 178 523.9899
Residual	47	5 716 091.030 28	121 618.958 09
Uncorrected total	50	198 251 663.000	
(Corrected total)	49	17 960 602.5000	

R squared $= 1 -$ Residual SS/Corrected $SS = 0.68174$

Parameter	Estimate	Asymptotic std error	Asymptotic 95% confidence interval	
			Lower	Upper
A	576.291 051 48	147.037 551 44	280.489 652 21	872.092 450 76
B	1.324 940 518	0.229 206 984	0.863 835 541	1.786 045 494
C	0.031 651 695	0.168 193 097	−0.306 709 171	0.370 012 562

Asymptotic Correlation Matrix of the Parameter Estimates

	A	B	C
A	1.0000	−0.2798	−0.2895
B	−0.2798	1.0000	−0.8174
C	−0.2895	−0.8174	1.0000

There is strong evidence that advertising in the same year is associated with sales but the estimate of c has a large standard error. The point estimate for c of 0.032 is rather small, though it might be considerably higher for monthly data. The high negative correlation between the estimators of b and c implies that if b is overestimated c will tend to be underestimated, and is to be expected with the products bc etc. in the model. Lambin (1997) proposes the model

$$\text{sales} = a(\text{sales\_1})^b (x\_0)^c$$

which is linearisable by taking logarithms. If this is fitted

$$\ln(\text{sales}) = 0.670 + 0.252(\text{sales\_1}) + 0.682(x\_0)$$

with $s = 0.1118$ and an R^2 of 89.9%.

5.4.2 Robust regression

Whilst unusual data should be investigated, and may lead to important findings, they must not be excluded from the analysis, unless they are found to be mistakes or are outside the range of values of predictor variables required for the application. Outlying points can be taken as evidence against a normal distribution of errors. If errors have a Laplace distribution (back-to-back exponentials), the maximum likelihood estimators of the parameters are obtained by minimising the sum of absolute deviations rather than the sum of squared deviations (Exercise 5.8). This can be done with a numerical optimisation routine. More generally, M-estimators minimise

$$\sum f(y_i - x_i^T \tilde{\boldsymbol{B}})$$

with respect to $\tilde{\boldsymbol{B}}$, where the function f depends on the likelihood function (hence the M) of the chosen error distribution. Hettmansperger and McKean (1998) and Montgomery and Peck (1992) give a detailed account of these and other robust regression methods.

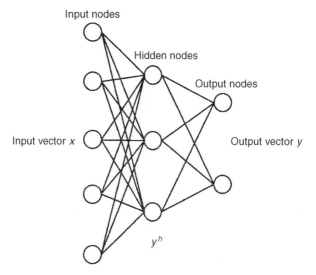

Input nodes

Hidden nodes

Output nodes

Input vector x

Output vector y

y^h

Figure 5.7   Schematic of a simple artificial neural network (ANN)

5.4.3   Artificial neural networks (ANN)

Artificial neural networks (ANN) relate outputs to inputs in a very general structure, but in most cases the results have very little physical interpretation. They seem to be particularly effective for classification problems (e.g. Michie *et al.*, 1994, and Hand and Jacka, 1998). The network in Figure 5.7 is referred to, in the neural network literature, as the architecture of a multilayer perception. It could, for example, be used to classify the output of an oil well (y) as negligible, poor, moderate or good, from geophysical data including production at neighbouring sites. The values of y_j^h at the hidden nodes are given by

$$y_j^h = f^h \left(\sum_{\text{inputs}} w_{ji} x_i \right)$$

A constant term can be included by setting one of the x_i equal to 1. The outputs are given by

$$y_k = f \left(\sum w_{kj} y_j^h \right)$$

where w_{ji} and w_{kj} are weights which are determined empirically. The functions f^h and f are usually chosen to limit the output to a finite range, often approximating a threshold function which is 0 if the sum is below some critical value and 1 otherwise. The tanh function and the logistic function ($f(x) = 1/(1 + e^{-x})$) are common choices. In this example there is only one output, and the function f might be the composition of a logistic function and a function which maps the ranges: (0, 0.25), [0.25, 0.5), [0.5, 0.75), [0.75, 1) into the four categories. The weights have to be estimated from a data set for which both x and y are known. A typical ANN will be set up with many parameters to be estimated and it is likely that a good fit can be obtained for the data. However, as in multiple regression analysis, an excessive number of parameters will give an excellent fit but lead to unreliable predictions. Therefore it is good practice to divide the data set into a training set and a set for testing the performance of the ANN fitted to the training set. The parameters are estimated by some non-linear least squares algorithm. With a large number of parameters to be estimated the optimisation is challenging and sophisticated algorithms have been developed. There is no guarantee of finding a global minimum, and stochastic optimisation procedures such as genetic algorithms (Chapter 7) and simulated annealing can be used.

5.5   Measurement error (ME) models

Let X_i and Y_i represent n measurements of moisture content of pitta bread to be made by forced air oven method and on-line moisture meter, respectively. It is assumed that they are subject to independent errors, E_i and H_i, respectively, of zero mean and constant variances σ_E^2 and σ_H^2. Let u_i and v_i represent the unobservable error-free measurements. It is supposed that

$$v_i = \alpha + \beta u_i$$

for some α and β, and one of the aims of the project is to see whether there is any evidence that these parameters differ from 0 and 1, respectively.

Substituting

$$u_i = X_i - E_i \quad \text{and} \quad v_i = Y_i - H_i$$

into the model relating v_i and u_i gives

$$Y_i = \alpha + \beta X_i + (H_i - \beta E_i)$$

Despite first appearances, this is not the standard regression model because the assumption that the errors are independent of the predictor variable is not satisfied:

$$\text{cov}(X_i, (H_i - \beta E_i)) = E[(X_i - u_i)(H_i - \beta E_i)]$$
$$= E[(E_i(H_i - \beta E_i))] = -\beta \sigma_E^2$$

Maximum likelihood solution

The usual approach to this problem is to assume the errors are normally distributed and that the ratio of σ_H^2 to σ_E^2, denoted by λ in the following, is known. In practice, some reasonable value has to be assumed for λ, preferably based on information from replicate measurements. The maximum likelihood estimates of α and β, using a notation S_{xy} for $\sum (x - \bar{x})(y - \bar{y})$, etc. are:

$$\hat{\beta} = \left\{ (S_{yy} - \lambda S_{xx}) + [(S_{yy} - \lambda S_{xx})^2 + 4\lambda S_{xy}^2]^{1/2} \right\} / (2S_{xy})$$
$$\hat{\alpha} = \bar{y} - \hat{\beta}\bar{x}$$

These follow from the likelihood function $\mathscr{L}$ which is proportional to

$$\sigma_E^{-1}\sigma_H^{-1}\exp\left\{-\tfrac{1}{2}\sigma_E^{-2}\sum_{i=1}^{n}(x_i-u_i)^2-\tfrac{1}{2}\sigma_H^{-2}\sum_{i=1}^{n}(y_i-(\alpha+\beta u_i))^2\right\}$$

if α, β, σ_E and u_i are treated as the unknown parameters.

Estimates of σ_E^2 and σ_H^2 are given by

$$\hat{\sigma}_H^2=\lambda\hat{\sigma}_E^2=(S_{yy}-\hat{\beta}S_{xy})/(n-2)$$

If $\lambda=1$, the values of $\hat{\alpha}$ and $\hat{\beta}$ are the slope and intercept of the line such that the sum of squared perpendicular distances from the plotted points to the line is a minimum.

Upper and lower points of the $(1-\alpha)\times100\%$ confidence interval for β are given by

$$\lambda^{1/2}\tan(\arctan(\hat{\beta}\lambda^{-1/2})\pm\tfrac{1}{2}\arcsin(2t_{\alpha/2}\theta))$$

where

$$\theta^2=\frac{\lambda(S_{xx}S_{yy}-S_{xy}^2)}{(n-2)((S_{yy}-\lambda S_{xx})^2+4\lambda S_{xy}^2)}$$

and $t_{\alpha/2}$ is the upper $(\alpha/2)\times100\%$ point of the t-distribution with $n-2$ degrees of freedom. An approximation to the standard deviation of $\hat{\beta}$ is given by one-quarter of the width of this interval. Since $\hat{\beta}$ is independent of $\bar{x}$

$$\mathrm{var}(\hat{\alpha})\simeq\hat{\sigma}_H^2/n+\bar{x}^2\mathrm{var}(\hat{\beta})+\hat{\beta}^2\hat{\sigma}_E^2/n$$

Cheng and Van Ness (1999) provide a general survey of measurement error models.

Table 5.4 ME regression

Oven	Meter
17.0	14.6
16.3	18.8
8.3	8.9
17.0	16.4
13.9	13.8
17.5	14.7
16.1	16.0
15.8	13.8
20.6	20.1
8.4	12.3
14.4	12.8
13.4	15.2
7.4	8.9
11.1	9.2
14.6	13.7
5.0	9.6
13.7	14.8
8.3	8.8
17.5	13.7
14.8	15.9

Example 5.8

The data in Table 5.4 are 20 pairs of measurements of moisture made with the forced air oven (x) and the on-line moisture meter (y). The standard deviation of the errors in the on-line meter measurements is half the standard deviation of the errors in the oven measurements. Hence λ is $(1/2)^2$. The formulae give:

$$S_{xx} = 329.97 \quad S_{xy} = 217.31 \quad S_{yy} = 199.40$$

$$\hat{\beta} = 0.840\,146 \quad \hat{\alpha} = 2.211\,83$$

$$\hat{\sigma}_H = 0.9669 \quad \hat{\sigma}_E = 1.9338$$

$$\theta^2 = 0.004\,038\,73$$

A 95% confidence interval for β is [0.629, 1.177] and its approximate standard deviation is 0.137.

5.6   Poisson regression

The records of machine breakdowns in TableWSmachbrk are from a large manufacturing company in the Middle East. The number of breakdowns was considered a serious problem and a manager was asked to find ways of improving the situation. He decided to involve the operators in routine preventative maintenance and allowed them time to do this. He also arranged a training programme, and encouraged all staff in the company to take more responsibility for their work. Absenteeism was reduced, and replies to questionnaires suggested that most people welcomed the changes. The success of the initiative, over 18 months at least, is clear from the graph in Figure 5.8(a). One question the manager hoped to answer from the data was whether employing additional operators to provide extra cover for absentees would help reduce breakdowns.

A Poisson regression model is a convenient abbreviation for a generalised linear model with Poisson errors. A routine for fitting it is available in the NAG Statistical Add-ins for Excel. Let Y be the number of breakdowns in a week. We assume that Y has a Poisson distribution with a mean μ, and that μ depends on several predictor variables which can include the number of absentees and time measured in weeks from January 1996. To keep the specification general we will refer to the predictor variables as $x_1, \ldots, x_k$. The dependence is assumed to have the form:

$$\mu = a(x_1)^{b_1}(x_2)^{b_2} \ldots (x_k)^{b_k}$$

which is equivalent to

$$\ln(\mu) = \ln(a) + b_1 \ln(x_1) + \ldots + b_k \ln(x_k) \tag{5.12}$$

This model is fitted by maximum likelihood. The principle is straightforward but implementing it requires numerical optimisation. Suppose we have data from n weeks; $(x_{1i}, \ldots, x_{ki}, y_i)$ for i from 1 to n. The likelihood of these data ($\mathcal{L}$) is the probability of obtaining the observed y_i, assuming Y_i are independently

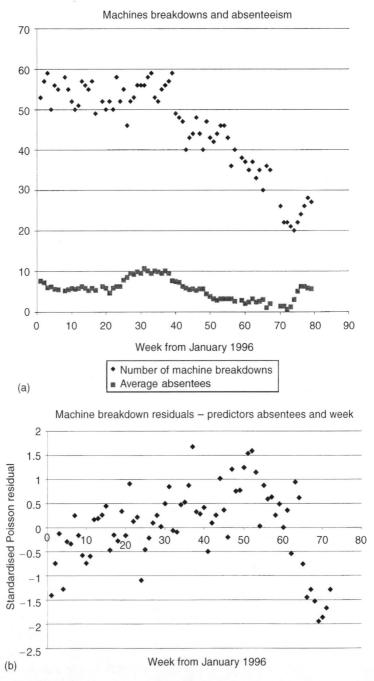

Figure 5.8 (a) Number of machine breakdowns and average numbers of absentees against week number; (b) residuals after fitting regression of breakdowns on absentees and week

distributed as Poisson variables with means μ_i where

$$\mu_i = E[Y_i] = a(x_{1i})^{b_1} \ldots (x_{ki})^{b_k}$$

That is,

$$\mathscr{L} = \prod_{i=1}^{n} e^{-\mu_i}(\mu_i)^{y_i}/y_i!$$

Since the y_i are observed values, the value of $\mathscr{L}$ depends on the parameters $a, b_1, \ldots, b_k$. The numerical values of these parameters which maximise $\mathscr{L}$ are the maximum likelihood estimates (MLE). Once installed in Excel the NAG Add-ins are found in: Insert ▷Function ▷ Engineering. The on-line help facility is in Nag ▷ Nag Extract ▷ Help. I started with the numbers of breakdowns in column B and the numbers of absentees and week numbers in columns C and D respectively. Select POISSON_GLM. The dialogue box should be completed as follows:

Link L   Mean M   X C2:D75   Isx {1, 1}   Y B2:B75

The 'L' is for the log link between μ and equation (5.12) which is linear in the parameters: $\ln(a)$, and the b_j. The 'M' is because we wish to include an intercept in the model, as an estimate of $\ln(a)$. The Isx is a convenient device for selecting which of the x_l we wish to include (1 for include, 0 for omit). The result is contained in ⟨list of 764⟩. To expand this: copy to at least 764 rows (but do not paste), region should be highlighted, click on the formula bar, then ctrl + shift + enter. The following information should now be displayed (in a column).

#G02GCF:0 |dev| 51.18622 |idf| 71 |irank| 3

|b(3)| 3.919 633   0.033 079   −0.007 979

|se(3)| 0.069 507   0.007 953   0.000 886

followed by some auxiliary details. The parameter estimates are given under |b(3)|. That is, the fitted model is

$$\ln(\mu) = 3.919\,63 + 0.033\,079 x_1 - 0.007\,979 x_2$$

where x_1 is the number of absentees and x_2 is the week number from January 1996. The deviance D is defined by,

$$D = 2\ln\left(\frac{\text{likelihood of maximal model with } n \text{ parameters}}{\text{likelihood of fitted model with } p \text{ parameters}}\right)$$

and is given under |dev|. The maximal model has the same number of parameters as there are data and is therefore an exact fit with $\mu_i = y_i$ for $i = 1, \ldots, n$. For a Poisson model, D has an approximate chi-square distribution with $(n - p)$ degrees of freedom.

The mean of a chi-square distribution is equal to its degrees of freedom and percentage points can be found from Excel (f_x ▷ Statistical ▷ CHIINV). In this case the deviance is 51.19 which is less than the mean of a chi-square distribution with $74 - 3 = 71$ degrees of freedom. There is therefore no evidence against the hypothesis that the data are from the fitted model. In general, large values of D

indicate that the model is not realistic and can be due to influential predictor variables which have not been included in the model and which may not have been monitored. Alternatively, a large value of D can arise because the response has a higher variance than a Poisson variable. This is known as overdispersion. We can make a more detailed assessment of the fit by investigating the residuals. Since the variance of a Poisson distribution is equal to its mean, standardised residuals are given by

$$(y_i - \mu_i)/\sqrt{\mu_i}$$

In this case there is a distinctive pattern (Figure 5.8(b)) and the first ten serial correlations are: 0.69, 0.57, 0.47, 0.39, 0.31, 0.26, 0.22, 0.22, 0.08, −0.01. One explanation for such high serial correlations is that machines are not repaired quickly, but remain out of use for weeks because of delays in obtaining spare parts. Another possible explanation is that breakdowns depend on the type of work, or the workload, which might persist over several weeks. The manager decided to investigate repair times, and asked for these to be recorded in addition to the breakdown data.

The estimated standard deviations of the parameter estimates are given under $|\text{se}(3)|$. These estimates are based on the assumptions that the response has a Poisson distribution, and that responses are independent. If the deviance D exceeds the degrees of freedom the estimates should be multiplied by $(D/\text{degrees of freedom})^{\frac{1}{2}}$ to allow for possible overdispersion. Approximate 95% confidence intervals are given by

$$b(j) \pm 2\text{se}(j)$$

In this case the 95% confidence interval for the coefficient of the number of absentees is calculated as $[0.017, 0.049]$, and the 95% confidence interval for the coefficient of the week number is $[0.0062, 0.0098]$. It seems there is strong evidence of a decreasing trend and evidence that the mean number of breakdowns increases if the number of absentees increases. However, there are high serial correlations and these confidence intervals are too narrow. Apart from the serial correlations, the results of any investigation over time are equivocal. For instance, it could be argued that absenteeism increases during the hot season, which it does, and the hot weather, rather than any reduced maintenance, makes machines more susceptible to breakdown. The manager needs to discuss this possibility with an engineer. The conclusions of the investigation are that there has been a substantial reduction in breakdowns over the past 18 months, that repair times need further investigation, that provision of extra cover for absentees is worth trying, and that the numbers of breakdowns, numbers of absentees and repair times should all be monitored.

The NAG algorithm allows predictor variables to be included as an offset. An offset x_i is raised to a power of 1 and hence the coefficient of $\ln(x_i)$ in the expression for $\ln(\mu)$ is also 1. In the machine breakdown example the number of working days, or better, machine hours worked, in each month would have been included as an offset if the data had been available.

If NAG add-ins are not available and the counts are quite high and fairly stable, then a standard multiple regression would suffice. If there is a substantial trend the variance could be stabilised by analysing square roots of the counts (Exercise 5.11) using standard multiple regression. However, the Poisson regression has several advantages. It avoids the possibility of negative estimates being included in the limits of prediction. It avoids the complication that arises with logarithmic regression, and also with a regression for any other function of Y (Section 5.2.2), because it models $\ln[E[Y]]$ rather than $E[\ln(Y)]$. A Poisson regression also has the advantage of being an explicit formulation for a Poisson variable rather than an approximation.

5.7 Logistic regression

A car manufacturer tests a prototype design of plastic fuel tank by subjecting tanks to high crushing impacts which exceed those expected in crashes. The results are given in Table 5.5. The manufacturer wishes to estimate the impact at which the probability of failure is 1 in 1000. It is not satisfactory to assume that the probability of failing (p) depends linearly on the impact (x), because this makes no allowance for the constraint that p must lie between 0 and 1. A more realistic assumption is that the logit of p, defined as the natural logarithm of the ratio of p to $1 - p$, depends linearly on x.

$$\text{logit}(p) = \ln(p/(1 - p))$$

The ratio of p to $(1 - p)$ is known as the odds, and the inverse relationship is

$$\text{logit}(p) = \theta \Leftrightarrow p = e^{\theta}/(1 + e^{\theta})$$

Notice then $\text{logit}(p)$ ranges from $-\infty$ to $+\infty$ as p moves from 0 to 1.

5.7.1 Binomial response

Let Y_i be the number of failures in n_i independent trials (each with two possible outcomes, 'failure' and 'success'), with k predictor variables taking values $x_{1i}, \ldots, x_{ki}$, and a probability of failure p_i. That is

$$E[Y_i/n_i] = p_i$$

The binary logistic regression model is

$$\text{logit}(p_i) = \beta_0 + \beta_1 x_{1i} + \ldots + \beta_k x_{ki}$$

The exact maximum likelihood solution is preferable to analysing $\text{logit}(\hat{p}_i)$, where $\hat{p}_i = Y_i/n_i$, using the standard model. One reason for this is that $\hat{p}_i$

Table 5.5 Results of tests of plastic fuel tanks

Impact (coded units) x_i	Number of tanks tested n_i	Number of tanks failing y_i
20	8	0
22	8	2
24	6	3
26	4	2
28	4	2
30	4	4

might equal 0 or 1, in which case logit $(\hat{p}_i)$ is not defined, and a second reason is that the variance of logit $(\hat{p}_i)$ depends on n_i and to a lesser extent p_i. The likelihood function is

$$\mathscr{L} = \prod_{i=1}^{m} p_i^{y_i}(1-p_i)^{n_i-y_i} {}_{n_i}C_{y_i}$$

where the p_i are defined in terms of the parameters $\beta_0, \ldots, \beta_k$ and the known values of the predictor variables. This has to be maximised with respect to the parameters, and Minitab or the NAG add-ins for Excel provide a convenient means of doing so. An important special case is when all the n_i equal 1, and the response Y_i is a binary variable, possibly representing passing or failing a specification.

The NAG BINOMIAL-GLM function is used in a similar way to the POISSON_GLM (Section 5.6) except that the position of the column of n_i has to be entered in a box which is labelled T (for total). The Minitab Binary Logistic Regression routine is straightforward to use. Either routine gives the following results for the data in Table 5.5.

$$\hat{\beta}_0 = -11.095 \quad \widehat{\text{stdev}}(\hat{\beta}_0) = 3.694$$

$$\hat{\beta}_1 = 0.4361 \quad \widehat{\text{stdev}}(\hat{\beta}_1) = 0.1498$$

with a deviance of 4.507 on 4 degrees of freedom. The estimated standard deviations are based on an assumption that the number of failures has a binomial distribution in which case the expected value of the deviance equals the degrees of freedom. In this case the deviance is reasonably close to its expected value. If there is evidence of overdispersion the standard deviations of estimators should be increased by multiplying by the square root of the ratio of the deviance to its degrees of freedom. The fitted model is

$$\ln(p/(1-p)) = -11.095 + 0.4361x$$

To answer the original question, if p is equal to 0.001 then logit$(p) = -6.907$ and the corresponding value of the impact x is 15.8. However, there is considerable uncertainty about this estimate and a parametric bootstrap could be used to construct a confidence interval for the impact. Minitab also provides an 'odds ratio' of 1.55, with a 95% confidence interval from 1.15 to 2.07. This odds ratio is the increase in the odds if the impact x is increased by 1 unit, i.e. exp(0.4361)

Survival of international corporations in China

Pan and Chi (1999) use binary logistic regression to investigate the impact of entry timing, mode of market entry, market focus and location on the survival of multinational companies in China. They interviewed the general managers of a stratified random sample of about 1000 foreign manufacturing enterprises. Their results are reproduced in Table 5.6. Most of the predictor variables are indicator variables to represent different categories. For example, there are three modes of market entry categories: cooperative operations between local and foreign firms, equity joint ventures (EJV) and wholly foreign-owned ventures. They used two indicator variables to compare the other categories

Table 5.6   Likelihood of survival in the second year (1993) of the two year study. Binary logistic regression: survival = 1/closure = 0. Reprinted from Pan and Chi (1999). Copyright © 1999 John Wiley & Sons Ltd

Determinants	No control variables	With control variables
I. Year in which production began		
Prior to 1988	−0.456 (0.289)	−0.459 (0.319)
Between 1988 and 1990	−	−
After 1990	0.323 (0.252)	0.492 (0.281)[*]
II. Mode of market entry		
Cooperative operations	−0.592 (0.302)[**]	−0.652 (0.330)[**]
Equity joint ventures	−	−
Wholly foreign-owned	−0.428 (0.287)	−0.219 (0.353)
III. Local sales vs exporting		
Sales in China	0.002 (0.003)	0.004 (0.003)
Chinese sale channels in China		
No	−	−
Yes	0.371 (0.349)	0.374 (0.377)
IV. Location-specific factors		
MNC location		
SEZ	−	−
National municipalities	1.323 (0.514)[**]	1.217 (0.565)[**]
Open coastal cities	0.516 (0.405)	0.435 (0.445)
Provincial capitals	0.250 (0.494)	0.118 (0.541)
Other cities	0.595 (0.341)	0.520 (0.369)
Preferential income tax treatment		
No	−	−
Yes	0.742 (0.250)[***]	0.652 (0.268)[**]
Annual railroad shipping capacity	−27.326 (18.211)	−31.066 (19.822)
V. Control variables		
Reached 80% production capacity		
No		−
Yes		0.867 (0.237)[***]
Amount of investment (in US$10,000 at time of survey)		0.014 (0.020)
Imported technology		0.251 (0.253)
Imported machinery and equipment		−0.470 (0.286)
Local working capital		−0.001 (0.003)
Chinese raw materials		−0.003 (0.003)
Manager compensation scale		−0.079 (0.044)[*]
Personnel recruiting		
Other means		−
Open recruitment		0.049 (0.277)
Recommended by Chinese firm		0.849 (0.278)[***]
Recommended by local labour department		0.127 (0.295)
Hired through employment service firms		0.188 (0.308)
Type of industry		
Electric and electronic		−
Food		0.510 (0.531)
Beverage and tobacco		−2.012 (1.001)[**]
Textile		0.561 (0.439)
Wearing apparel		−0.081 (0.369)
Leather, fur products		0.022 (0.480)
Wood, bamboo products		−0.223 (0.654)
Paper and printing		−0.698 (0.603)
Chemical materials		0.614 (1.113)

Table 5.6   Continued

Determinants	No control variables	With control variables
Type of industry (continued)		
Chemical products		0.062 (0.538)
Petroleum and coal products		N/A
Rubber products		0.369 (1.106)
Plastic products		C.077 (0.466)
Non-metallic mineral products		1.168 (1.089)
Basic metals		N/A
Fabricated metals		−0.587 (0.527)
Machinery and equipment		−0.327 (0.559)
Transport equipment		−0.218 (0.753)
Precision instruments		0.357 (0.694)
Miscellaneous industrial products		0.161 (0.628)
Country of origin of foreign MNC		
Hong Kong (Macao)		–
Taiwan		−0.060 (0.301)
USA		−0.202 (0.494)
Japan		1.081 (0.782)
Other Asian countries		0.030 (0.598)
EU countries		−0.417 (0.732)
All other countries		N/A
$-2 \log L$	36.78***	88.77***
Concordant	66.7%	74.9%
N	859	859

Note: There were 122 operations that were closed down by the second year of our survey. The dependent variable is continuing business or closedown in 1993. The independent measures were survey information for the previous year, i.e. 1992. The analysis used SAS logistic procedure. Numbers in parentheses are standard errors, and test of significance is based on chi-square statistics. *** $p < 0.01$; ** $p < 0.05$; * $p < 0.1$.

with equity joint ventures: x_3 is 1 if the mode of market entry was a cooperative operation and 0 otherwise, and x_4 is 1 for wholly foreign-owned ventures and 0 otherwise. In the model with control variables included the coefficients of x_3 and x_4 are −0.652 and −0.219 respectively. The odds of a cooperative operation surviving are a factor of $\exp(-0.652)$, which equals 0.52, less than the odds of an EJV company surviving. The estimated standard deviation of this co-efficient is given as 0.330, so an approximate 95% confidence interval for the factor is given by

$$[\exp(-0.652 - 2 \times 0.330), \exp(-0.652 + 2 \times 0.330)]$$

which is [0.27, 1.01]. Hence the finding is not quite statistically significant at a 5% level although it is at a 10% level. In this context the control variables were not of primary interest to the investigators, but were included because they might have affected the probability of survival. Then the estimates of the effects of different modes of market entry, for example, are made after allowing for differences in control variables as well as differences in the other variables of primary interest. The comparison between the cooperative operations and EJV is not very sensitive to the inclusion of the control variables. The value of the estimate of the constant $(\hat{\beta}_0)$ is not given in their table but it is irrelevant to the comparisons. Elsewhere they state that 12% of EJVs went out of business in 1993, compared to 20% of wholly foreign-owned ventures and 24% of cooperative operations.

5.7.2 Ordinal logistic regression

Pan and Chi (1999) also investigated the effect of entry timing, mode of market entry, market focus and location on the profitability of multinational companies in China. The level of profitability was categorised on an ordinal scale: incurred a heavy loss (0); incurred a slight loss (1), made no profit (2), profit was less than 3% (3), profit was between 3% and 8% (4), profit was between 8% and 15% (5), and profit was greater than 15% (6). The ordinal logistic regression model has the form

$$\text{logit}(\Pr(\text{category} > j)) = \beta_{0j} + \beta_1 x_1 + \ldots + \beta_k x_k$$

where $\beta_{00} > \beta_{01} > \ldots > \beta_{06}$. Their results, for the model without control variables, are reproduced in Table 5.7. This table does not include $\hat{\beta}_{0j}$, but approximate values, based on the 1992 overall results, are: $\hat{\beta}_{00} = 2.20$, $\hat{\beta}_{01} = 0.41$, $\hat{\beta}_{02} = 0.12$, $\hat{\beta}_{03} = -0.53$, $\hat{\beta}_{04} = -1.27$, $\hat{\beta}_{05} = -2.44$.

Minitab has an Ordinal Logistic Regression routine which can be used to fit the model.

Table 5.7 Profitability in 1992 and 1993: ordered logistic regression. Reprinted from Pan and Chi (1999) with permission. Copyright © 1999 John Wiley & Sons Ltd

Determinants	No control variables
I. Year in which production began	
Prior to 1988	0.556 (0.144) ***
Between 1988 and 1990	–
After 1990	−0.713 (0.106) ***
II. Mode of market entry	
Cooperative operations	−0.318 (0.157) **
Equity joint ventures	–
Wholly foreign-owned	−0.312 (0.140) **
III. Local sales vs. exporting	
Sales in China	−0.000 (0.001)
Chinese sales channels in China	
No	–
Yes	0.052 (0.133)
IV. Location-specific factors	
MNC location	
Special economic zone	–
National municipalities	0.493 (0.221) **
Open coastal cities	0.378 (0.210) *
Provincial capitals	0.060 (0.247)
Other cities	0.082 (0.185)
Preferential income tax treatment	
No	–
Yes	−0.000 (0.090)
Annual railroad shipping capacity	−22.221 (6.639)***

Note. Numbers in parentheses are standard errors, and test of significance is based on chi-square statistics. *** $p < 0.01$; ** $p < 0.05$; * $p < 0.1$.

Nominal logistic regression

Another variant on logistic regression supported by Minitab is nominal logistic regression. The model has the form

$$\Pr(\text{category } j) = \exp(\boldsymbol{\nu}_j) / \left(1 + \sum_{i=1}^{J-1} \exp(\boldsymbol{\nu}_i) \right)$$

for $j = 1, 2, \ldots, (J-1)$ and

$$\Pr(\text{category } J) = 1 - \sum_{i=1}^{J-1} \Pr(\text{category } j)$$

The $\boldsymbol{\nu}_j$ have the form

$$\boldsymbol{\nu}_j = \beta_{0j} + \sum_{i=1}^{k} \beta_{ij} x_i$$

5.8   Principal components analysis

In chemical manufacturing processes, and power generation, variables such as flows, temperatures and pressures are measured continuously at many locations to help ensure safe operation and as part of the process control. These measurements will typically be highly correlated. Although a policy of monitoring individual variables should pick up dramatic changes, such as those arising from burst pipes or pump failures or burner failures, it is likely to miss more subtle signs of insidious changes such as leaks or small temperature rises. Particular linear combinations of the variables may be much more effective at doing this. For example, suppose that measurements from three flow meters are positively correlated. Two measurements well above target and one well below target could signal a problem even if none of the measurements considered on its own would do so. The technique of principal components can be used to construct promising linear combinations (e.g. Tong and Crowe, 1995), and it is common to concentrate on monitoring a few of these.

  Another potential application of principal components is in the manufacture of complex mechanical components. Bank Bottom Engineering manufacture industrial robot arms. They routinely measure critical features at 14 positions on the arm. There can be up to six physically distinct measurements, lengths and angles, made at each position. Data from 206 robot arms manufactured during March of 1995 are given on the book website. There are 29 variables, and the objective is to find a smaller number of linear combinations of these variables which could be used to monitor the quality of the product.

Principal components algorithm

Suppose x is a random variable with k components:

$$x^T = [X_1, \ldots, X_k]$$

and a variance–covariance matrix V. The first principal component W_1 is a linear combination of the X_i,

$$W_1 = a_{11}X_1 + \ldots + a_{1k}X_k$$

which has a maximum variance, subject to the constraint that $\sum a_{1i}^2 = 1$. If we define

$$a_1^T = [a_{11}, \ldots, a_{1k}]$$

then the variance of W_1 is

$$\text{var}(W_1) = \text{var}(a_1^T x)$$
$$= a_1^T V a_1$$

and the constraint is that $a_1^T a_1 = 1$. In vector terms the length of a_1 is 1, and this is known as the normalisation condition. The method of Lagrange multipliers can now be used to find the a_1 which maximises

$$G = a_1^T V a_1 + \lambda(1 - a_1^T a_1)$$

in which λ is the Lagrange multiplier. The derivative of G with respect to a_1 is

$$\frac{\partial G}{\partial a_1} = 2V a_1 - 2\lambda a_1$$

and setting it to zero to find the stationary point gives

$$(V - \lambda I)a_1 = 0 \tag{5.13}$$

This is an eigenvalue problem, and for a non-zero solution we require

$$\det(V - \lambda I) = 0$$

There will be k roots of this equation, but the following argument demonstrates that we need the greatest: multiply equation (5.13) by a_1^T to obtain

$$a_1^T V a_1 = \lambda$$

and remember that the objective is to maximise $a_1^T V a_1$, which is the variance of W_1. It is convenient to denote this greatest eigenvalue by λ_1, the corresponding eigenvalue being a_1. The second principal component W_2 is a linear combination of the X_i

$$W_2 = a_{21}X_1 + \ldots + a_{2k}X_k$$

which has a maximum variance, subject to the constraints that the length of a_2 as a vector equals 1, i.e. the normalisation criterion $\sum a_{2i}^2 = 1$, and a_1 and a_2 are orthogonal, i.e. $a_1^T a_2 = \sum a_{1i}a_{2i} = 0$. We now need to introduce two Lagrange multipliers λ_2 and θ. The objective is to maximise

$$G = a_2^T V a_2 + \lambda_2(1 - a_2^T a_2) + \theta a_1^T a_2$$

Setting the partial derivative with respect to a_2 equal to 0 gives

$$2(V - \lambda_2 I)a_2 + \theta a_1 = 0 \tag{5.14}$$

Pre-multiply by a_1^T and use the orthogonality and normalisation conditions to obtain:

$$2a_1^T V a_2 + \theta = 0$$

Now pre-multiply equation (5.13) by a_2^T and use the orthogonality condition to demonstrate that $a_1^T V a_2$ equals zero. It follows that $\theta = 0$ and we are left with a second eigenvalue problem

$$(V - \lambda_2 I)a_2 = 0$$

so λ_2 is the second greatest eigenvalue and a_2 is the corresponding eigenvector. The construction is continued until all k principal components have been obtained. Two properties of the eigenvalues of any square matrix A are that their sum is tr(A) and their product is det(A) (Exercise 5.12). In this application the sum of the eigenvalues will equal the sum of the variances of the X_i. That is:

$$\sum \lambda_i = \sum \text{var}(W_i) = \sum \text{var}(X_i) = \text{tr}(V)$$

and λ_j represents the proportion of the total variance of the X_i which is accounted for by the jth principal component. A PCA can also be applied to the correlation matrix, which is the variance–covariance matrix of the standardised variables.

Example 5.9
The data in TableWSrobot are measurements made on 206 robot arms manufactured by Bank Bottom Engineering during March of 1995. To demonstrate the algorithm we will apply a PCA to the three measurements made at point 4. The three deviations are lengths in the x, y and z directions, and as they are measured in the same unit there is some justification for applying PCA to the covariance matrix. Some summary statistics are:

	Mean	Standard deviation
x	0.005 31	0.011 80
y	0.012 82	0.008 73
z	0.036 86	0.015 40

The covariance matrix is

		x	y	z
	x	13 921		
$10^{-8} \times$	y	2859	7619	
	z	11 342	9262	23 706

where the upper half follows from symmetry. The eigenanalysis of the variance–covariance matrix is

Eigenvalue λ, $10^{-8} \times$	34 430	8156	2660
Proportion of variance	0.761	0.180	0.059
Cumulative proportion	0.761	0.941	1.000

and the principal components are

	PC1	PC2	PC3
x	0.492	0.807	−0.326
y	0.331	−0.520	−0.788
z	0.806	−0.279	0.522

All the coefficients for PC1 are positive, so it can be thought of as a measure of how much the component is over-size. The second principal component can be interpreted as a measure of compression in the y- and z-directions accompanied by a lengthening in the x-direction, as if it were being squashed. Together the first two components account for 94% of the sum of the variances of the three original variables.

We will now compare the above results with those obtained by applying a PCA to the correlation matrix. This is:

	x	y	z
x	1		
y	0.278	1	
z	0.624	0.689	1

The eigenanalysis of the correlation matrix is

Eigenvalue λ	2.0786	0.7241	0.1974
Proportion of variance	0.693	0.241	0.066
Cumulative proportion	0.693	0.934	1.000

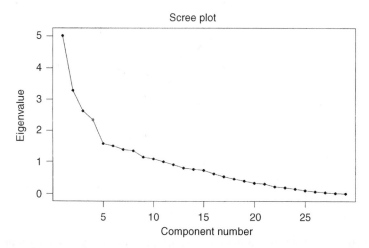

Figure 5.9   Scree plot for principal components calculated from robot arm data

and the principal components are

	PC1	PC2	PC3
$(x - \bar{x})/s_x$	−0.520	0.748	0.413
$(y - \bar{y})/s_y$	−0.551	−0.663	0.507
$(z - \bar{z})/s_z$	−0.653	−0.036	−0.757

The interpretation of PC1 is qualitatively similar to the previous analysis but PC2 is now the difference between the standardised x deviation and the standardised y deviation.

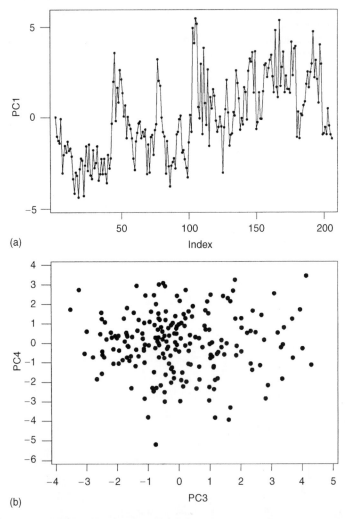

(a)

(b)

Figure 5.10   (a) Plot of first principal component for robot arms against order of manufacture; (b) plot of fourth principal component against third principal component for robot arms

The choice between using the variance–covariance matrix and the correlation matrix depends on whether or not it makes more sense to standardise the variables. If it does, then the correlation matrix should be used.

If we now apply a principal component analysis to all 29 variables, excluding 7(2) which is constant at 0, the correlation matrix should be used because some measurements are lengths and others are angles. A plot of eigenvalue against component number is known as a scree plot (Figure 5.9). The first four principal components account for 45.8% of the variance.

The component scores are calculated for each arm by adding the products of standardised variables with the PC coefficients. A plot of PC1 against order of manufacture and a plot of PC4 against PC3 are shown in Figure 5.10(a) and (b). PC1 shows an upward trend over the period, but there are no striking anomalies in the plot of PC4 against PC3. In this case the production manager might consider monitoring production by keeping charts of the first four principal components, on the basis that the corresponding eigenvalues are noticeably larger than the remainder.

5.9   Factor analysis

A researcher, Marzuki Bakar, was asked to investigate communications within a company that manufactures industrial machinery. He asked employees to complete the questionnaire given in Appendix 8. There are 70 statements, and respondents are asked to indicate their level of agreement with each statement on a four-point scale (four points exclude an indifferent position). The statements are grouped under 14 categories (as shown on the score sheet), and the total score for each category lies between 0 and 15. The categories can be described as follows.

OBJ	Communication of company's objectives
ORG	Organisation within the company
CON	Control within the company
TEAM	Teamwork
RES	Availability of resources
PEOP	Staff development
EFF	Efficiency
MKT	Sensitivity to changing market conditions
PSV	Persuasiveness of management
INT	Integration within the company
DOWN	Downward flow of information
TRUS	Trust
UP	Upward flow of information
ADMIN	Administration

In all cases, high scores are desirable. It seems reasonable to suggest that successful communication depends on employees' motivation as well as on ready access to clear information. One of the objectives of the research was to investigate the extent to which a respondent's scores for each category could be accounted for by two underlying factors which were termed 'information availability' and 'social relations'. All employees of the company were asked to complete the questionnaire and 97, which was most of them, did so. The results are given on the book website.

Common factor model

Suppose x is a random variable with k components

$$x^T = [X_1, \ldots, X_k]$$

zero mean, and a variance–covariance matrix V. For instance $X_1, \ldots, X_k$ could be scores, relative to the population means, on k tests which a company uses as part of its recruitment procedures. The model for the factor structure assumes that there are m common factors $F_1, \ldots, F_m$, with m substantially less than k, which between them account for appreciable proportions of the scores. They might be presentation skills, verbal ability and mathematical ability. The remainder of the scores is attributed to specific factors $E_1, \ldots, E_k$. So

$$X_1 = b_{11}F_1 + \ldots + b_{1m}F_m + E_1$$
$$\vdots$$
$$X_k = b_{k1}F_1 + \ldots + b_{km}F_m + E_k$$

The coefficient b_{ij} is the loading of factor j on variable i. The model can be written in matrix form as

$$x = Bf + e$$

The F_i are defined to be independently distributed with zero means and unit variances, and the E_i are independently distributed with zero means and variances ϕ_i^2. Hence, the variance of x can be partitioned as

$$V = BB^T + G$$

where G is a diagonal matrix with elements ϕ_i^2. In component form,

$$\text{var}(X_i) = b_{i1}^2 + \ldots + b_{im}^2 + \phi_i^2$$

The sum, $b_{i1}^2 + \ldots + b_{im}^2$, is known as the communality. It is that part of the variance of X_i which is related to the common factors. The variance of the E_i, ϕ_i^2, is known as the specificity. It follows from the definition of the model that

$$\text{cov}(X_i, X_j) = \sum_{l=1}^{m} b_{il}b_{jl}$$

Fitting the model

If we have scores on k tests for a sample of n people we can calculate a sample variance–covariance matrix $\hat{V}$. We then have to choose a number of factors (m), after which the estimation problem is to find a matrix $\hat{B}$ and diagonal matrix $\hat{G}$ such that

$$\hat{V} = \hat{B}\hat{B}^T + \hat{G}$$

This cannot be achieved exactly because the number of free parameters is less than the number of scalar equations in any interesting case $(k > 2m)$.

However, any approximate solution $\hat{B}$ is not unique because

$$\hat{B}\hat{B}^T = (\hat{B}U)(\hat{B}U)^T$$

for any orthogonal matrix U, i.e. $UU^T = I$. In geometric terms orthogonal matrices represent a rotation about the origin. The simplest approach to the problem is to begin with a principal components analysis and use the first few principal components as unrotated factors (Manly, 1994). A full principal components analysis is of the form

$$W_1 = a_{11}X_1 + \ldots + a_{1k}X_k$$
$$\vdots$$
$$W_k = a_{k1}X_1 + \ldots + a_{kk}X_k$$

The transformation from x to w is orthogonal so the inverse relationship is simply

$$X_1 = a_{11}W_1 + \ldots + a_{k1}W_k$$
$$\vdots$$
$$X_k = a_{1k}W_1 + \ldots + a_{kk}W_k$$

(See Exercise 5.10.)

The first m principal components can be used as factors if they are normalised by dividing by their standard deviations, $\sqrt{\lambda_j}$. Hence

$$b_{ij} = \sqrt{\lambda_j}\, a_{ji}$$

The specific factors are not uncorrelated with each other but all approaches to the problem give some residual correlations. Minitab version 12 offers the principal component method of extraction and a more sophisticated method which assumes the data are from a multivariate normal and applies the maximum likelihood principle (e.g. Morrison, 1967, and Everitt and Dunn, 1991). This gives smaller residual correlations and also provides the basis for a test of the hypothesis that m common factors are sufficient to explain the observed correlations. A factor analysis can be applied to either the covariance matrix or the correlation matrix, which is just the variance–covariance matrix of standardised test scores.

Rotation of factors

If there are just two factors, the effect of rotation can be seen graphically (Exercise 5.10). The varimax criterion is to rotate the axes so as to maximise the variance of the squared loadings within each column of the matrix $\hat{B}$.

Factor score coefficients

In some applications we may wish to estimate a score on each factor for each person taking the test. There are again several different methods, which give slightly different results, but Bartlett's (1937, 1938) is relatively straightforward. The scores for a person are found by minimising the sum of squares of the specific factors divided by their standard deviations. The expression to be minimised is

$$\sum_{i=1}^{k} E_i^2 / \phi_i^2 = (x - Bf)G^{-1}(x - Bf)$$

If the derivative with respect to the factor scores, f, is set equal to zero

$$f = (B^T G^{-1} B)^{-1} B^T G^{-1} x$$

The estimated factor scores for a person, $\hat{f}$, are given by premultiplying that person's vector of test scores x by

$$(\hat{B}^T \hat{G}^{-1} \hat{B})^{-1} \hat{B}^T \hat{G}^{-1}$$

Results

The means, standard deviations and correlations between the 14 variables are given in Table 5.8. The results of a Minitab factor analysis, using the principal

Table 5.8   Communications questionnaire: summary statistics

Variable	Mean	StDev
OBJ	11.216	1.589
ORG	6.237	1.778
CON	9.959	1.040
TEAM	6.093	1.860
RES	7.175	1.233
PEOP	9.1649	0.9539
EFF	7.144	1.974
MKT	8.268	1.490
PER	5.412	1.281
INT	7.969	1.365
DOWN	7.113	1.658
TRUS	7.567	1.561
UP	4.072	1.602
ADMN	9.031	1.104

Correlations

	OBJ	ORG	CON	TEAM	RES	PEOP	EFF	MKT	PER	INT	DOWN	TRUS	UP	ADMN
ORG	0.53													
CON	0.52	0.53												
TEAM	0.54	0.44	0.35											
RES	0.26	0.24	0.40	0.16										
PEOP	0.50	0.59	0.31	0.40	0.19									
EFF	0.48	0.37	0.49	0.40	0.40	0.37								
MKT	0.31	0.07	0.28	0.01	0.47	0.13	0.47							
PER	0.61	0.59	0.47	0.47	0.38	0.52	0.50	0.30						
INT	0.63	0.44	0.45	0.51	0.39	0.52	0.60	0.32	0.53					
DOWN	0.56	0.42	0.35	0.49	0.30	0.44	0.51	0.16	0.44	0.48				
TRUS	0.65	0.69	0.41	0.57	0.23	0.61	0.47	0.09	0.70	0.52	0.57			
UP	0.64	0.56	0.45	0.51	0.24	0.60	0.52	0.16	0.65	0.62	0.49	0.77		
ADMN	0.61	0.43	0.59	0.33	0.39	0.32	0.51	0.37	0.54	0.48	0.52	0.45	0.45	

Table 5.9 PCA for the communications questionnaire

Principal component factor analysis of the correlation matrix

Unrotated factor loadings and communalities

Variable	Factor 1	Factor 2	Communality
OBJ	−0.815	−0.040	0.667
ORG	−0.724	−0.283	0.604
CON	−0.669	0.226	0.499
TEAM	−0.650	−0.313	0.520
RES	−0.474	0.585	0.567
PEOP	−0.682	−0.308	0.559
EFF	−0.716	0.318	0.613
MKT	−0.370	0.749	0.699
PER	−0.800	−0.035	0.642
INT	−0.771	0.085	0.602
DOWN	−0.695	−0.065	0.488
TRUS	−0.820	−0.349	0.793
UP	−0.809	−0.238	0.711
ADMN	−0.710	0.287	0.586
Variance	6.9467	1.6040	8.5508
% Var	0.496	0.115	0.611

Rotated factor loadings and communalities – varimax rotation

Variable	Factor 1	Factor 2	Communality
OBJ	0.718	−0.389	0.667
ORG	0.766	−0.134	0.604
CON	0.455	−0.541	0.499
TEAM	0.718	−0.069	0.520
RES	0.101	−0.746	0.567
PEOP	0.742	−0.091	0.559
EFF	0.447	−0.643	0.613
MKT	−0.073	−0.833	0.699
PER	0.702	−0.386	0.642
INT	0.615	−0.473	0.602
DOWN	0.628	−0.305	0.488
TRUS	0.882	−0.127	0.793
UP	0.815	−0.216	0.711
ADMN	0.458	−0.614	0.586
Variance	5.5081	3.0427	8.5508
% Var	0.393	0.217	0.611

Factor score coefficients

Variable	Factor1	Factor2
OBJ	0.113	−0.039
ORG	0.181	0.097
CON	0.009	−0.171
TEAM	0.181	0.118
RES	−0.131	−0.347
PEOP	0.183	0.113
EFF	−0.015	−0.223
MKT	−0.197	−0.427
PER	0.110	−0.041
INT	0.067	−0.103
DOWN	0.107	−0.017
TRUS	0.214	0.125
UP	0.177	0.067
ADMN	−0.006	−0.206

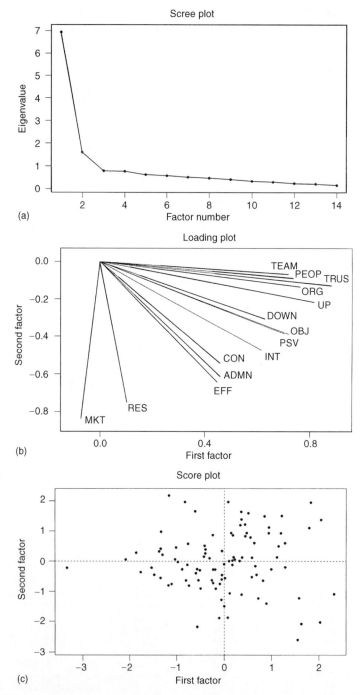

Figure 5.11   Minitab factor analysis of questionnaire data; (a) variance contributions; (b) loadings; (c) score plot for respondents

components method of extraction on the correlation matrix followed by varimax rotation, are shown in Table 5.9. The rotation matrix, U, is

$$\begin{pmatrix} 0.8548 & -0.5189 \\ 0.5189 & 0.8548 \end{pmatrix}$$

which corresponds to an anti-clockwise rotation of 149°. The scree plot is shown in Figure 5.11(a). The plot of the rotated factor loadings shown in Figure 5.11(b) is a great help for explaining the results, which are quite consistent with an interpretation that the first factor represents 'information availability' whilst the second represents 'social relations'. For example, both 'Teamwork' and 'Communication of the company's objectives' are high on information availability, but the former is higher on the social relations scale. It is plausible that 'Sensitivity to changing market conditions' and 'Availability of resources', which are lowest on both factors, are less dependent on good communications within the company than the other variables.

The factor scores for each respondent are calculated by multiplying that respondent's standardised scores by the factor score coefficients. For example, the first person's original scores are

$$(15, 7, 11, 7, 8, 10, 8, 7, 6, 10, 8, 9, 6, 10)$$

which become

$$(2.381, 0.429, 1.001, 0.488, 0.669, 0.875, 0.434, -0.851,$$
$$0.459, 1.488, 0.535, 0.918, 1.203, 0.878)$$

when they are standardised. If these standardised scores are multiplied by the factor score coefficients, we obtain the following factor scores for the first person

$$(1.291, -0.199)$$

Pairs of scores can be calculated for all 97 respondents, and these are shown in Figure 5.11(c). The follow up is to see whether there is a tendency for some jobs, or some departments, to be associated with respondents having low scores.

5.10 Discriminant function analysis

A consultant specialises in advising companies how to implement management systems which comply with ISO9000, and which are easy to adapt in response to changing requirements from customers. The consultant intends sending promotional material to companies which are likely to be interested in this service and would appreciate a method for selecting such companies.

One of the questions on the survey sent to Malaysian companies was whether they were likely to seek registration to ISO9000 in the future if they were not already registered to it. There were three categories of response; 'no', 'maybe' and 'yes'. The companies can be grouped according to this response, and a discriminant function can be used to predict which group a company is in from other variables. For the discriminant function to be useful, these variables will have to be restricted to those that are either known, or can easily be found out, for all companies, whether or not they took part in the survey. As an example, I have selected: the number of employees; whether or not the company's business is based on natural resources; and the production technology classed as small batch, large batch and continuous process.

Although discriminant function analysis is a widely used technique, nominal logistic regression (Section 5.7.2) and ANN (Section 5.4.3) are alternative methods for achieving a similar end. If the categories are ordered, as they are in this case, ordinal logistic regression (Section 5.7.2) might be preferable with a small data set because there are fewer parameters to estimate. Discriminant function analysis is based on distances of multivariate data points from the group means.

Denote the annual sales revenue by x_1, the percentage of annual sales revenue from foreign sales by x_2, and the natural logarithm of the number of employees by x_3. Let x_4 be 1 for a small batch company and 0 otherwise, and let x_5 be 1 for continuous production and 0 otherwise. Then x_i is a multivariate observation $(x_1 x_2 x_3 x_4 x_5)^T$ for the ith company. The first step is to calculate a sample mean $\bar{x}_k$ and variance–covariance matrix $\hat{V}_k$ from the data for the companies in each group. Then a pooled variance–covariance matrix $\hat{V}$ can be calculated as a weighted average of the $\hat{V}_k$, with weights proportional to sample sizes less one. The Mahalanobis distance is used to measure the distance of an observation from the centre of each group:

$$d_k^2(x_i) = (x_i - \bar{x}_k)^T \hat{V}^{-1} (x_i - \bar{x}_k)$$

If the covariances happened to be zero this would be the Euclidean distance, calculated from measurements made in standardised units. In general, an adjustment is made for covariances, so that the distance is a measure of the distance from the centre of a multivariate normal distribution. This is apparent if $d_k^2(x)$ is compared with the pdf of the multivariate normal distribution for a p-dimensional variable:

$$f(x) = (2\pi)^{-p/2} |V|^{-\frac{1}{2}} \exp(-\tfrac{1}{2}(x - \mu)^T V^{-1} (x - \mu))$$

The Mahalanobis distance can be rewritten as

$$d_k^2(x_i) = -2[\bar{x}_k^T V^{-1} x_i - 0.5\bar{x}_k^T V^{-1} \bar{x}_k] + x_i^T V^{-1} x_i$$

This result is easy to check. An observation is classified into group k, if the Mahalanobis distance $d_k^2(x_i)$ is the smallest. The advantage of using the second form for $d_k^2(x_i)$ is that the term in the square brackets, being the only term to depend on the group (k), is a linear function of x known as the linear discriminant function. Since it is multiplied by -2, a point x_i will be nearest the group for which the linear discriminant function is largest.

Example 5.10
Ninety of the Malaysian companies were not registered to ISO9000 and answered the question about their plans for future registration. The response has been coded 0 for 'no', 1 for 'maybe' and 2 for 'yes'. The predictor variables are: the natural logarithm of the number of employees in a company (x_1); whether or not a company is resource based (x_2) coded as 0 for 'no' and 1 for 'yes'; and two indicator variables which serve to compare small batch

production and continuous processes with large batch production, x_3 being 1 for small batch and 0 otherwise, and x_4 being 1 for a continuous process and 0 otherwise. The means for the different groups are:

	Group		
Variable	0	1	2
x_1	2.85	4.37	4.48
x_2	0.80	0.64	0.43
x_3	0.40	0.44	0.45
x_4	0.00	0.08	0.17
Group size	5	25	60

The lower half of the symmetric pooled covariance matrix is:

$$\begin{array}{rrrr} 1.506 & & & \\ -0.053 & 0.245 & & \\ -0.117 & -0.015 & 0.255 & \\ -0.010 & 0.027 & -0.062 & 0.117 \end{array}$$

The linear discriminant functions for the three groups are:

	Group		
Variable	0	1	2
Constant	−5.390	−9.350	−9.4216
x_1	2.271	3.345	3.415
x_2	3.848	3.329	2.392
x_3	3.061	4.019	4.304
x_4	0.902	2.303	3.421

The Mahalanobis distances between groups j and k can be calculated as $d_k^2(\bar{x}_j)$ or equivalently $d_j^2(\bar{x}_k)$. These are referred to as the squared distance between groups in the Minitab results: between groups 0 and 1, 1.86; between groups 0 and 2, 2.88; and between groups 1 and 2, 0.30. Before using the discriminant function to predict group membership, its performance can be assessed by predicting group membership for the companies used to calculate it.

		True group		
		0	1	2
Predicted group	0	4	5	8
	1	0	11	12
	2	1	9	40

The discriminant functions indicate that large companies using continuous processes are most likely to seek ISO9000 registration, but it is not particularly reliable with only 61% correct classifications. The consultant has asked how likely it is that a continuous process company with 500 employees will intend obtaining ISO9000 registration. Since ln(500) is 6.215 the vector x is (6.215, 0, 0, 1) and the discriminant function values for groups 0, 1 and 2 are 9.63, 13.74 and 15.22 respectively. The prediction is that it belongs to group 2 and it is likely that the company will be interested in the consultant's advice. Minitab calculates the Mahalanobis distances from the point x to the different group means and the probability of memberships: 20.724 from group 0 with probability 0.003, 12.499 from group 1 with probability 0.185, and 9.538 from group 2 with probability 0.812.

A common application of discriminant function analysis is to assess credit worthiness. It may be useful as a means of identifying a group of people who are at risk of defaulting on payments but it is not a substitute for skilled judgement. People who fall in a group which is perceived as risky should have their cases reviewed by a manager.

Prior probabilities can be incorporated into a discriminant function analysis, and you are asked to check the result in Exercise 5.13. It is possible to relax the assumption that the group variance–covariance matrices are equal, which justifies taking their weighted mean, and to define

$$d_k^2(x_i) = (x_i - \bar{x}_k)\hat{V}_k^{-1}(x_i - \bar{x}_k)$$

However, this does not simplify into a linear function and the method is known as quadratic discriminant analysis. Notice that $d_k^2(\bar{x}_j)$ no longer equals $d_j^2(\bar{x}_k)$.

5.11   Cluster analysis

Discriminant function analysis is used for classifying observations into specific groups which have been defined at the beginning of the analysis. In contrast to this, cluster analysis looks for natural groupings arising from the data. It is based on some measure of similarity of multivariate observations $x_i^T = (x_{i1}, \ldots, x_{ik})$ and $x_j^T = (x_{j1}, \ldots, x_{jk})$. There are many possible measures but the simplest, and the Minitab default, is the Euclidean distance

$$d_{ij} = [(x_{i1} - x_{j1})^2 + \ldots + (x_{ik} - x_{jk})^2]^{\frac{1}{2}}$$

An example from the Minitab Release 12 User's Guide and Johnson and Wichern (1992) is the comparison of 12 breakfast cereal brands in terms of five nutritional characteristics: protein, carbohydrate and fat contents; calories and percentage of the daily allowance of Vitamin A per 37.5 g serving. The data are given in Table 5.10. The Minitab cluster analysis commands are:

Multivariate ▷ Cluster Observations
Variables: Protein–Vitamin A
Linkage Method: Complete
Distance Measure: Euclidean
⊠ Standardise variables
• Number of clusters 4

The dendogram is shown in Figure 5.12. Everitt and Dunn (1991) give a detailed account of cluster analysis.

Table 5.10 Cereal nutritional information (Johnson and Wichern, 1992)

Row	Brand	Protein	Carbo	Fat	Calories	Vitamin A
1	Life	6	19	1	110	0
2	Grape Nuts	3	23	0	100	25
3	Super Sugar Crisp	2	26	0	110	25
4	Special K	6	21	0	110	25
5	Rice Krispies	2	25	0	110	25
6	Raisin Bran	3	28	1	120	25
7	Product 19	2	24	0	110	100
8	Wheaties	3	23	1	110	25
9	Total	3	23	1	110	100
10	Puffed Rice	1	13	0	50	0
11	Sugar Corn Pops	1	26	0	110	25
12	Sugar Smacks	2	25	0	110	25

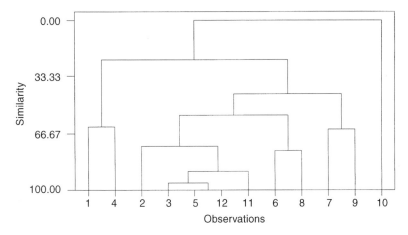

Figure 5.12 Results of Minitab cluster analysis for cereal data

5.12 Summary

1. Multiple regression is a technique for predicting some continuous variable (y) from a set of predictor variables (x_j). There is no restriction that the x_j be uncorrelated although interpretation of results is easier if they are. If they can be chosen in advance it is desirable that they be uncorrelated, or nearly so. The predictor variables can be continuous or discrete and discrete variables can be used as indicator variables for different categories.
2. Excel has a multiple regression routine. It requires that the values of the predictor variables are in k consecutive columns. The NAG add-ins for Excel include most of the more specialist analyses described in this chapter. However, I find a specialist statistics package, especially Minitab, easier to use. It is also quite easy to move worksheets between Excel and Minitab. Minitab Release 12 does not include a Poisson regression routine

or a macro which can be used to fit models by non-linear least squares. If SPSS is available, its non-linear least squares routine is straightforward to use.
3. Factor analysis is a technique for describing a multivariate data set in terms of a small number of common factors and independent variation.

Exercises

5.1 X has a log-normal distribution if $\ln X$ has a normal distribution. Let $Y = \ln X$ and $Y \sim N(a, b^2)$. Then

$$f(y) = \frac{1}{b\sqrt{2\pi}} \exp\{-[(y-a)/b]^2\}$$

(i)   Start with $F(y)$, deduce an expression for $F(x)$ and differentiate to obtain

$$f(x) = \frac{1}{xb\sqrt{2\pi}} \exp\{-[(\ln x - a)/b]^2\} \quad \text{for } 0 \le x$$

(ii)   Find the mean of the distribution for X, $E[X] = E[e^Y]$, from either

$$\int xf(x)\,dx \quad \text{or} \quad \int e^y f(y)\,dy$$

The mean, standard deviation, and skewness of X are

$$\mu = \exp(a + b^2/2)$$
$$\sigma = \mu[\exp(b^2) - 1]$$
$$\gamma = [\exp(b^2) - 1]^3 + 3[\exp(b^2) - 1]$$

(iii) In general, suppose $\phi(X)$ is some function of a variable X which has a mean μ and variance σ^2. A Taylor series expansion for $\phi(X)$ about μ is:

$$\phi(X) \simeq \phi(\mu) + \phi'(\mu)(X - \mu) + \phi''(\mu)(X - \mu)^2/2$$

Deduce that $E[\phi(X)]$ is approximately $\phi(\mu) + \phi''(\mu)\sigma^2/2$. Apply this result to Y with ϕ equal to $\exp(\ )$ and obtain an approximation to μ_X.

5.2   A manager in a company which sells electronic equipment thinks that sales may depend on the number of special offers displayed in the window. The manager intends trying $0, 1, 2, \ldots, 6$ offers in a random order during seven weeks in February and March, and will record sales (y). If the number of offers is x, calculate the leverages.

5.3   Check from its definition that H is an idempotent matrix, i.e.

$$H^2 = H$$

Show that if H is idempotent then $I - H$ is also idempotent.

5.4 Find the least squares estimator of the slope of a line constrained to go through the origin if the model is

$$Y_i = bx_i + E_i$$

and the E_i satisfy the usual assumptions.

5.5 The diagram below represents a junction box in an oil pipeline. The oil enters at A and leaves at either B or C, and there is no leakage. There are three meters 1, 2 and 3 which give readings of flow which are subject to independent measurement errors with zero mean and the same variance σ^2. Let the flows be θ_1, θ_2 and θ_3. Because there is no leakage θ_1 equals the sum of θ_2 and θ_3 and we can arbitrarily choose any two of the flows to be the unknown parameters. The measurements of the

flows are Y_1, Y_2 and Y_3. Choosing θ_2 and θ_3 as the unknown parameters leads to an error sum of squares,

$$\psi = (Y_1 - (\theta_2 + \theta_3))^2 + (Y_2 - \theta_2)^2 + (Y_3 - \theta_3)^2$$

(i) Find the least squares estimators of θ_1, θ_2 and θ_3.
(ii) What is the variance of these estimators?

5.6 Prove that

$$\sum (y_i - \hat{y}_i)(\hat{y}_i - \bar{y}) = 0$$

for a regression with one predictor variable x_i. A hint is to substitute

$$\hat{y}_i = \bar{y} + \hat{\beta}(x_i - \bar{x})$$

where, in the notation of Section 5.5, $\hat{\beta} = S_{xy}/S_{xx}$. Generalise the result to an arbitrary number of predictor variables.

5.7 Take the six pairs

x	1	2	3	4	5	6
y	1	4	9	16	25	40

(i) Regress y on x and find R^2, R^2-adjusted and the mean and standard deviation of the errors defined as $y_i - \hat{y}_i$. Note that the latter is $\sqrt{4s^2/5}$ rather than s for comparability with part (ii).
(ii) Regress $\ln y$ on x and find R^2, R^2-adjusted and the mean and standard deviation of the errors defined as $y_i - \exp(\ln \hat{y}_i)$

5.8 Consider the model

$$Y_i = \beta_0 + \beta_1 x_i + E_i$$

(a) Suppose that the E_i are independent $N(0, \sigma^2)$. Show that the least squares estimators of β_0 and β_1 are the same as the maximum likelihood estimators.

(b) Suppose the E_i have a Laplace distribution with parameter θ. That is, the Laplace distribution, with mean 0, has a pdf

$$f(w) = \frac{1}{2\theta}\exp\left(\frac{-|w|}{\theta}\right)$$

(i) Show that the variance of E_i is $2\theta^2$.

(ii) Show that the ML estimators of β_0 and β_1 are found by minimising the sum of absolute deviations.

5.9 Useful matrix results for statistical analysis.

(i) Let $A = (a_{ij})$ be a square $n \times n$ matrix. The *trace* of A is defined by

$$\text{tr}A = \sum_{i=1}^{n} a_{ii}$$

If B is another $n \times n$ matrix, it follows from the definition that

$$\text{tr}(A + B) = \text{tr}\,A + \text{tr}\,B$$

Also, if P and Q are $m \times n$ and $n \times m$ matrices, respectively, then

$$\text{tr}(PQ) = \text{tr}(QP)$$

since

$$\sum_{i=1}^{m}\left\{\sum_{\alpha=1}^{n} p_{i\alpha}q_{\alpha i}\right\} = \sum_{i=1}^{n}\left\{\sum_{\alpha=1}^{m} q_{i\alpha}p_{\alpha i}\right\}$$

Verify this result for $m = 1$ and $n = 2$.

(ii) Let A be a square $n \times n$ matrix and c any $n \times 1$ matrix which is not identically equal to a column of zeros. The scalar expression $c^T A c$ is known as a *quadratic form* in the elements of c. The matrix A is *positive definite* if and only if

$$c^T A c > 0$$

and *positive semi-definite* if and only if

$$c^T A c \geq 0$$

For example, let Y_1 and Y_2 be random variables with variances σ_1^2 and σ_2^2 and covariance σ_{12}. The covariance matrix of Y_1 and Y_2 is

$$C = E[(Y - E(Y))(Y - E(Y))^T] = \begin{pmatrix} \sigma_1^2 & \sigma_{12} \\ \sigma_{12} & \sigma_2^2 \end{pmatrix}$$

where $Y^T = (Y_1 Y_2)$. Show that C is positive semi-definite. Also show that

$$\text{var}(c^T Y) = c^T C c$$

This is a generalisation of the result for a single random variable. It is true for any number of random variables Y_i in an $n \times 1$ matrix Y. The essential steps in the proof are demonstrated by the 2×1 case considered above.

(iii) If an $n \times n$ matrix A is positive semi-definite and has an inverse, its inverse is also positive semi-definite. To prove this, set $c = A^{-1}v$ in the definition of positive semi-definiteness.

(iv) If Y is an $n \times 1$ matrix of random variables Y_i and A is a constant $n \times n$ matrix, then

$$E[Y^T A Y] = E[Y]^T A E[Y] + \text{tr}(AC)$$

where C is the covariance matrix of Y. This is a generalisation of the result,

$$E[aY_i^2] = aE[Y_i]^2 + a\,\text{var}[Y_i]$$

for a single random variable. Verify the matrix result for $n = 2$; the general proof follows a similar argument.

5.10 In two dimensions an anti-clockwise rotation of the vector w through an angle θ about 0 is given by Uw where

$$U = \begin{pmatrix} \cos\theta & -\sin\theta \\ \sin\theta & \cos\theta \end{pmatrix}$$

(i) By taking v equal to $(1 \quad 0)^T$ and then $(0 \quad 1)^T$, verify that U is the rotation described.

(ii) What is the geometric interpretation of U^T?

(iii) If $v = Uw$ show that $w = U^T v$.

5.11 Use a Taylor series expansion about μ to show that if $X \sim \text{Poisson}(\mu)$ then $\sqrt{X}$ has an approximately constant variance.

5.12 Let A be any 2×2 matrix. Show that the sum and product of the eigenvalues equals $\text{tr}(A)$ and $\det(A)$ respectively.

5.13 From the Minitab Release 11 Reference Manual.
Suppose we have prior probabilities for group membership in a discriminant function analysis p_k and let $f_k(x)$ be the pdf for group k. The posterior probability that x is in group k is

$$p_k f_k(x) \Big/ \sum p_j f_j(x)$$

The largest posterior probability is equivalent to the largest value of $\ln(p_k f_k(x))$. Show that if $f_k(x)$ are normal with a common variance–covariance matrix then this is equivalent to finding the maximum of the values of

$$-2[\mu_k^T V^{-1} x - 0.5\mu_k^T V^{-1}\mu_k + \ln p_k]$$

6

Control and improvement of the process

6.1 Introduction

6.1.1 Process control

The objective of process control is to maintain the process, or system or sub-system, in some desirable steady state. This can be achieved automatically, for example when thermostats in an office are used to control the central heating, or it may involve human intervention, for example when machines in a workshop are reset at the beginning of a shift. It can involve continuous feed-back control, or at least some control action whenever the process is sampled, or just monitoring with occasional action when something untoward happens. The former strategy tends to be associated with 'process industries' that continu-ously manufacture some product, a common means of operation in the chemical sector, and with the terms automatic process control (APC) and engineering process control (EPC). The monitoring strategy is often adopted for batch processing, and by the 'parts' industry. The results of monitoring are usually displayed on control charts, which are an important aspect of statistical process control (SPC). However, it is the natural behaviour of the system and the cost of intervention, rather than the product, which should dictate control strategy.

A process is said to be in a state of statistical control if deviations from target (errors) are randomly and independently distributed with a mean of zero and a constant variance, such deviations often being referred to as common cause variation. The paradigm for applying control charts is a process that is naturally in a state of statistical control, but occasionally affected by special cause variation. The presence of special cause variation is signalled by the control chart, and action is then taken to either remove it or to allow for it. Deming (1986) emphasises that there is no rationale for adjusting a process which is in statistical control because errors are independent, so past errors give no information about the next error, and warns that tampering will increase the process variability.

Stationarity is a weaker requirement than statistical control. The errors are not defined as independent. The mean and autocorrelation structure of the distribution of the errors are assumed to be constant over time. Stability is a closely related concept. A process is stable if it returns to an equilibrium after it is given a perturbation. With this definition, a stable process could include some cyclical variation in the mean, for example, and need not satisfy the definition of stationarity. Alwan and Roberts (1995) estimated that over 85% of applications of standard control charts are to processes that provide evidence of substantial medium-term variation, such as cycles. Medium-term variation may be acceptable if the product remains within speci-fication. However, control limits based on a variance which has been calcu-lated as an average of variances of small sub-groups of consecutive observations will lead to an excessive number of action signals. This is because the variation within sub-groups will be less than that for the process as a whole (e.g. Caulcutt, 1995). An alternative to removing medium-term variation, or any other cause of autocorrelation amongst the errors, is to make use of the information and apply a feedback control (e.g. Box and Kramer, 1992, and Metcalfe, 1992). Jones *et al.* (1996) adapted Deming's funnel experiment (Deming, 1986), which was originally constructed to illus-trate the follies of tampering, to provide a very convincing demonstration of the use of feedback. The drawback is that the cost of applying regular feed-back control action might be prohibitive if it involves stopping the production line every time.

Box and Luceño (1997) point out that many processes, if they were not appropriately controlled, would permanently drift away from the target – possibly with disastrous results. Such processes would be non-stationary, and unstable, if left in their natural state. In most cases feedback control will be essential, but it is possible that the trend in the mean is sufficiently predictable for allowance to be made in advance. This is known as feedforward control, and wear of machine tools is an example. The machine might be set up to remove slightly too much metal, on average, at the beginning of the shift, knowing that the tool will wear during the shift and end up removing too little metal.

6.1.2 Feedback control and feedforward control

Feedback control relies on regular process measurements to detect any differ-ence between the actual state and desired state, the difference (error) being used as a control signal. Inventions of devices for achieving automatic feedback control can be traced back to ancient Egyptian water clocks (Mayr, 1970). Mechanical devices for automatic control are still in common use. If the temperature in my office drops below 60°F the thermostats, which are bi-metallic strips connected to radiator valves, open the valves more. Another thermostat is used to monitor the temperature of the circulating water and increase the heat supply if the water temperature falls below the stipulated value. In contrast to passive mechanical systems, automatic control on modern chemical plant will typically use transducers to convert process measurements into electrical signals which are fed into a microprocessor. The output of the microprocessor is directed to an actuator which converts the electrical signal into a physical change in the process inputs. However, feedback control does not have to be automatic. I recently visited a factory which makes car tyres. The width of the rubber sheet, which is cut for the side walls, is monitored every half hour. The process operator can adjust the width at a console, without stopping the process, and easily implement feedback control if it is thought to be appropriate. Box and Luceño give several examples of 'feedback control' by manual adjustment. It is useful to reserve the term feedback control for situations in which control action is at least anticipated after every process measurement.

A common example of feedforward control is the setting of timing clocks for central heating systems. The disadvantage is that the system cannot automatically respond to unexpectedly warm or cold weather. Another example of feedforward control is the setting of traffic light dwell times. It can be argued that there are feedback loops in the overall traffic control system because settings will be changed if the traffic flow is unsatisfactory. However, this feedback is over a much longer time period than the daily process. It is also possible to control the dwell times by automatic traffic counters, in which case the feedback is explicit.

6.1.3   Process improvement

This chapter concentrates on maintaining processes in some acceptable steady state. This is an important short-term aim, but in many cases there will be scope to improve the process by, for example, increasing the average yield or reducing the variability. Control charts help process managers identify potential improvements. The next stage is to carry out well-designed experiments to test these ideas. Experiments can be carried out off-line, or incorporated into the day-to-day running of the plant. The latter strategy, which is sometimes referred to as evolutionary operation (EVOP), requires considerable commitment from the team of operators. There will be more of a delay before the results from successful off-line experiments can be used to modify the process. One of Taguchi's main contributions has been to emphasise the need for experimental conditions to approximate the range of environmental conditions under which the process is required to operate. These methods are covered in Chapter 7.

6.2   Process capability

6.2.1   Process capability indices

Over-zealous use of process capability indices has attracted considerable criticism. Kotz and Lovelace (1998) give a useful summary, but the underlying principle of comparing process variability with specification limits is crucial to managing a business. The standard indices are defined in terms of the process mean (μ), the process standard deviation (σ), and the upper and lower specification limits (*USL, LSL*). If the process has upper and lower specification limits, dimensions of components for example, the two most common indices are:

$$\text{process capability, } C_p = \frac{(USL - LSL)}{6\sigma}$$

and the

$$\text{process performance ratio, } C_{pk} = \min \left\{ \frac{USL - \mu}{3\sigma}, \frac{\mu - LSL}{3\sigma} \right\}$$

If there is only an upper limit, an appropriate index is:

$$C_{pU} = \frac{USL - \mu}{3\sigma}$$

Typical examples are the run-out of brake discs or bicycle wheels (i.e. the distance a pointer near the circumference, and perpendicular to the plane of the disc, moves during a revolution). If there is only a lower limit, breaking load for hand brake cables for example, the corresponding index is

$$C_{pL} = \frac{\mu - LSL}{3\sigma}$$

The rationale behind these definitions is that if the variable is normally distributed, and the process mean is at the centre of the specification, then a C_p of 1 corresponds to about 2 in 1000 out-of-spec items. The precise value is found from tables of the normal distribution as $2 \times (1 - \Phi(3))$ which equals 0.0027. A C_p of 1.33 corresponds to about 60 parts per million out-of-spec. The process performance ratio allows for the process mean to be offset from the centre of the specification. This might be because it cannot be changed easily or because of different costs of being out-of-spec. In ship-building a box section can be shortened by cutting away the excess, whereas if it is already too short it will have to be scrapped or have a fillet inserted, and this may not be acceptable to certification bodies. Volvo (1992) required a C_{pk} of at least 1.33 on all key processes, and Tioxide currently expect a value of 1.2.

Example 6.1
Isao Ohno (1990) reported that Sumi-Auto welding had a mean tack weld length (μ) of 2.23 mm and a standard deviation of tack length (σ) of 0.63 mm. The specification is that the length should be between 1.5 and 3.0.

$$C_{pk} = \min((3.30 - 2.23)/(3 \times 0.63), (2.23 - 1.5)/(3 \times 0.63))$$

$$= \min(0.407, 0.386) = 0.39$$

This was far too low and Isao Ohno described how the yard had reduced the standard deviation to about 0.25 mm. If the process mean was adjusted to the middle of the specification, 2.25 mm, this corresponds to

$$C_p = (3.0 - 1.5)/(6 \times 0.25) = 1.0$$

A C_p of 1.0 would be quite acceptable, in this context, provided it can be maintained.

There are three issues that arise in the practical use of these indices. The first is that different people may have different views about what constitutes the process.

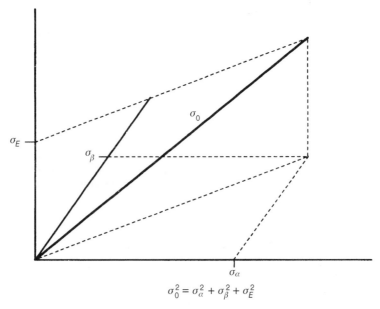

$$\sigma_0^2 = \sigma_\alpha^2 + \sigma_\beta^2 + \sigma_E^2$$

Figure 6.1　Independent components of variance are additive. In the context of Example 6.2; α, β and E are m, s and e respectively

Example 6.2

A company (A) specialises in fitting built-in wardrobes and cupboards, and buys wooden doors from a manufacturer (B). A specifies that the widths of doors should be within plus or minus 2 mm of the requested width. B uses a machine supplied by C. Once the machine is set up the standard deviation of widths of doors (σ_m) is 0.3 mm. However, the error in setting up the machine for a requested width has a standard deviation (σ_s) of 0.6 mm. The widths of doors also have a natural variability due to changes in the environment (e.g. changes in temperature and humidity), and for typical widths of doors this has a standard deviation (σ_e) of 0.4 mm. These sources of error are independent, so variances are additive (Appendix 2), and standard deviations are combined according to Pythagoras' theorem (Figure 6.1). C might claim the inherent capability of the machine for this process is $4/6\sigma_m$, which equals 2.22. The standard deviation of the widths of doors leaving B is,

$$\sigma_B = (\sigma_m^2 + \sigma_s^2)^{\frac{1}{2}} = 0.67$$

and B could claim a capability of $4/(6 \times 0.67)$ which equals 0.99. The standard deviation of doors fitted by A is,

$$\sigma_A = (\sigma_m^2 + \sigma_s^2 + \sigma_e^2)^{\frac{1}{2}} = 0.78$$

and A might respond that the capability is only 0.85. A and B need to agree whether the process capability applies to goods leaving the factory or goods received by the customer. B and C need to establish whether the setting-up errors are because of poor training of B's employees or because the machine itself is awkward to set up precisely. There is little point in reducing σ_m while the other sources of variation are substantially larger.

The second issue is that the variable may not be approximately normally distributed. The probabilistic interpretation relates to individual components and is therefore sensitive to the form of the distribution. Run-out measurements might be nearer to a folded normal distribution than a normal distribution (Exercise 6.1). Weld lengths are likely to have a positively skewed distribution.

Example 6.3
A car manufacturer buys in exhaust systems. The manufacturer offers a comprehensive three-year guarantee on all cars sold. Some electronic components have lifetimes that are well approximated by an exponential distribution, i.e. they are no more likely to fail as they get older. However, this would be very unrealistic for anything subject to obvious physical wear and tear, such as exhaust systems. Even so, a normal distribution, which is symmetric, is not necessarily a good model for component lifetimes. The Weibull distribution (Exercise 6.3) can take shapes ranging from close to a normal distribution, to more extreme than an exponential distribution. It is therefore often used in reliability analyses to model component lifetimes. There is some further justification for using the Weibull distribution. If we imagine that the exhaust system consists of a large number of pieces of metal, and that it fails when the first piece becomes perforated, then the distribution of lifetimes will be well approximated by a Weibull distribution (Exercise 6.3). Table 6.1 gives the skewness of the distribution, and tail areas, for values of the shape parameter from 1 to 10. Suppose distribution of lifetimes is well approximated by a Weibull distribution with a negative skewness of −0.50, and $\mu - 3\sigma$ is set at 3 years. Then an expected 5 out of 1000 exhaust systems would need replacing under guarantee.

Table 6.1 Skewness and tail areas of Weibull distributions

α	Skewness	Below				Above			
		$\mu - 4\sigma$	$\mu - 3\sigma$	$\mu - 2\sigma$	$\mu - \sigma$	$\mu + \sigma$	$\mu + 2\sigma$	$\mu + 3\sigma$	$\mu + 4\sigma$
1.0	2.00	0	0	0	0	0.1353	0.0498	0.0183	0.0069
1.5	1.07	0	0	0	0.1445	0.1547	0.0448	0.0107	0.0021
2.0	0.63	0	0	0.0016	0.1638	0.1618	0.0374	0.0056	0.0006
2.5	0.36	0	0	0.0058	0.1677	0.1642	0.0308	0.0029	0.0001
3.0	0.17	0	0	0.0144	0.1678	0.1645	0.0256	0.0015	0
3.5	0.03	0	0	0.0205	0.1668	0.1639	0.0214	0.0008	0
4.0	−0.09	0	0.0004	0.0247	0.1654	0.1628	0.0182	0.0004	0
4.5	−0.18	0	0.0012	0.0278	0.1641	0.1616	0.0156	0.0002	0
5.0	−0.25	0	0.0020	0.0300	0.1628	0.1604	0.0136	0.0001	0
5.5	−0.32	0	0.0027	0.0317	0.1617	0.1592	0.0119	0.0001	0
6.0	−0.37	0.0001	0.0034	0.0331	0.1606	0.1580	0.0106	0	0
6.5	−0.42	0.0002	0.0041	0.0342	0.1597	0.1569	0.0095	0	0
7.0	−0.46	0.0003	0.0046	0.0350	0.1588	0.1558	0.0085	0	0
7.5	−0.50	0.0004	0.0051	0.0357	0.1580	0.1549	0.0078	0	0
8.0	−0.53	0.0005	0.0055	0.0363	0.1573	0.1539	0.0071	0	0
8.5	−0.56	0.0006	0.0059	0.0368	0.1567	0.1531	0.0065	0	0
9.0	−0.59	0.0007	0.0063	0.0373	0.1561	0.1523	0.0061	0	0
9.5	−0.62	0.0008	0.0066	0.0377	0.1556	0.1516	0.0056	0	0
10.0	−0.64	0.0009	0.0069	0.0380	0.1551	0.1509	0.0053	0	0

The third issue is that indices are estimated from samples and demonstrating reasonable confidence that a requested value is exceeded will require a large sample or a process which has a higher capability than required. If reducing variability involves buying more expensive machinery, this last option is unlikely to appeal to most companies. Kotz and Lovelace (1998) give a detailed account of constructing confidence intervals for process indices. A useful result is the following approximation for a confidence interval for C_{pk} (Kushler and Hurley, 1992, and Exercise 6.4)

$$\hat{C}_{pk} = \min \left\{ \frac{USL - \bar{x}}{3s}, \frac{\bar{x} - LSL}{3s} \right\}$$

and approximate $(1 - \alpha) \times 100\%$ confidence limits for C_{pk} are given by

$$\hat{C}_{pk}[1 \pm z_{\alpha/2}/(2n - 2)^{\frac{1}{2}}]$$

where n, $\bar{x}$ and s are the sample size and its mean and standard deviation. The result can also be used to construct confidence intervals for C_{pU} and C_{pL}.

Example 6.4
The diameters of a random sample of 50 pistons were measured and $\bar{x} = 100.93$ mm and $s = 0.17$ mm. The specification is that the diameter should be between 100.3 and 101.5.

(a) Calculate an approximate 80% confidence interval for C_{pk}.

$$\hat{C}_{pk} = 1.12$$

An 80% confidence interval for C_{pk} is given by

$$1.12 \times [1 \pm 1.282/(2 \times 50 - 2)^{\frac{1}{2}}]$$

which equals $[0.98, 1.26]$
(b) Calculate an approximate lower limit for a one-sided 95% confidence interval for C_{pk}.

$$1.12 \times [1 - 1.645/(2 \times 50 - 2)^{\frac{1}{2}}] = 0.93$$

(c) How large a sample do we need for the lower limit of an approximate 90% confidence interval for C_{pk} to be $0.9\hat{C}_{pk}$? We require

$$0.9 = [1 - 1.282/(2n - 2)^{\frac{1}{2}}]$$

which is equivalent to

$$2n - 2 = [1.282/(1 - 0.9)]^2$$

and has the solution $n = 83$.

6.2.2   Components of variance

Two levels

If we are to reduce variation we need first to identify its sources. A long-established company makes a range of burners for domestic gas appliances. The process starts with steel plates that are then perforated, rolled into tubes

Table 6.2   Distances from base of gas burner to first outlet hole. Random samples of five burners from six runs

	Run 1	Run 2	Run 3	Run 4	Run 5	Run 6
	68.20	68.09	68.32	67.93	68.14	68.19
	68.10	68.24	68.19	67.89	68.09	68.30
	68.10	67.92	68.33	67.95	67.94	68.25
	68.13	68.06	68.21	67.94	67.92	68.19
	68.01	68.06	68.26	67.97	68.01	67.95
Mean	68.108	68.074	68.262	67.936	68.020	68.176
Std dev.	0.068	0.114	0.063	0.030	0.095	0.134

and flanged at the ends. The last two operations are carried out by a single remarkable and rather old machine which needs careful setting at the start of each run of burners of a particular type. The distance from the base of the finished burner to the first gas outlet hole is critical for burner performance. The specified distance for a particular type of burner is between 67.50 mm and 68.50 mm. The manufacturer needs to demonstrate that the machine is capable of meeting this requirement if it is set up to give a mean of exactly 68.00 mm, and also wishes to know how much variation about this target mean of 68.00 mm to expect when the machine is set up. Some variation is inevitable because we cannot be sure about the exact setting and even if we could be, it might not be practical to attempt making fine adjustments. Initially, random samples of five burners were taken from six short runs of the process. The data are given in Table 6.2. The analysis will be based on the following model for the process generating the data. Let Y_{ij} represent the distance for burner j in the sample from run i. Then

$$Y_{ij} = \mu + \alpha_i + E_{ij}$$

where μ is the overall mean, α_i is the deviation from this overall mean in run i, and E_{ij} are independent random errors. The α_i are independently distributed with a mean of 0 and a variance σ_b^2, where the b stands for between runs. The E_{ij} are independently distributed with a mean 0 and a variance σ_w^2, where the w stands for within runs. We wish to estimate μ, σ_w^2 and σ_b^2. The estimators for the first two of these parameters are the overall mean, and the average of the variances of the samples. To generalise the formulae suppose that there are m runs, and that the sample size from each run is n. The estimator of μ is,

$$\hat{\mu} = \sum_{i=1}^{m} \sum_{j=1}^{n} Y_{ij}/mn = \bar{Y}_{..}$$

where the dots designate the subscripts that have been averaged over, and the estimator of σ_w^2 is

$$S_w^2 = \sum_{i=1}^{m} \left[\sum_{j=1}^{n} (Y_{ij} - \bar{Y}_{i.})^2/(n-1) \right] \Big/ m$$

and has $(n-1)m$ degrees of freedom. The estimator of σ_b^2 is obtained by noticing that the variance of the sample means ($\bar{Y}_{i.}$) will be

$$\sigma_b^2 + \sigma_w^2/n$$

The first term is the variance of the actual run means, and the second term allows for the variance of the sample means about the run means. Formally,

$$\bar{Y}_{i.} = \mu + \alpha_i + \bar{E}_{i.}$$

$$\text{var}(\bar{Y}_{i.}) = \text{var}(\alpha_i) + \text{var}(\bar{E}_{i.})$$

$$= \sigma_b^2 + \sigma_w^2/n$$

The sample variance of the $\bar{Y}_{i.}$ is $S_{\bar{Y}}^2$:

$$S_{\bar{Y}}^2 = \sum_{i=1}^{m} (\bar{Y}_{i.} - \bar{Y}_{..})^2/(m-1)$$

and the estimator of σ_b^2, denoted by S_b^2, is found from

$$S_{\bar{Y}}^2 = S_b^2 + S_w^2/n$$

If the process were to continue with more runs, in the same fashion, the variance of the dimension would be

$$\text{var}(Y_{ij}) = \text{var}(\alpha_i) + \text{var}(E_{ij})$$

$$= \sigma_b^2 + \sigma_w^2$$

For the data in Table 6.2

$$\bar{y}_{..} = (68.108 + \ldots + 68.176)/6 = 68.096$$

$$s_w^2 = [(0.068)^2 + \ldots + (0.134)^2]/6 = 0.008\,26$$

$$s_{\bar{y}}^2 = [(68.108 - 68.111)^2 + \ldots + (68.176 - 68.111)^2]/(6-1) = 0.013\,19$$

and hence

$$s_b^2 = 0.013\,19 - 0.008\,26/5 = 0.011\,54$$

An estimate of the overall variance is

$$\widehat{\text{var}}(Y_{ij}) = s_b^2 + s_w^2 = 0.0198$$

A more straightforward estimate is the variance of the 30 data in Table 6.2, which equals 0.0182. The two estimates should be close. A Minitab boxplot of the data in Table 6.2 is shown in Figure 6.2, and there are noticeable differences between the runs. An example of a similar analysis in a different context, inter-laboratory trials, is given in Exercise 6.5.

Three levels

Graham and Martin (1946) described the method they used to ensure the concrete paving used for the main runway at Heathrow airport met the specification. A short excerpt from the published data is given in Table 6.3, and relates to six deliveries of cement. Each delivery of cement was used to

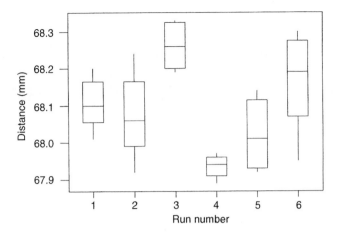

Figure 6.2   Distances from base of gas burners to first outlet hole

make many batches of concrete, of which two were selected at random. Four concrete test cubes were made from each batch, and tested for compressive strength after 28 days. The arrangement is shown schematically in Figure 6.3.

Let Y_{ijk} be the strength of test cube k from batch of concrete j from delivery of cement i. A model for this strength is

$$Y_{ijk} = \mu + \alpha_i + \beta_{ij} + E_{ijk}$$

where μ is an overall mean, and α_i for i from 1 to 6 is the discrepancy from μ for delivery i. The β_{ij} represents the discrepancy from $\mu + \alpha_i$ for batch j, where j is 1 or 2. Finally E_{ijk} is random variation for cube k about the mean $\mu + \alpha_i + \beta_{ij}$. We assume

$$\alpha_i \sim \text{ mean 0 and variance } \sigma_\alpha^2$$

$$\beta_{ij} \sim \text{ mean 0 and variance } \sigma_\beta^2$$

$$E_{ijk} \sim \text{ mean 0 and variance } \sigma_E^2$$

Table 6.3   Compressive strengths of 28-day concrete cubes (MPa). (Source: Graham and Martin, 1946)

35.6	38.6	30.7	31.7	30.0	27.9	34.3	38.7	33.2	35.8	39.5	38.7
33.6	41.6	30.5	30.0	35.0	27.7	36.4	38.5	35.2	37.1	42.1	36.1
34.1	40.7	27.2	33.8	35.0	29.0	33.4	43.3	37.8	37.1	38.5	35.9
34.5	39.9	26.8	29.6	32.6	32.8	33.4	36.7	35.4	39.5	40.2	42.8

Batch means

34.35	40.20	28.20	31.28	33.15	29.35	34.38	39.30	35.40	37.38	40.08	38.38

Cement delivery means

37.32		30.04		31.25		36.84		36.39		39.23	

Overall mean

35.18

Figure 6.3 Hierarchical structure for concrete test cubes

and that all components of variance are independent. It follows that

$$\mathrm{var}(Y_{ijk}) = \sigma_\alpha^2 + \sigma_\beta^2 + \sigma_E^2$$

and hence

$$\mathrm{stdev}(Y_{ijk}) = \sqrt{\sigma_\alpha^2 + \sigma_\beta^2 + \sigma_E^2}$$

The result for the standard deviation is an application of Pythagoras' theorem in three dimensions (Figure 6.1).

Our objective is to estimate the three components of variance. Define:

$\bar{Y}_{ij\cdot}$ as the mean of the four strength measurements within a batch, the batch mean;
$\bar{Y}_{i\cdot\cdot}$ as the mean of the two batch means within a cement delivery, the delivery mean;
$\bar{Y}_{\cdots}$ as the mean of the six delivery means, the overall mean.

In order to generalise the formulae we now assume n cubes within a batch, b batches of concrete made from a delivery of cement and a deliveries of cement:

$$i = 1, \ldots, a; \quad j = 1, \ldots, b; \quad \text{and} \quad k = i, \ldots, n$$

The estimator, s_E^2 of σ_E^2 is the average of the within-batch estimators of the test variance.

$$s_E^2 = \sum_{i=1}^{a} \sum_{j=1}^{b} \left\{ \sum_{k=1}^{n} (Y_{ijk} - \bar{Y}_{ij\cdot}) / (n-1) \right\} \Big/ (ab)$$

The next step is to estimate σ_β^2 from the standard deviation of the batch means, but we have to take into account the variation of the estimators of the within-batch means. That is, for a fixed value of i,

$$\mathrm{var}_{j|i}(\bar{Y}_{ij\cdot}) = \sigma_\beta^2 + \sigma_E^2/n$$

where the subtext $j|i$ is to emphasise that it is the variance over batches (j) given a fixed delivery of cement (i). The estimator of the variance of batch means within cement deliveries is,

$$\widehat{\mathrm{var}}_{j|i}(\bar{Y}_{ij\cdot}) = \sum_{i=1}^{a} \left\{ \sum_{j=1}^{b} (\bar{Y}_{ij\cdot} - \bar{Y}_{i\cdot\cdot})^2 / (b-1) \right\} \Big/ a$$

so it follows that the estimator of σ_β^2 is,

$$s_\beta^2 = \widehat{\text{var}}_{j|i}(\bar{Y}_{ij\cdot}) - s_E^2/n$$

The estimator of σ_α^2 follows from the variance of the cement delivery means by a similar argument:

$$\text{var}(\bar{Y}_{i\cdot\cdot}) = \sigma_\alpha^2 + \sigma_\beta^2/b + \sigma_E^2/(nb)$$

The estimator of $\text{var}(\bar{Y}_{i\cdot\cdot})$ is

$$\widehat{\text{var}}(\bar{Y}_{i\cdot\cdot}) = \sum_{i=1}^{a} (\bar{Y}_{i\cdot\cdot} - \bar{Y}_{\cdots})^2/(a-1)$$

and hence

$$s_\alpha^2 = \widehat{\text{var}}(\bar{Y}_{i\cdot\cdot}) - s_\beta^2/b - s_E^2/(nb)$$

If we apply these formulae to the data in Table 6.3 we obtain: $s_E^2 = 4.1797$, $\widehat{\text{var}}_j(\bar{y}_{ij\cdot}) = 7.0562$, $\widehat{\text{var}}(\bar{y}_{i\cdot\cdot}) = 13.3956$. Hence

$$s_E^2 = (2.044)^2$$

$$s_\beta^2 = 7.0562 - (2.044)^2/4 = (2.452)^2$$

$$s_\alpha^2 = 13.3956 - (2.452)^2/2 - (2.044)^2/8 = (3.141)^2$$

The best strategy for reducing overall variability is to concentrate on the factors that make the biggest contributions to the sum of variances. The most important factor appears to be variation in cement between deliveries, but the variation between batches is also quite substantial. Also, our estimators of the variances are not very precise: s_E^2 is based on $(4-1) \times 12 = 36$ degrees of freedom; but s_β^2 is based only on $(2-1) \times 6 = 6$ degrees of freedom; and s_α^2 is based only on $6-1=5$ degrees of freedom. A 90% confidence interval for the ratio of σ_β^2 to σ_α^2 is given by

$$\left[\frac{s_\beta^2}{s_\alpha^2} F_{5,6,0.95}, \frac{s_\beta^2}{s_\alpha^2} F_{5,6,0.05} \right]$$

(see Appendix A2.13) which is

$$[0.123, 2.67]$$

The corresponding 90% confidence interval for the ratio of σ_β to σ_α is $[0.35, 1.63]$, and since this includes 1 we cannot be confident that the between-batch variability is less important. It would be prudent to try and reduce both the variation between deliveries of cement, and the variation between batches made from the same delivery.

The calculations can be done using Minitab. Set up a text data column headed 'deliveries' with 8 As, 8 Bs, ..., 8 Fs. Set up another text data column headed 'batches' with 4 A1s, 4 A2s, 4 B1s, ..., 4 F2s. Set up a third column with the compressive strengths. The worksheet should look something like the following.

C1-T deliveries	C2-T batches	C3 strength ·
A	A1	35.6
A	A1	33.6
A	A1	34.1
A	A1	34.5
A	A2	38.6
⋮	⋮	⋮
F	F2	42.8

Now select:

Statistics ▷ ANOVA ▷ Fully Nested ANOVA
Response: strength
Factors: deliveries batches

You can get similar information from the more general routine:

Statistics ▷ ANOVA ▷ General Linear Model
Response: strength
Model: deliveries batches (deliveries)
Random factors: deliveries batches

6.2.3   Autocovariance, correlogram and variogram

Shipyards paint steel plates before welding. The depth of the paint finish must be sufficient to protect the metal, but surplus paint adds unnecessary cost and interferes with the welding process. The data in TableWSpaintcap are the results of a process capability study in which the operators measured the depth of the paint film on 629 consecutive plates (Figure 6.4). The specification was that the depth of paint should be within plus or minus 86 (coded units) of a target value. The operators noticed a tendency for the depth to increase slightly over time, and made adjustments once they noticed that the upper specification limit had been exceeded. These adjustments were made after plate numbers 102, 146, 240, 320 and 497, giving six sequences of paint depths between adjustments. The regression equations for depth of paint against plate number, within each sequence, are summarised in Table 6.4. Overall, there is evidence to support the operators' claim that the depth tends to increase. The standard deviation of all 629 measurements of paint depth is 29.36. We can compare this with the standard deviation of errors in the regressions and hence estimate the reduction we might achieve by removing the cause of the increasing trend. It is better to average the variances of the errors than to average the standard deviations because the variances are unbiased, and if the variances are weighted in proportion to the corresponding degrees of freedom the average is 758. Taking square root gives an estimate of the standard deviation of errors about the trend of 27.53. Although the operators are adroit at compensating for the trend, removing it would result in a decrease of almost 6% in the process standard deviation and make their job easier.

Another objective of the capability study is to investigate whether errors in depth can reasonably be considered as independent. The autocorrelation

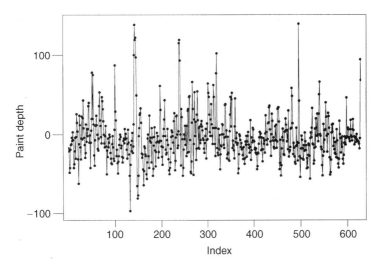

Figure 6.4   Depth of paint films on 629 consecutive plates (deviations from target in coded units)

function or correlogram provides a useful summary of the correlation structure. Let $\{Y_t\}$ be a sequence of random variables defined at equally spaced time intervals, $t = 1, 2, \ldots$ with a constant mean (μ) and variance (σ^2). The auto-covariance function (acvf), $\gamma(k)$, is defined as the covariance of variables separated by the time difference (k), according to the formula:

$$\gamma(k) = E[(Y_t - \mu)(Y_{t+k} - \mu)] \quad \text{for } k = 0, 1, \ldots$$

where the expectation is thought of as being over all possible sequences that might arise. The time difference k is known as the 'lag', and the acvf at lag 0 is the variance. The unit of the acvf is the square of the unit of Y_t, and for many purposes it is more convenient to have a dimensionless form known as the autocorrelation function (acf), $\rho(k)$, defined by

$$\rho(k) = \gamma(k)/\gamma(0) \quad \text{for } k = 0, 1$$

Like any correlation, $\rho(k)$ can only take values in the range $[-1, 1]$. The acvf and acf can be defined for negative integer k, but as they are symmetric about $k = 0$,

$$\gamma(k) = \gamma(-k)$$

Table 6.4   Regressions of paint depth against plate number in six runs of the process

Number of plates	Plate numbers	$\hat{\beta}_0$	$\hat{\beta}_1$	s	R^2	P-value
102	1:102	−8.65	0.130 (0.091)	26.92	2%	0.154
44	103:146	−256.36	2.044 (0.478)	40.23	30%	0.000
94	147:240	−101.73	0.472 (0.120)	31.61	14%	0.000
80	241:320	−106.12	0.400 (0.144)	29.69	9%	0.007
177	321:497	−28.50	0.054 (0.035)	23.76	1%	0.122
132	498:629	−91.87	0.150 (0.052)	22.81	6%	0.005

and it suffices to consider non-negative k. The usual sample estimates of $\gamma(k)$ and $\rho(k)$ are:

$$c(k) = \sum_{t=1}^{n-k} (y_t - \bar{y})(y_{t+k} - \bar{y})/n$$

where

$$\bar{y} = \sum_{t=1}^{n} y_t/n$$

and

$$r(k) = c(k)/c(0)$$

respectively. With these definitions $c(0)$ is the variance of $\{y_t\}$ calculated using a divisor n rather than $(n-1)$. A plot of the sample autocorrelation function $r(k)$ against k is sometimes known as the correlogram. The estimation is discussed in more detail in Chapter 8, but it is essential that $\bar{y}$ is calculated from a sufficiently long record for n to exceed, considerably, the largest lag at which there is any noticeable correlation.

A more general measure of correlation, which can be used usefully when trends are present, is the expected value of the squared difference at lag k,

$$V_k = E[(Y_{t+k} - Y_t)^2] \quad \text{known as the variogram}$$

The non-dimensional ratio V_k/V_1 is the standardised variogram. If $\{Y_t\}$ is stationary, that is Y_t have constant mean μ and variance σ^2, then

$$V_k = 2\sigma^2 - 2\gamma(k)$$

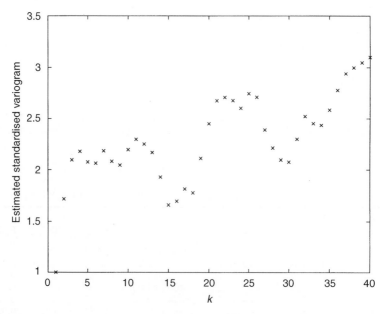

Figure 6.5 Variogram for depths of paint film on plates 498 to 629

and V_k approaches an upper limit $2\sigma^2$. The variogram can be estimated by

$$\hat{V}_k = \sum_{t=1}^{n-k} (y_t - y_{t+k})^2/(n-k)$$

Division by $(n-k)$ rather than n which was used for the correlogram helps to emphasise any trend.

Example 6.5
The sixth sequence of paint depths was measured on plates 498 up to 629. The estimated standardised variogram is shown in Figure 6.5. There is a clear increase in variance when the separation (k) increases beyond 1, and then there is a tendency for the variance to increase despite the considerable sampling variation in $\hat{V}$. This increase is a consequence of the slight trend in the sequence which was identified by the regression analysis (Table 6.4). The residuals from the regression have no trend and the correlogram is shown in Figure 6.6. There is clear evidence of autocorrelation at, at least, lag one. The practical consequences of this correlation are that we should avoid taking consecutive plates when drawing samples for control charts, and that we might be able to reduce variability by applying an automatic feedback controller. This is quite feasible as several contactless devices for measuring paint depth are available.

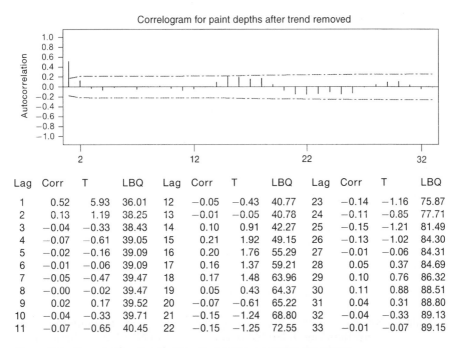

Correlogram for paint depths after trend removed

Lag	Corr	T	LBQ	Lag	Corr	T	LBQ	Lag	Corr	T	LBQ
1	0.52	5.93	36.01	12	−0.05	−0.43	40.77	23	−0.14	−1.16	75.87
2	0.13	1.19	38.25	13	−0.01	−0.05	40.78	24	−0.11	−0.85	77.71
3	−0.04	−0.33	38.43	14	0.10	0.91	42.27	25	−0.15	−1.21	81.49
4	−0.07	−0.61	39.05	15	0.21	1.92	49.15	26	−0.13	−1.02	84.30
5	−0.02	−0.16	39.09	16	0.20	1.76	55.29	27	−0.01	−0.06	84.31
6	−0.01	−0.06	39.09	17	0.16	1.37	59.21	28	0.05	0.37	84.69
7	−0.05	−0.47	39.47	18	0.17	1.48	63.96	29	0.10	0.76	86.32
8	−0.00	−0.02	39.47	19	0.05	0.43	64.37	30	0.11	0.88	88.51
9	0.02	0.17	39.52	20	−0.07	−0.61	65.22	31	0.04	0.31	88.80
10	−0.04	−0.33	39.71	21	−0.15	−1.24	68.80	32	−0.04	−0.33	89.13
11	−0.07	−0.65	40.45	22	−0.15	−1.25	72.55	33	−0.01	−0.07	89.15

Figure 6.6 Correlogram for depths of paint film on plates 498 to 629

6.3   SPC charts

6.3.1   Shewhart mean and range charts

A company manufactures and markets shampoo products. Some years ago a new line was set up to dispense a new shampoo product into plastic bottles with nominal contents of 110 ml. The supplier of the filling machine had agreed to provide plant to fill 48 bottles a minute with a standard deviation of volume dispensed of less than 1.1 ml. The Code of Practical Guidance for Packers and Importers (HMSO, 1985), issued under the UK Weights and Measures Act 1985, was referred to when setting up the control charts. The machine had 12 filling heads and 12 bottles were filled simultaneously every 15 seconds. During a test period bottles were weighed, marked and passed down the line, interspersed with a few unmarked bottles. The unmarked bottles were placed between marked bottles to mimic a sampling procedure of selecting 4 bottles from about 24 consecutive bottles which pass the sampling point in 30 seconds. The marked bottles were weighed when full and the weight of shampoo was deduced by subtracting the weight of the empty bottle. The density of the shampoo was known, to considerable accuracy, and weights were converted to volumes. Variances were calculated within groups of four consecutive marked bottles. The average of 80 within-group variances (s_w^2) was $(0.94)^2$. The variance of the volumes in all the bottles (s_0^2) was $(1.02)^2$. The between-groups variance can be estimated by s_b^2, in the formula

$$s_0^2 = s_b^2 + s_w^2$$

and equals 0.40. This is an alternative to calculating the variance of the group means (Section 6.2.2). It does not lead, explicitly, to an F-test but the Code (HMSO, 1985) gives critical values of the ratio s_0/s_w for testing a hypothesis that σ_b equals 0. In this case the ratio was 1.08. This is statistically significantly greater than 1, and of practical significance. A likely explanation for the between-groups variability is that the machine has 12 filling heads, and errors during a single filling cycle are not independent. The overall standard deviation (1.02) was less than the upper limit quoted by the supplier of the filling machine, and it would not have been practical to reduce the between-groups standard deviation.

In normal operation, selected bottles would not be weighed before filling. The average weight of empty bottles would be subtracted from the weights of sampled bottles. Therefore, allowance also had to be made for the variance of the weights of empty bottles. A random sample of 80 empty bottles were weighed (tare weight) and the mean and standard deviation of tare weight were 25.71 g and 0.23 g respectively.

Before setting up the Shewhart mean chart (Shewhart, 1931), the company had to decide on a target volume. The Code (HMSO, 1985) can be summarised as:

- Rule 1   Actual contents shall be not less, on average, than the nominal quantity.
- Rule 2   Not more than 2.5% of the packages may be non-standard, i.e. be more than one tolerable negative error (TNE) below the nominal quantity. (For a 110 ml package the TNE is 4.95 ml.)

- Rule 3   No package may be inadequate, i.e. be more than 2 TNE below the nominal quantity. (In practice none is usually interpreted as fewer than 1 in 10 000.)

If the actual quantity is normally distributed with a mean equal to the nominal quantity and a standard deviation of half the TNE, these rules would just be followed. The standard deviation of the volume of shampoo dispensed was estimated as 1.02 which is considerably less than half the TNE. The production manager suggested a target volume of 1% above the nominal contents, i.e. 111.1 ml, but the sales manager thought the target should be lower and recommended 110.5 ml.

Shewhart mean chart

Suppose that observations of a process have a mean μ, a standard deviation σ, and are independent. Samples of n observations are taken at convenient intervals. Let $\bar{X}$ represent a sample mean. Then

$$\bar{X} \sim N(\mu, \sigma^2/n)$$

provided the observations are near enough normally distributed for the mean of n of them to have a distribution which is nearly normal. We aim to set the process up at some target value T. The Shewhart mean chart is a plot of sample means against time with the target value and action lines (also referred to as control lines) drawn parallel to the time axis. The action lines are at

$$T \pm 3.09\sigma/\sqrt{n}$$

The rationale for these lines is that if the process is on target they will only be crossed once in 500 samples, whereas if the process is off target the closer action line is more likely to be crossed. The benefits of the ensuing corrective action outweigh the costs of occasional detrimental adjustments. It is quite common to add warning lines at plus or minus $1.96\sigma/\sqrt{n}$ and to take action if two consecutive means lie outside the same warning line.

Example 6.6
If the company had implemented its original strategy, of taking samples of size 4 from 24 consecutive bottles passing the sample point during 30 seconds, the sample means would have had an estimated standard deviation of

$$(s_b^2 + s_w^2/4 + s_T^2/4)^{\frac{1}{2}}$$

where s_T is the estimated standard deviation of tare, expressed in units of equivalent volume, which was about 0.25 ml. The standard deviation of sample means would have been 0.63.

   Instead, the company decided to take samples of 4 bottles, with approximate 30-second intervals between bottles. The 30-second separation was chosen so that the bottles were not filled during the same filling cycle. The volumes of shampoo dispensed were then found to be approximately independent. In normal operation the standard deviation of estimated volumes (σ) was estimated by

$$\hat{\sigma} = (s_0^2 + s_T^2)^{\frac{1}{2}} = 1.05$$

Hence, the standard deviation of sample means was $1.05/\sqrt{4}$, which equals 0.525. Sampling takes an extra 90 seconds but this is justified by the decrease in the standard deviation of sample means. Action lines would be set at

$$\text{target} \pm 3.09 \times 1.05/\sqrt{4}$$

which is target plus or minus 1.62. However, the target value had still to be set.

Average run length

If $\bar{X} \sim N(\mu, \sigma^2/n)$ we can calculate the probability (p) that $\bar{X}$ falls below the lower action line if μ is less than the target value T. The probability of $\bar{X}$ falling above the upper action line if $\mu < T$ is small enough to be neglected. Let R be the number of samples until the first mean lies below the lower action line. Then R has a geometric distribution, which has the probability mass function, $\Pr(R = r)$:

$$P(r) = (1 - p)^{r-1} p \quad \text{for } r = 1, 2, \ldots$$

$$E[R] = 1/p \quad \text{and} \quad \text{var}(R) = (1 - p)/p^2$$

The average run length (ARL) is $E[R]$
If $\bar{X} \sim N(\mu, \sigma^2/n)$

$$\Pr(\bar{X} < T - 3.09\sigma/\sqrt{n}) = \Pr(Z < [(T - 3.09\sigma/\sqrt{n}) - \mu]/(\sigma\sqrt{n}))$$

$$= \Pr(Z < (T - \mu)\sqrt{n}/\sigma - 3.09)$$

Hence the ARL is given by

$$[\Phi((T - \mu)\sqrt{n}/\sigma - 3.09)]^{-1}$$

The ARL formula can be set up as a formula in an Excel spreadsheet, which is very convenient for investigating the effect of changing T and n.

Example 6.7
For the shampoo filling line, $\sigma = 1.05$ and $n = 4$. If the sales manager's suggestion is followed, T will be 110.5 ml. Suppose μ is 110.0 ml, which is the lowest it can be without infringing the code. The average run length is 61.4. This corresponds to an average of 30 hours before the process is adjusted, if samples are taken every half hour. The production manager's suggestion of 111.1 ml for the target corresponds to an ARL of 6.3, when μ equals 110.0 ml. A compromise target value of 110.9 ml, which corresponds to an ARL of 11.8, was agreed as an interim measure (Figure 6.7(a) drawn from data in TableWSshampoo). The sales manager still thought the savings that would accrue from setting a lower target would be worthwhile. She asked what sample sizes would give an ARL of about 15 when μ equals 110.0 ml if the target was set at 110.7 ml and if it was set at 110.5 ml. If the target is 110.7 ml, sample sizes of 5 and 6 give ARL of 18.2 and 13.8 respectively. If the target is 110.5 ml, sample sizes of 11 and 12 give ARL of 15.3 and 13.4 respectively.

Average run lengths are reduced if action is taken when two consecutive means fall outside the same warning line (see Table 6.5).

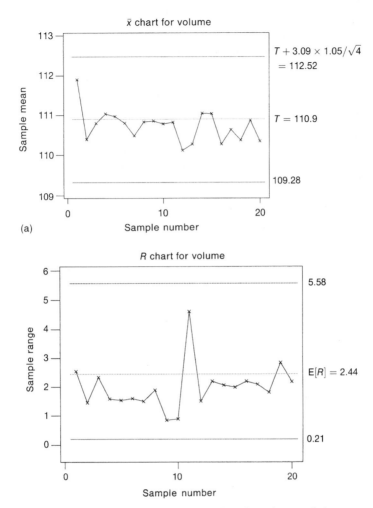

Figure 6.7   (a) Shewhart means chart; (b) range chart for volumes of shampoo

Table 6.5   Comparison of average run lengths using action lines only and using action lines with warning lines. The probability of means lying outside the lines on the opposite side to the non-zero deviations has been ignored

Deviation $(T - \mu)\sqrt{n}/\sigma$	Action lines only	Action lines and warning lines
0	500	320
0.5	208.4	108.0
1.0	54.6	26.4
1.5	17.9	8.9
2.0	7.3	4.1
2.5	3.6	2.5

Shewhart mean charts are not very sensitive to an assumption that the variable is normally distributed if the sample size is four or more, provided the observations are not substantially serially correlated. This is a consequence of the central limit theorem.

Range chart

It is as important to detect changes in variability of the process as it is to detect changes in the mean. Increased variability may be attributable to machine wear, lack of maintenance or inexperienced operators. A sudden decrease in variability is likely to be due to faults in measuring equipment. Table AST.6 gives factors of the standard deviation from which to calculate upper (1 in 1000) and lower (1 in 1000) action lines for the range of samples of size n. They are calculated from the distribution of the range of n independent variables from a normal distribution. In contrast to the mean chart, the range chart is sensitive to the assumption that the variable is normally distributed. Appropriate action lines for the range of samples from other distributions could be obtained by simulation or set empirically if a sufficiently long record from the process, when it has been working satisfactorily, is available. For the shampoo filling line, $\sigma = 1.05$ and $n = 4$, the upper and lower action lines are at 0.2×1.05 and 5.31×1.05 respectively, Figure 6.7(b). The lines that are shown by default in Minitab are set at the mean range plus or minus three standard deviations of the distribution of the range. Since the distribution of the range is positively skewed these are not the same as the lines shown in Figure 6.7(b).

Mean chart designed from specification limits

In Example 6.1 the specification for the tack weld length was between 1.5 mm and 3.0 mm. A control chart could be set up about a target value of 2.25 mm. The purpose of the chart is to detect changes in the process mean from the target value, which is a reasonable strategy if deviations from target lead to poor quality. However, the main concern may be just to keep within specification, especially for one-sided specification limits such as run-out of brake discs and purity of laboratory reagents.

The specification for the distance to the first gas outlet hole of the domestic gas burner, described in Section 6.2.2, is between 67.50 mm and 68.50 mm. Once the machine is set up the standard deviation (σ) of this distance is 0.1 mm, when the machine is running properly. Setting up the machine is an intricate process. The distances appear to have a near normal distribution. The upper 0.2% point of a normal distribution is 2.88, and the manufacturer will be satisfied if the process mean is within 2.88σ mm of the specification limits. However, it would not be satisfactory for the mean to be any closer to the specification limit, and the manufacturer thinks it would be prudent to set a probability of 80% of resetting the machine if the mean is within 2.88σ of the limit. This requirement will be met if an action line is set a distance

$$2.88\sigma + z_{0.20}\sigma/\sqrt{n}$$

where n is the sample size, in from the specification limit. The $z_{0.20}$ point is 0.842, and substituting 0.1 for σ and taking a sample size of 4 gives a distance of 0.330. We must now check that this action line is not nearer to the middle of the specified range than $3.09\sigma/\sqrt{n}$, which is 0.154 and corresponds to a distance of 0.346 from the specification limit. For this application the action limits based on the construction which moves in from the specification are slightly further apart than they would be for a standard Shewhart chart with a target value set at the middle of the specification. It is unusual to have such a wide specification relative to the process standard deviation, and the manufacturer has set exacting criteria that could not be met by a process with a C_p of 1.33. The probability of resetting the machine if the process mean is on the action line needs to be greater than 50% if we are to avoid the odd effect of action lines becoming closer to the specification limits as the sample size decreases (Hill, 1956). The construction can also be used if there is medium-term variation in the mean or if the specification is one-sided.

Effect of departures from assumptions

Suppose X_i, for $i = 1, \ldots, n$, are independent variables from a distribution with mean μ, variance σ^2, skewness γ and excess kurtosis $(\kappa - 3)$. Then the skewness and excess kurtosis of $\bar{X}$ are $\gamma/\sqrt{n}$ and $(\kappa - 3)/n$ respectively (Exercise 6.6). The skewness and kurtosis have no effect on the results that

$$E[\bar{X}] = \mu \qquad E[S^2] = \sigma^2$$

but excess kurtosis does considerably inflate the variance of S^2

$$\text{var}(S^2) = \frac{2\sigma^4}{(n-1)}\left(1 + \frac{(\kappa - 3)(n - 1)}{2n}\right)$$

and this effect does not disappear as the sample size increases (Wetherill and Brown, 1991). So mean charts are insensitive to the assumption that X_i are from a normal distribution, but the standard limits for the range chart will lead to too many action signals if the sampled distribution has a higher kurtosis than the normal distribution. If successive observations, only, are assumed to have a correlation ρ, it is easy to show (Exercise 6.7) that

$$\text{var}(\bar{X}) = \sigma^2(1 + 2\rho(1 - 1/n))/n$$

Also the expected value of the sample variance is reduced

$$E[S^2] = \sigma^2(1 - 2\rho/n)$$

A positive autocorrelation will lead to too many false positives on a mean chart, and too few on a range chart, if the limits are based on an assumption of independence.

6.3.2   One at a time data

Runs chart

Many products in process or batch chemical industries will just be sampled once at each sampling time. A plot of the sample values against time or batch number is known as a runs chart. This is a simple but effective display and it is useful to

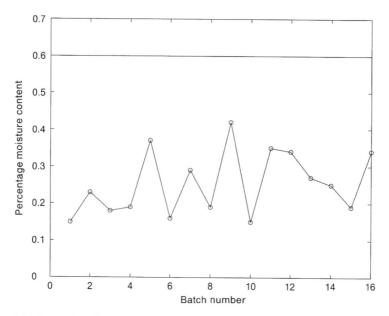

Figure 6.8 Runs chart for moisture content of batches of zirconium silicate

show the specification limits. A company sells milled zirconium silicate to the ceramics industry as a refractory and glaze opacifer. The raw material is milled in a continuous process but packed in one-tonne batches. The product specification is in terms of chemical composition, particle size and moisture content. The upper limit for moisture is 0.6%. Following a customer complaint, a runs chart for moisture content has been kept and a typical sequence is shown in Figure 6.8.

Moving average chart

A manager in a motoring organisation, which provides roadside assistance to members whose cars break down, keeps a record of the number of letters of complaint received each month. Most of the complaints are about the length of time until help arrives and the organisation has been recruiting new patrol staff to improve the situation. Complaints also tend to follow a seasonal pattern and peak during January. Let Y_t be the complaints at the end of month t. Instead of plotting Y_t, the manager plots the average of the past 12 months every month. That is

$$M_t = (Y_t + Y_{t-1} + \ldots + Y_{t-12})/12$$

is plotted against t. Data for four years are given in Table 6.6 and M_t is plotted against t for $t = 12, \ldots, 48$ in Figure 6.9. The moving average can be of any number that is relevant to the business. For instance, weekly sales of a product for which there is a fairly consistent demand might be averaged in fours to give a (lunar) monthly moving average. Action limits can be set up in the usual way for a means chart. Notice that if Y_t are independent with

Table 6.6 Number of letters of complaint received by a motoring organisation over four years, 1996–99, and moving average, of length 12, for 1999

Month	1996	1997	1998	1999	Moving average (1999)
January	27	31	28	18	17.58
February	34	31	21	20	17.50
March	31	19	18	21	17.75
April	24	17	23	6	16.33
May	18	10	19	9	15.50
June	19	25	16	29	16.58
July	17	26	10	12	16.75
August	12	18	24	19	16.33
September	26	18	11	14	16.58
October	14	10	11	19	17.25
November	18	4	13	16	17.50
December	33	20	27	23	17.17

variance σ^2, then a moving average of m terms M_t will have variance σ^2/m. Successive M_t will not be independent, but this does not affect the result that an average of 998 in 1000 M_t will lie within $3.09\sigma/\sqrt{m}$ of some target value if the process remains on target. However, it does imply that two consecutive M_t are much more likely to lie outside the same action line than are two consecutive independent means. The moving average chart is more effective than the usual Shewhart mean chart for detecting small process shifts (Montgomery, 1991), and a moving average can be kept of consecutive means.

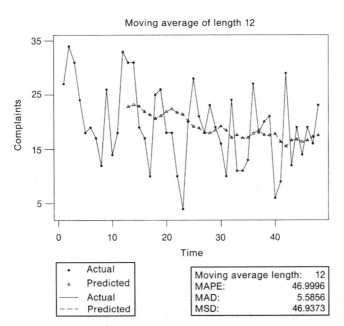

		Moving average length: 12
•	Actual	MAPE: 46.9996
▲	Predicted	MAD: 5.5856
——	Actual	MSD: 46.9373
– – –	Predicted	

Figure 6.9 Moving average chart for number of letters of complaint per month for motoring organisation

Exponentially weighted moving average

A manager in Albatross Airways keeps a record of the numbers of unsold seats and overbooked seats each week. Rather than plot the number of unsold seats, Y_t, each week, the manager calculates an average, $\tilde{Y}_t$, of this week's number and last week's average $\tilde{Y}_{t-1}$. That is:

$$\tilde{Y}_t = (1 - \theta) Y_t + \theta \tilde{Y}_{t-1}$$

for some constant θ which lies between 0 and 1. Repeated back substitution gives:

$$\tilde{Y}_t = (1 - \theta) Y_t + (1 - \theta)\theta Y_{t-1} + \ldots + (1 - \theta)\theta^t Y_0$$

and hence the name 'exponentially weighted moving average' (EWMA), the exponent being the power of θ. Since the weights form a geometric progression, the EWMA is sometimes called a geometric moving average. If the Y_t are independent with variance σ^2, the variance of $\tilde{Y}_t$ is given by

$$\text{var}(\tilde{Y}_t) = (1 - \theta)^2 (1 + \theta^2 + \ldots + \theta^{2t})\sigma^2$$

$$= \frac{(1 - \theta)^2 (1 - \theta^{2t})}{1 - \theta^2} \sigma^2$$

Upper and lower action limits can be set at

$$\text{target value} \pm 3.09\sigma\sqrt{(1 - \theta)/(1 + \theta)}$$

for moderately large t. Albatross Airways data are given in Table 6.7 and the EWMA is plotted in Figure 6.10. The Minitab 'weight' is $(1 - \theta)$, and the mean of the first six observations is used to start the procedure.

Cusum charts

Laboratory personnel employed by a pharmaceuticals company, who work with radioactive chemicals, are asked to wear personnel dosimeters which are replaced each week. An average dose of 2 millirem/week is considered safe. Let x_t be the weekly dose for a particular employee. The cumulative sum (cusum) chart is a plot of

$$S_t = \sum_{i=1}^{t} (x_i - L)$$

where L is the safe level or, more generally, some target value. Steep slopes indicate a change in the process mean. A V-mask can be used to decide whether action is justified (Figure 6.11 is drawn for the data in Table 6.8). The width of the front of the mask is 10 standard deviations of the variable being plotted when the process is in statistical control, 0.1 millirem/week in this case. The gradients of the two arms are plus or minus 5 standard deviations per 10 sampling units. The front of the mask is centred on the current observation and action is indicated if the plot crosses an arm. The false alarm rate is about 1 in 440, if the variable has a normal distribution. In this example the only concern is with the lower arm. If this is crossed the employee would be moved to some other task. The advantage of a cusum chart over a Shewhart chart for the mean is that, on average, the cusum leads to quicker control action with the same chance of false alarms.

Table 6.7 Albatross Airways unsold seats, and EWMA with theta equal to 0.8

Week	Unsold seats	EWMA
1	164	150.8
2	90	138.6
3	157	142.3
4	157	145.2
5	181	152.4
6	136	149.1
7	116	142.5
8	167	147.4
9	183	154.5
10	217	167.0
11	129	159.4
12	146	156.7
13	192	163.8
14	152	161.4
15	108	150.7
16	135	147.6
17	130	144.1
18	161	147.5
19	179	153.8
20	105	144.0
21	201	155.4
22	106	145.5
23	171	150.6
24	192	158.9
25	156	158.3

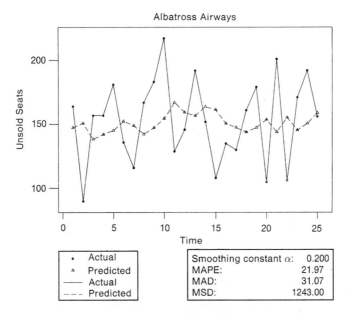

Figure 6.10 EWMA for unsold seats on Albatross Airways flights

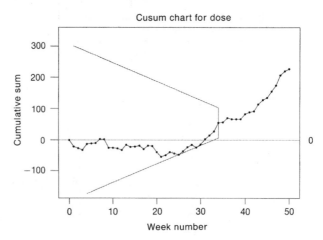

Figure 6.11   Cusum chart for weekly radiation doses for a laboratory employee. Action should have been taken after week 34.

6.3.3   Control charts for attributes

c-Chart

Every day, a team leader at a car manufacturing plant inspects a randomly selected finished car and records the number of imperfections. These are not

Table 6.8   Weekly radiation dose for an employee working in a laboratory (deviations from 2 millirem/week in units of 0.01 millirem/week)

1	−21	−21	26	11	−36
2	−5	−26	27	13	−23
3	−6	−32	28	8	−15
4	19	−13	29	−9	−24
5	2	−11	30	10	−14
6	1	−10	31	17	3
7	13	3	32	12	15
8	−1	2	33	13	28
9	−27	−25	34	27	55
10	0	−25	35	2	57
11	−2	−27	36	13	70
12	−5	−32	37	−3	67
13	17	−15	38	0	67
14	−7	−22	39	0	67
15	1	−21	40	16	83
16	3	−18	41	6	89
17	−11	−29	42	3	92
18	11	−18	43	22	114
19	−1	−19	44	14	128
20	−20	−39	45	7	135
21	−15	−54	46	20	155
22	5	−49	47	19	174
23	10	−39	48	33	207
24	−4	−43	49	13	220
25	−4	−47	50	7	227

functional and most purchasers would not notice them. Examples might include a small paint run inside the engine compartment or a slight variation in the width of a trim. An average count of about 8 is considered acceptable and occasional comparisons are made with hire car vehicles from other manufacturers. The NTSB in the USA records details of all reported aircraft incidents. An incident is an event other than an accident associated with the operation of an aircraft which affects, or could affect, the safety of operations. A manager plots the number of incidents each month to see whether there is any evidence of changes in the underlying rate of occurrence. In both these cases the number being recorded, Y_t, is a count, and it is possible that Y_t has a Poisson distribution. If Y_t has a Poisson distribution with a mean, μ, which exceeds about 6, then a reasonable approximation is that

$$Y_t \sim N(\mu, \mu)$$

A count chart (c-chart) can be set up with a centre line at μ and action lines at μ plus or minus $3.09\sqrt{\mu}$.

Example 6.8

A manager in a bakery which specialises in decorated cakes has kept a record of the number of complaints over the past 20 months: 6, 7, 3, 9, 3, 7, 9, 7, 4, 13, 5, 7, 8, 2, 6, 8, 6, 4, 11, 7. Complaints are usually about the decoration rather than the quality of the cakes and the manager thinks it is reasonable to assume complaints are random and independent. The mean number of complaints is 6.6 per month, the standard deviation is 2.72, and the variance, which is 7.40, is reasonably close to the mean, as expected for a sample from a Poisson distribution. The manager decides to set up a c-chart. If this is done with a mean of 6.6, and the data are plotted retrospectively, none lies outside the upper action line at:

$$6.6 + 3.09 \times \sqrt{6.6} = 15.54$$

The normal distribution is not satisfactory for setting the lower action limit, because -1.34 is not physically possible. If $Y \sim \text{Poisson}(6.6)$ then $\Pr(Y = 0)$ is 0.0014, but a lower action limit is not relevant in this context. However, a normal approximation is reasonable for setting the upper limit, and the exact $\Pr(Y > 15)$ is 0.0014. However, the manager aims to reduce the mean number of complaints to 3 per month by providing a clearer description of the decoration customers should expect. The normal approximation would give an upper limit of,

$$3 + 3.09 \times \sqrt{3} = 8.35$$

but this is unlikely to be a satisfactory approximation, because the mean is substantially less than 6. If $Y \sim \text{Poisson}(3)$ then

$$\Pr(Y > 8) = 0.0038 \quad \text{and} \quad \Pr(Y > 9) = 0.0011$$

An action line at 10 or more complaints would be near 1 in 1000.

In fact there were 28 complaints in the next six months. An approximate 95% confidence interval for the difference in mean monthly claims before and after the change is

$$(6.60 - 4.67) \pm 1.96 \times \sqrt{\frac{132}{20^2} + \frac{28}{6^2}}$$

which gives

$$1.93 \pm 2.31$$

Although the trend is in the right direction, the reduction is not statistically significant at a 5% level because the confidence interval includes 0.

p-Charts

The proportions chart (p-chart) is also usually based on a normal approximation. It is useful whenever we classify a sample as good or defective. Let X be the number of defectives in a random sample of size n, and suppose that a target proportion is p. If the process in on target, the normal approximation is

$$X \sim N(np, np(1-p))$$

or equivalently

$$\hat{p} \sim N(p, p(1-p)/n)$$

where $\hat{p} = X/n$ is the sample proportion. Upper and lower warning lines can be set at

$$p \pm 3.09 \times \sqrt{p(1-p)/n}$$

The approximation is reasonable provided the smaller of np and $n(1-p)$ exceeds 5. It is convenient if n has the same size for each sample.

Example 6.9

A company manufactures a range of filters for the removal of oil and other contaminants from compressed air systems. There are several hand operations involved in the assembly of the filter elements. Each batch is sampled, according to British Standard 6001, at the end of each operation and the sample is tested for mechanical properties. All filter elements are checked for end cap alignment and overall length using an electronic Go–No Go gauge. Also, all filter elements are inspected for visual defects before they are stamped with the company logo and packed. Ten years ago, the proportion of filters classed as defective at this stage was about 10%. Although several of the defects, such as a missing O-ring or dirty end caps, could be rectified immediately, and most could be remedied by re-work, the company wished to reduce the proportion of defects. The main problem area seemed to be the socking operation, in which a foam sleeve (a micro-pore anti-re-entrainment barrier) is cut to length, glued along its seam, pulled over the element, and glued to the end caps. The company employed a postgraduate student to work on the problem with the operators. They started by keeping a p-chart of the results from a sample inspection after the socking operation over 26 days. The data are given in Table 6.9 (Pritchard *et al.*, 1993), and the proportions of defectives are plotted in Figure 6.12. The upper control limit (UCL) has been calculated from the formula

$$\text{UCL} = \bar{p} + 3\sqrt{(\bar{p}(1-\bar{p})/n)}$$

Table 6.9 Results of sample inspection of batches of filter elements

Date	Batch size	Sample size	Number defectives	$\hat{p}$	Defect codes
24/9	1345	109	7	0.064	4Y 2Z 1AA
25	1326	117	9	0.077	1Z 8AA
26	1431	148	13	0.088	7Y 2Z 4AA
27	800	70	8	0.114	4Y 1Z 3O
28/9	230	16	0	0.000	-
2/10	100	8	6	0.75	1Y 5AA
3	600	60	5	0.083	4Y 1BP
4	600	60	0	0.000	−
5	214	21	5	0.238	1BV 4Z
8/10	1505	107	33	0.308	25Y 6Z 2O
9	2100	170	31	0.182	25Y 5Z 1O
10	915	89	3	0.034	2Y 1BV
11	1005	89	6	0.067	4Y 2Z
12	450	43	5	0.116	4Y 1BP
15/10	1200	103	10	0.097	6Y 1Z 1BV
16	610	62	7	0.101	2Y 5Z
18	550	44	4	0.091	4Y
19	1112	88	4	0.045	4Y
22/10	536	53	5	0.094	3Y 1V 1BT
23	1920	140	8	0.057	2BP 1V 5T
24	1050	63	3	0.047	3Y
25	968	73	3	0.041	2Y 1BP
26	400	28	11	0.39	3Y 8BV
29/10	1423	100	15	0.15	8Y 2Z 2V 3BV
30	1224	84	6	0.071	1Y 1BV 4AA
31/10	1008	93	5	0.053	2Y 1Z 2BV

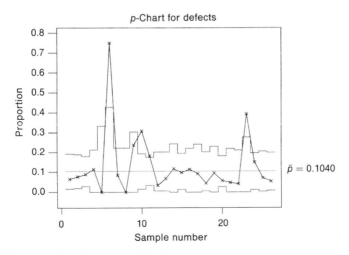

Figure 6.12 *p*-chart for filter elements

where $\bar{p}$ is the total number of defectives divided by the total of the sample sizes and n is the sample size. The lower control limit (LCL) is given by

$$\text{LCL} = \bar{p} - 3\sqrt{\bar{p}(1 - \bar{p})/n}$$

or 0 if this is negative.

The chart shows a value for $\bar{p}$ of over 0.1. There were 4 samples with a $\hat{p}$-value above the UCL. On 2nd October 1990, 6 out of 8 elements were defective, 5 of the 6 with code AA, dirty end caps. On 8th and 9th October 1990 the defect was mostly code Y, insufficient adhesive on the sock, and 25 such elements were found on both days. On 26th October 1990 there were 11 defectives out of a sample of 28, with most of them having a code BV which is a missing O-ring. The defects AA and BV could be prevented by taking more care, and the Y defect suggests that the operators are not paying enough attention to the requirement of attaching the sock to the end cap around the circumference. The $\bar{p}$-value for the socking operation was too high. Although the faults could usually be rectified, the time taken to re-work the items and re-inspect them was an unnecessary cost. Producing consistently correct items should not take up any more time, and probably less, than was being wasted in rectification.

The UCL was exceeded on 4 out of 26 days. If defectives occur according to a binomial distribution with a value of p of 0.1 the probability of 4 or more exceedances in 26 days would be negligible. Therefore the data were investigated in more detail by fitting a logistic regression model with a constant term only, i.e.

$$\ln(\hat{p}/(1 - \hat{p})) = \ln(p/1 - p) + \text{random variation}$$

The scaled deviance was 130 on 25 degrees of freedom. If the data were from the binomial distribution the expected value of the scaled deviance would equal the degrees of freedom, and the distribution of the scaled deviance would be approximately chi-squared with 25 degrees of freedom. The data are clearly 'over-dispersed' by an estimated factor of over 5. There are two explanations for this increased variance. Firstly, occurrences of defectives were not independent. A mistake such as missing out an O-ring is likely to occur for a consecutive sequence of elements. Secondly, the samples were made up from inspections of different product lines which were likely to have differing underlying proportions of defectives, miniature elements being particularly difficult to handle. This will increase the variance (Box et al., 1978). If these 26 days had been considered an acceptable level of operation the UCL would have been increased by a factor of 2.3 (approximately $\sqrt{130/25}$) for future use. However, further training resulted in a reduction in the proportion of defectives and helped counter the tendency to repeat the same mistake.

The elements are made to an exacting specification and a 100% dimensional and mechanical check, to determine that no part is missing, is made at the end of production. At the time of the investigation monthly targets were set. It was suggested that this policy might result in increasing proportions of defectives towards the end of the month as there was sometimes increased pressure on all involved to achieve the targets. The model

$$\ln(\hat{p}/(1 - \hat{p})) = \ln(p/(1 - p) + \beta \times (\text{day of month}) + \text{random variation}$$

was fitted to the 100% inspection data. The estimate of β was -0.01 with a standard deviation of 0.02, after allowing for over-dispersion. There was no evidence to support the suggestion of an increase in proportions of defectives during a calendar month. Nevertheless, the policy of setting monthly targets was reviewed as the results of setting up syndicates (quality circles) became available.

A proposal to set up control charts for a hand assembly operation is bound to be viewed with suspicion. Although it should lead to useful information which will allow the process to be improved, a benefit to all concerned, it could also be used to single out individual operators for poor work. It would not be sensible to make an issue over occasional individual lapses. Furthermore, even if all operators were identical in their working practice, natural variation in the process will always lead to someone being perceived as worst, during any particular period. It is a common exercise in statistics classes for management students, to ask them to draw samples from a bag containing beads of two colours. The person with the least 'defectives' (defined by one of the colours) is rewarded as 'best operator' for that round. The operators at the company agreed to a trial period of charting provided it was restricted to the process as a whole. The involvement of an independent investigator, who was studying for a higher degree, was appreciated by the operators. They thought this demonstrated that the investigation was an unbiased attempt to improve procedures for all concerned. It was generally agreed that the control charts helped identify areas where extra training would be useful. However, it was thought that it would be better to implement charting as an occasional audit exercise than as a permanent system.

6.4   Multivariate process control

6.4.1   Introduction

It is not always sufficient to monitor a single process variable. Mechanical components will usually have several critical dimensions, 28 in the case of the robot arms manufactured by Bank Bottom Engineering. Another example is a cough mixture with two active ingredients, codeine phosphate hemihydrate 3 mg/10 ml and creosote 0.015 ml/10 ml, and 12 other ingredients apart from water. A case from the service sector is that of a car repair workshop. Each day a randomly selected customer is asked to complete a brief questionnaire. The questions ask for the respondent's opinion of the cleanliness of the returned vehicle, the time spent in the reception area when delivering and collecting the vehicle, the actual cost relative to the estimated cost, and whether the repair was completed on time. The weekly average results are recorded. Chemical manufacturing processes have provided an incentive for much of the work on multivariate process control. It is common to monitor up to 100 variables at one-minute intervals, using automatic measuring equipment, and it would not be helpful to plot a means chart of hourly averages for every one of the variables, and take action whenever one of the means falls outside an action line. This would lead to far too many unnecessary adjustments. Conversely, relying on individual means charts could result in missed warning signs when many of the variables are simultaneously close to, but do not fall beyond, the action lines.

6.4.2   The T^2-chart

Suppose a process is characterised by k variables. The target value is μ and the variance–covariance matrix is V. A sample of size n is taken at approximately, but not exactly, half hour intervals. The sample mean is $\bar{x}_t$. Statistical process control (SPC) can be based on Hotelling's statistic:

$$T^2 = n(\bar{x}_t - \mu)^T V^{-1}(\bar{x}_t - \mu)$$

If the process is on target, T^2 has a chi-square distribution with k degrees of freedom and an action limit can be set at $\chi^2_{k,0.002}$. As with univariate SPC, the variance–covariance matrix will have to be estimated from data obtained during some stable period.

Example 6.10
The variables 6(1), 6(2) and 6(3) measured on the 206 Bank Bottom robot arms (TableWSrobotarms) have sample means of $-0.020\,97$, $0.020\,39$, and $0.070\,21$ respectively. The sample correlations are:

$$
\begin{array}{ccc}
 & 6(1) & 6(2) \\
6(2) & 0.243 & \\
6(3) & -0.606 & 0.257
\end{array}
$$

and the sample covariance matrix is

$$\hat{V} = \begin{pmatrix} 26\,982 & 5945 & -20\,994 \\ 5945 & 22\,161 & 8061 \\ -20\,994 & 8061 & 44\,479 \end{pmatrix} \times 10^{-8}$$

For illustration a T^2-chart for samples of size 4 with

$$\mu = \begin{pmatrix} -0.02 \\ 0.02 \\ 0.07 \end{pmatrix} \quad \text{and} \quad V = \begin{pmatrix} 26 & 6 & -20 \\ 6 & 22 & 8 \\ -20 & 8 & 44 \end{pmatrix} \times 10^{-5}$$

is shown in Figure 6.13. The plotted points are for samples made up from the data for the following arm numbers: $\{1, 6, 11, 16\}$; $\{21, 26, 31, 36\}$; $\{41, 46, 51, 56\}$; ...; $\{181, 186, 191, 196\}$.

   The limitation of the T^2-chart is that it does not indicate which variables are primarily responsible for the out-of-control signal, and there have been many suggestions for augmenting it (e.g. Kourti and MacGregor, 1996). A good proportion of these rely on the first few principal components. Although such methods often seem to be effective, there are potential drawbacks. The first is that they indicate which principal components are out of control rather than which of the original variables are out of control. This need not be a disadvantage if there are useful physical interpretations of the principal components, but there will not always be any. A second possibility is that one of the last principal

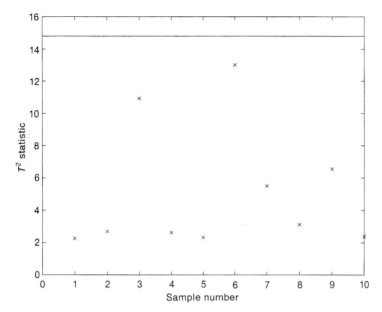

Figure 6.13 T^2-chart for variables 6(1), 6(2), 6(3) on robot arms. The mean vector for the first sample of 4 is (−0.0248 0.0282 0.0760), and the value of T^2 is 2.56. The 0.2% control line is shown at 14.80

components, which might be ignored, is associated with some critical quality of the final product. It is easy to construct such examples and Hadi and Ling (1998) discuss the consequences. Jones (1995, 1999) gives an ingenious alternative to principal components, for which it is convenient to use the following standard results for the multivariate normal distribution.

6.4.3   Multivariate normal distribution

1. If X_i, for i from 1 to n, is a random sample of size n from a multivariate normal distribution, $N(\mu, V)$, then $\bar{X} \sim N(\mu, V/n)$
2. If $X \sim N(\mu, V)$ and X is partitioned as

$$X = \begin{pmatrix} X_1 \\ X_2 \end{pmatrix}$$

with

$$V = \begin{pmatrix} V_{11} & V_{12} \\ V_{21} & V_{22} \end{pmatrix}$$

$$X_1 \mid X_2 = x_2 \sim N(\mu_1 + V_{12} V_{22}^{-1}(x_2 - \mu_2), V_{11} - V_{12} V_{22}^{-1} V_{21})$$

The multiple regression result is easy to verify for a bivariate normal distribution and Graybill (1976) gives a general proof.

6.4.4 Decomposition of T^2

Jones's (1995) description of his algorithm, for the case of k equal to 4, shows the general principle.

- Step 0   Calculate T^2, and if it exceeds $\chi^2_{4,0.002}$ proceed to Step 1. In the following a statistic is significant if its absolute value exceeds $z_{0.001}$, i.e. 3.09.
- Step 1   For $i = 1$ to 4 calculate

$$(\bar{x}_i - \mu_i)/(\sigma_i/\sqrt{n})$$

   If none is significant, let p be the subscript corresponding to the variable for which the absolute value of this statistic is a minimum. If at least one is significant, let p be the subscript corresponding to the variable for which the absolute value of this statistic is a maximum, and identify X_p as being out of control. In any event proceed to Step 2, but notice that the criteria for selecting the value of p are different.

- Step 2   Determine the conditional distributions of $\bar{X}_i$, given that $\bar{X}_p = \bar{x}_p$, and calculate

$$\frac{\bar{x}_i - \text{mean}(\bar{X}_i \,|\, \bar{X}_p = \bar{x}_p)}{\text{standard deviation}(\bar{X}_i \,|\, \bar{X}_p = \bar{x}_p)}$$

   If none is significant let q correspond to the smallest in absolute magnitude. If at least one is significant let q correspond to the largest in absolute magnitude and identify X_q as out of control given the value of $\bar{X}_p$. Therefore, variables X_p and X_q may both need attention.

- Step 3   Determine the conditional distributions of $\bar{X}_i$, given that $\bar{X}_p = \bar{x}_p$ and $\bar{X}_q = \bar{x}_q$, and calculate

$$\frac{\bar{x}_i - \text{mean}(\bar{X}_i \,|\, \bar{X}_p = \bar{x}_p \text{ and } \bar{X}_q = \bar{x}_q)}{\text{standard deviation}(\bar{X}_i \,|\, \bar{X}_p = \bar{x}_p \text{ and } \bar{X}_q = \bar{x}_q)}$$

   If none is significant let r correspond to the smallest in absolute magnitude. If at least one is significant let r correspond to the largest in absolute magnitude and identify X_r as out of control given the values of $\bar{X}_p$ and $\bar{X}_q$. It is probably best to investigate all three variables.

- Step 4   Determine the distribution of $\bar{X}_i$, given that $\bar{X}_p = \bar{x}_p, \bar{X}_q = \bar{x}_q$ and $\bar{X}_r = \bar{x}_r$ and calculate

$$\frac{\bar{x}_i - \text{mean}(\bar{X}_i \,|\, \bar{X}_p = \bar{x}_p, \bar{X}_q = \bar{x}_q, \bar{X}_r = \bar{x}_r)}{\text{standard deviation}(\bar{X}_i \,|\, \bar{X}_p = \bar{x}_p, \bar{X}_q = \bar{x}_q, \bar{X}_r = \bar{x}_r)}$$

   This is the final step, and if it is significant X_i is identified as out of control given the values of $\bar{X}_p, \bar{X}_q$ and $\bar{X}_r$. It is probably best to investigate all the variables. An example is given in Exercise 6.8.

6.4.5   Elliptical control charts

In some applications there is a good physical reason for considering a specific combination of variables. An example is the deviation of the centre of a cylinder bore in an engine block from the specified point. If X and Y are the errors across and along the engine block, it would be more appropriate to monitor the offset

$$R = \sqrt{X^2 + Y^2}$$

than both X and Y. The specification would be a circle based on the centre point. In some applications errors in one direction may be more serious, and the specification limit might be an ellipse. If control limits are set in from the specification, a principle discussed towards the end of Section 6.3.1, they would be approximately elliptical. A mathematical justification would depend on the assumed form of the bivariate distribution, but the construction does have an intuitive appeal. It is justified if X and Y are independent normal variables, because the contours of the bivariate pdf are ellipses with major axes along the major axis of the specification, and an assumption of independence may often be reasonable in a machining context.

Papayannopoulos *et al.* (1999) set up a chart with elliptical control limits for a process in the manufacture of television tubes. The joining of the neck supporting the electron gun to the rest of the television tube is a critical process, which directly affects the quality of the picture. The ends of the neck and tube are nominally circles of the same diameter, and they should be concentric when joined in order to provide good support and alignment of the electron gun. One consequence of the principle used for the design of these colour television tubes is that errors in the x-direction are more critical than those in the y-direction. It was anticipated that a control chart with elliptical limits would lead to a reduction in the number of unessential adjustments of the machinery, and hence an increase in production and a reduction of the variability of the process in the critical x-direction. The reason for the reduction in variability is that it is not possible to adjust in the y-direction without resetting in the x-direction, and frequent unnecessary resetting increases process variability.

The process was monitored by measuring four tubes every few hours. The tubes were not selected consecutively, in case of any serial correlation. Once the machine had been set up, the standard deviation (σ) of errors in the x- and y-directions was known to be about 13 units. The specification for the centre of the neck relative to the centre of the tube was plus or minus 50. The original control limits had been set as plus or minus $3.09\sigma/\sqrt{n}$, which equals 20 units in both the x- and y-directions.

The strategy of resetting the head if the eccentricity exceeds plus or minus 20 in either the x- or the y-direction corresponds to the square acceptance region shown in Figure 6.14. The modified method that they investigated relaxes the limits in the y-direction to plus or minus 25, and replaces the rectangular region by an elliptical region (see Figure 6.14). This was based on the idea that a deviation in the process mean by up to 5 units in the y-direction was acceptable. It resulted in a C_{pk} of 1.15 rather than a C_p of 1.28, but the former was quite sufficient for the y-direction. The chart could be a plot of the deviations $(\bar{x}, \bar{y})$ as a sequence of numbered points in a

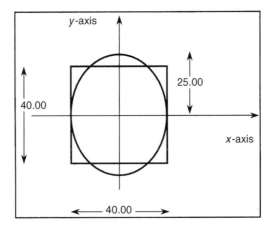

Figure 6.14   Elliptical control area for neck joining of TV tubes

predrawn ellipse or a plot of the statistic d given by:

$$d = \frac{\bar{x}^2}{20^2} + \frac{\bar{y}^2}{25^2}$$

on ordinary graph paper (Figure 6.15). The action line at 1 corresponds to points being outside the ellipse.

The acceptance region was loosely based on intuition that a locus of points giving equal picture quality was an ellipse with its major axis aligned in the y-direction. The area of the square region was 0.40×0.40, which equals 0.16. The area of the ellipse was given by $\pi \times 0.25 \times 0.20$, which equals 0.157. Although there is little change in the size of the acceptance region the elliptical shape will contain more of a bivariate normal distribution of deviations $(\bar{x}, \bar{y})$, particularly if the standard deviation of $\bar{y}$ exceeds that of $\bar{x}$. The researchers claimed that the elliptical shape should result in improved picture quality because the region outside the ellipse but within the square was less desirable than the region outside the square but within the ellipse. During the trial period the number of adjustments would have been reduced by 23%. This was explained by points being more likely to lie at the ends of the major axis of the ellipse than in the corners of the square, perhaps because the machine operators are more careful with the x-direction setting.

The strategy of relating control charts to specification limits, rather than concentrating on the concept of statistical control, is described by Wetherill and Brown (1991). They state that, in general, this allows the process mean to wander and that statistical control is no longer achieved. However, this may not be a concern and does not necessarily apply if errors arise during the setting-up of the process. The combination of a chart based on specification limits in the x-direction and a standard chart in the y-direction would be a rectangle. A heuristic argument was used to replace this by an ellipse, which seems more compatible with elliptical specification limits.

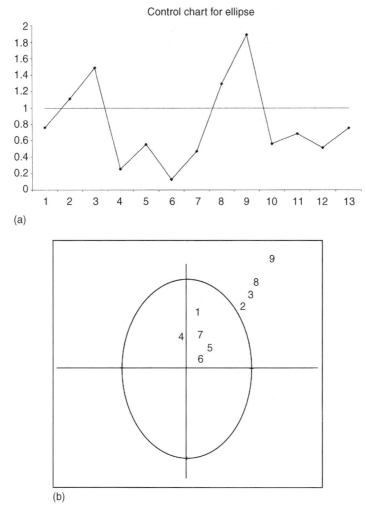

Figure 6.15   Elliptical control charts for neck joining of TV tubes process: (a) scaled distance to perimeter; (b) position within the elliptical control area with points shown by sample number

6.5   Optimal control of linear systems

Flour milled from wheat grown in countries within the European Community may not contain enough protein for baking large loaves of bread. Fearn and Maris (1991) described a feedback control system for maintaining the percentage of protein in flour at a target value by adding dried gluten. The target value will depend on the grade of flour being produced but is typically about 12%. Gluten is made by washing most of the starch out of wheat flour. A control system is necessary because variation from the target value will make the flour less suitable for baking and gluten is expensive. Measurements of the protein content of the flour were made at one-minute intervals using a near infra-red spectrophotometer (NIROS). They defined the state of the system at time t, x_t, as the difference in protein content from the target value. The following general results for discrete time linear systems are used to design a control algorithm for the flour mill, in Section 6.5.5.

6.5.1   State space representation

A small island has two newspapers, *The Quest* and a new title *The Signal*. Let q_t and s_t be sales of the *Quest* and *Signal* during week t. Suppose sales are governed by the following equations:

$$q_t = a_{11}q_{t-1} + a_{12}s_{t-1}$$
$$s_t = a_{21}q_{t-1} + a_{22}q_{t-1}$$

If the state, x_t, at time t is defined as $(q_t \quad s_t)^T$, then the equations can be written in the form

$$x_t = Ax_{t-1}$$

where the matrix A has elements (a_{ij}). This form would include a model which has sales during a week depending on sales one week ago and two weeks ago. For example, if

$$q_t = a_{11}q_{t-1} + a_{12}q_{t-2} + a_{13}s_{t-1}$$
$$s_t = a_{31}q_{t-1} + a_{33}s_{t-1} + a_{34}s_{t-2}$$

the state x_t could be defined as $(q_t q_{t-1} s_t s_{t-1})^T$ and

$$A = \begin{pmatrix} a_{11} & a_{12} & a_{13} & 0 \\ 1 & 0 & 0 & 0 \\ a_{31} & 0 & a_{33} & a_{34} \\ 0 & 0 & 1 & 0 \end{pmatrix}$$

If the initial state is x_0 then

$$x_t = A^t x_0$$

and the system is stable if $A^t \to 0$ as $t \to \infty$. It has a fixed point if A^t tends to some constant value as t tends to infinity (see Markov chains in Chapter 8). The system is unstable if A^t increases as t increases, and a fundamental requirement of any control system is that the closed loop system be stable. In general, if there are n states the square $n \times n$ matrix A will have n eigenvalues (λ_i). If the associated eigenvectors are linearly independent then A can be diagonalised, i.e. there exists an invertible matrix M and a diagonal matrix Λ such that

$$M^{-1}AM = \Lambda$$

and

$$A^k = M\Lambda^k M^{-1}$$

where Λ^k is the diagonal matrix with entries λ_i^k (e.g. Allenby, 1995, or Kreyszig, 1993). It follows that the system will be stable if A has eigenvalues, which may be complex, with modulus less than 1, i.e. within the unit circle centred at the origin in the complex plane.

A general state space representation of a linear system is

$$x_{t+1} = Ax_t + Bu_t + v_t$$

$$y_{t+1} = Cx_{t+1} + w_t$$

where x_t is an $n \times 1$ state vector, A is an $n \times n$ state transition matrix, u is an $l \times 1$ control vector, B is an $n \times l$ matrix, v_t is an $n \times 1$ vector of system noise, y_t is a $m \times 1$ observation vector, C is an $m \times n$ matrix and w_t is observation noise. The system and observation noise have variance–covariance matrices V and W respectively. Although the components of v_t and the components of w_t can be correlated, v_t and w_t are independent of each other and of past values.

6.5.2   Optimal control

The state x_t will typically be defined as deviation from target. One approach to control system design is to set $u_t = -Lx_t$ and choose L so that the matrix $(A - BL)$ has eigenvalues which are suitably close to zero. A formal optimal solution is to minimise the performance indicator

$$\sum_0^\infty (x_t^T Qx_t + u_t^T Ru_t)$$

where Q is chosen to represent the costs of being off target, and R is chosen to represent the costs of applying the control. The solution is:

$$L = (R + B^T PB)^{-1} B^T PA$$

where P is a solution of an algebraic Ricatti equation

$$P = A^T \{P - PB(R + B^T PB)^{-1} B^T P\}A + Q$$

In general, this equation has to be solved numerically. Barnett and Cameron (1985) and Kwakernaak and Sivan (1972) cover the details of the theory. Craven (1995) gives more theory for non-linear problems, including some that arise in management. Erickson and Hedrick (1999) cover the practical design of controllers for chemical plants.

6.5.3   Kalman filter

The optimal control assumes x_t is known, but in practice we cannot measure x_t and have to rely on y_t. The Kalman filter, or optimal observer, is an algorithm for estimating x_t from y_t. Let $\hat{x}_t$ be the estimate of x_t made from y_t and define the estimation error e_t as $(x_t - \hat{x}_t)$. The general principle is that the Kalman filter has the structure

$$\hat{x}_{t+1} = (A - KC)\hat{x}_t + Bu_t + Ky_t$$

Ignoring, for the moment, the noise terms it follows from the equation $y_t = Cx_t$ that

$$e_{t+1} = (A - KC)e_t$$

We need to choose K so that $(A - KC)$ has eigenvalues that are reasonably close to zero. However, when we introduce the noise terms the solution that minimises $\sum e_t^2$ is to set

$$K = HC^T(CHC^T + W)^{-1}$$

where

$$H = A\{H - HC^T(W + CHC^T)^{-1}CH\}A^T + V$$

6.5.4   IMA model

Suppose we have a sequence of measurements $\{Y_t\}$, at equally spaced time intervals, from some system, e.g. deviations of percentage protein or depth of paint from their target values. Box and Luceño (1997) have found that the dynamic behaviour of such systems is often approximated quite well by the model

$$Y_{t+1} = Y_t + E_t - \theta E_{t-1}$$

where $\{E_t\}$ is a sequence of independent random variables with mean 0 and variance σ^2. An equivalent form of the model is in terms of first differences

$$Y_{t+1} - Y_t = E_t - \theta E_{t-1}$$

and it is from this that the name 'integrated moving average' (IMA), of order 1, arises. It is 'integrated' because adding is the inverse operation to the differencing and 'moving average' because the right-hand side is a moving average of order 1. Differences at lag 1 have a correlation

$$E[(Y_{t+1} - Y_t)(Y_t - Y_{t-1})] = E[(E_t - \theta E_{t-1})(E_{t-1} - \theta E_{t-2})] = -\theta\sigma^2$$

Differences at any higher lag are independent. Since

$$\text{var}(Y_{t+1} - Y_t) = \text{var}(E_t - \theta E_{t-1}) = \sigma^2 + \theta^2\sigma^2$$

the autocorrelation is $-\theta/(1 + \theta^2)$ at lag 1 and 0 at higher lags. Repeated substitution gives

$$Y_{t+m} - Y_t = E_{t+m} + (1 - \theta)E_{t+m-1} + \ldots + (1 - \theta)E_{t+1} - \theta E_t$$

and hence

$$\text{var}(Y_{t+m} - Y_t) = \sigma^2 + (m - 1)(1 - \theta)^2\sigma^2 + \theta^2\sigma^2$$
$$= 2\theta\sigma^2 + m(1 - \theta)^2\sigma^2$$

The IMA model is not in the standard state space form, but the random walk

$$x_{t+1} = x_t + v_t$$

with a noisy observation equation

$$y_{t+1} = x_{t+1} + w_t$$

where v_t and w_t are sequences of independent random variables with zero mean and variances σ_v^2, σ_w^2 respectively, is in the standard form and has the same variogram as the IMA model:

$$\text{var}(y_{t+m} - y_t) = \text{var}(v_{t+m-1} + \ldots + v_{t-1} + w_{t+m} - w_t)$$

$$= m\sigma_v^2 + 2\sigma_w^2$$

The identification between the IMA and the random walk with the noisy observation (sticky innovation model of Box and Luceño, 1997) is that

$$\sigma_v^2 = (1 - \theta)^2 \sigma^2$$

$$\sigma_w^2 = \theta\sigma^2$$

If a control term is added to the random walk

$$x_{t+1} = x_t + bu_t + v_t$$

$$y_{t+1} = x_{t+1} + w_t$$

and we have the standard state space form with $A = 1$, $B = b$, $V = \sigma_v^2$, $C = 1$ and $W = \sigma_w^2$. If we take the cost of control as negligible ($R = 0$) the optimal control reduces to

$$u_t = -x_t/b$$

However, we cannot measure x_t and have to estimate it from y_t. The estimation equation has the form

$$\hat{x}_t = (1 - K)\hat{x}_{t-1} + bu_{t-1} + Ky_{t-1}$$

and since $u_{t-1} = -\hat{x}_{t-1}/b$

$$\hat{x}_t = K(y_{t-1} - \hat{x}_{t-1})$$

This is a proportion of the difference between the measurement at time $t - 1 (y_{t-1})$ and the estimate of x_{t-1} made at time $t - 2$. The Kalman filter simplifies considerably

$$K = H(H + W)^{-1}$$

where

$$H = H - H(W + H)^{-1}H + V$$

which is equivalent to

$$H^2 - VH - VW = 0$$

Hence

$$K = \frac{2\sigma_v^2}{\sigma_v^2 + \sqrt{\sigma_v^4 + 4\sigma_v^2\sigma_w^2}}$$

where $K > 0$ implies that only the positive square root in the solution of the quadratic equation need be considered. If σ_v^2 is much greater than σ_w^2, K will be close to 1 as y_t is a relatively precise estimate of x_t. If σ_w^2 is much greater than σ_v^2, the estimate of x_t is very poor and K is reduced towards 0.

6.5.5   Time delays

Fearn and Maris (1991) found that their process was well modelled by an IMA process with a θ of 0.75. However, the response of flow protein content at the sampling point to a change in the setting of the gluten feeder is delayed for two reasons. The first is that the sampling point is downstream of the gluten feeder and the second is the time taken for the feeder mechanism to adjust to the control signal. They investigated the system dynamics by making some large step changes to the gluten addition and monitoring the effects. No change was noted in the sample taken 1 minute later but the full effect was noted after 2 minutes. Time delays, also known as dead time, are a significant feature of processes and make control systems less effective and more prone to instability. We now need to introduce a second state in the model

$$\begin{bmatrix} x_{t+1} \\ x_t \end{bmatrix} = \begin{bmatrix} 1 & 0 \\ 1 & 0 \end{bmatrix} \begin{bmatrix} x_t \\ x_{t-1} \end{bmatrix} + \begin{bmatrix} b \\ 0 \end{bmatrix} u_{t-1} + \begin{bmatrix} v_t \\ 0 \end{bmatrix}$$

where u_{t-1} is restricted to be of the form:

$$u_{t-1} = -Lx_{t-1}$$

The feedback controlled system will be stable provided $0 < bL < 1$ (Exercise 6.9a). The control will be implemented with x_{t-1} replaced by an estimate given by

$$\tilde{x}_{t-1} = K(y_{t-1} - \hat{x}_{t-1})$$

where $\hat{x}_{t-1} = \tilde{x}_{t-2}$. It would be reasonable to try a value of bL equal to 0.5 to begin with and then vary it slightly to see if the performance of the controller improves. Ray (1981) covers in detail the design of controllers for systems with time delays. You are asked to try applying the optimal control algorithm in Exercise 6.9b.

6.5.6   Control of a spring coiling machine

Readman *et al.* (1998) compare the performance of fixed and adaptive control laws for controlling the free length of helical springs coiled on an automatic coiling machine. The length is measured using a contactless capacitive sensor and controlled by varying the spring pitch angle. A feedback control system was designed to reject disturbances caused by variation of the wire properties and periodic disturbances generated by the coiler. For the adaptive control law, they made recursive estimates of the parameters in the following model. The variables e_t and u_t are the error in the spring length and the control input respectively.

$$e_t = \alpha_1 e_{t-1} + \alpha_2 e_{t-2} + \ldots + \alpha_p e_{t-p} + g_1 u_{t-1} + \ldots + g_q u_{t-q}$$

The estimated parameters are then used to generate a control signal (Exercise 6.10)

$$u_t = -(\hat{g}_2 u_{t-1} + \ldots + \hat{g}_q u_{t-q+1} + \hat{\alpha}_1 e_t + \ldots + \hat{\alpha}_p e_{t-p+1})/\hat{g}_1$$

6.6 Summary

1. The process capability index, C_p, is defined as the ratio of the width of the specification interval to six standard deviations of the variable. If the variable has a normal distribution, then a C_p of 1.0 corresponds to 2 in 1000 items being outside specification. Other indices are also used, e.g. C_{pU}, which is appropriate when there is only an upper specification limit, is defined as the ratio of the difference between this limit and the mean of the variable to three standard deviations.
2. Control charts can be used to monitor the process and, provided the reasons for points crossing action lines are investigated, they contribute to improvements in the process.
3. Consecutive items from a process may be relatively similar, and this feature can be detected from a variogram or a correlogram. In principle, the correlation might be used to set up automatic process control and thereby reduce variability. In practice this may not be feasible, within sensible cost limits, and control charts may be used. In this case there should be an appropriate gap between taking items for samples.
4. Automatic process control requires a system of sensors and actuators and associated electronics. However, the cost of these can be balanced against the benefits of reduced variability.

Exercises

6.1 **The folded normal distribution**
A steel surface is ground until it appears smooth, but under a powerful microscope it is seen as pitted. The depths of pits are approximately normally distributed, with a mean of δ below the surface and a standard deviation σ. The surface is now highly polished and a layer of material of depth δ is removed. The remaining pits have depths Y which have a folded normal distribution with pdf:

$$f(y) = \frac{2}{\sigma\sqrt{2\pi}}e^{-y^2/\sigma^2} \quad \text{for } 0 \leq y$$

Show that Y has a mean value of

$$\sigma\sqrt{2/\pi}$$

and a standard deviation of

$$\sqrt{(1 - 2/\pi)}\sigma$$

What proportion of the distribution lies more than 3.09 standard deviations above the mean?

6.2 **Cauchy distribution**
A radioactive source is shielded so that it emits α-particles into a half plane at angles θ which are uniformly distributed over $[-\pi/2, \pi/2]$. Let X be the intercept a particle makes with a vertical line a distance a from the source.

(i)   Show that the intercepts have a distribution with pdf:

$$f(x) = \frac{a}{\pi(a^2 + x^2)} \quad \text{for } -\infty < x < \infty$$

known as the Cauchy distribution.

(ii) Demonstrate that X has an infinite variance if the mean, which is not properly defined, is taken as 0. It follows that the central limit theorem is not applicable. In particular, the mean of n independent variables from a Cauchy distribution has the same distribution as a single variable.

(iii) Find the upper 0.1% point of the distribution in terms of a.
      [The ratio of two independent standard normal variables has a Cauchy distribution.]

6.3   **Weibull distribution**
Imagine a long piece of material consists of N independent links, whose weakest member determines its failure. Assume the probability a link survives a load x is

$$e^{-(x/\theta)^a}$$

Let X be the load at which the piece of material fails. Show that the cdf of X has the form,

$$F(x) = 1 - e^{-(x/b)^a}$$

This is known as the Weibull distribution, and a method of fitting it to a set of data is given in Exercise 8.12.

6.4   Assume a random sample of size n from a normal distribution. A rough approximation to the sampling distribution of S, e.g. Metcalfe (1994), is:

$$S \sim N(\sigma, \sigma^2/(2n))$$

Deduce the approximate confidence limits for C_{pk}:

$$\hat{C}_{pk}[1 \pm z_{\alpha/2}/(2n)^{\frac{1}{2}}]$$

Using $(n-1)$ in place of n gives a slight improvement.

6.5   In order to investigate the accuracy and precision of a colourmetric method for estimating lead content of spring water, 15 litres of a carefully prepared 12 ppm solution were made up. The solution was thoroughly shaken and poured into 15 one-litre jars. Sets of three jars were chosen at random and sent to five randomly selected laboratories. The results are given below.

Lab A	Lab B	Lab C	Lab D	Lab E
9	12	13	9	13
11	14	13	12	11
10	10	16	12	12

Estimate the within-laboratory and between-laboratories variances, and show that the estimated between-laboratories standard deviation is 1.18 (to 2 decimal places).

Tests on the same material by the same operator using the same equipment in the same laboratory are said to be carried out under repeatability conditions, whereas tests by different operators in different laboratories are said to be carried out under reproducibility conditions. The repeatability value (r) is the value below which the absolute difference between two single test results obtained under repeatability conditions may be expected to lie with probability of 95% (BS5497: Part 1 1987). Calculate the repeatability defined as,

$$r = 1.96\sqrt{s_w^2 + s_w^2}$$

Calculate the reproducibility value (R) defined as,

$$R = 1.96\sqrt{s_b^2 + s_w^2 + s_b^2 + s_w^2}$$

6.6 Use the moment-generating function, up to the fourth power, to show that the skewness and excess kurtosis of $\bar{X}$ are $\gamma/\sqrt{n}$ and κ/n.

6.7 $\bar{X}$ is the mean of samples of size 4, from a distribution with variance σ^2, in which consecutive observations have a correlation ρ, but observations separated by more than one sampling interval are uncorrelated. Verify the result,

$$\mathrm{var}(\bar{X}) = \sigma^2(1 + 2\rho(1 - 1/n))/n$$

Prove the general result.

6.8 Assume that $n = 4$, $\mu = 0$, and

$$V/n = \begin{pmatrix} 1.0 & 0.1 & 0.5 & 0.1 \\ 0.1 & 1.0 & 0.5 & 0.6 \\ 0.5 & 0.5 & 1.0 & 0.8 \\ 0.1 & 0.6 & 0.8 & 1.0 \end{pmatrix}$$

Use the procedure of Section 6.4.4 for a datum (0.5735, 1.2044, −0.4298, 1.9924). This is the example Jones (1995) works through in detail.

6.9 (a) Show that the matrix

$$\begin{pmatrix} 1 & -bL \\ 1 & 0 \end{pmatrix}$$

has eigenvalues within the unit circle if $0 \le bL < 1$. You should consider the cases of $bL < 1/4$ and $1/4 < bL$ separately.

(b) Refer to the model for the gluten feeder in Section 6.5.5. Replace the entry in the second row and second column of the A matrix by a smaller number ε and try applying the optimal control algorithm.

6.10 Investigate the form of the controller described in Section 6.5.6 for p equal to 2 and q equal to 2.

7

Design of experiments

7.1 Introduction

The purpose of an experiment is to find out how variables which are important to the business, the response variables, depend on other variables that can either be set to specific values, control variables, or be monitored, concomitant variables. The first stage in planning an experiment is to identify these variables by encouraging everyone with experience of the process to contribute to the discussion. Once this is done the research team can propose an experimental procedure and consider how the variables can be measured. Singh *et al.* (1994) investigated the effect of message spacing on people's recall of television commercials. The response variables were whether people remembered the brand of product being advertised and whether they remembered the claims that were being made for it. The research team decided to investigate the effect on the response of the number of commercials between replicates of the test commercial and how the response depended on a subject's age and the time until recall was tested.

It is not good practice to investigate the effects of variables one at a time. This is always inefficient, and often misleading because of interactions between the variables. Reports of earlier research work mentioned instances of an interaction between the number of intervening items and time lapse until testing recall. The items were words and, if memory was tested soon after the second occurrence of the repeated word, recall was better if there were fewer intervening items. The opposite effect was noted when memory was measured after a long delay. Then, recall improved when there were more intervening items. If interactions are to be detected it is necessary to carry out factorial experiments. In a factorial experiment several different values (levels) are chosen for each of the control variables (factors), and if it is a full factorial design the experiment is run at every possible factor combination. The number of levels for each factor is often restricted to two or three.

7.2 Factorial designs

7.2.1 2^k designs

It is common to restrict the number of levels of each of k factors to two, typically termed low (L) and high (H). The drawback of this procedure is that it will miss any curvature in the response. For example, memory recall could be poor with a low number of intervening items and poor with a high number of intervening items, yet good with a mid-range number of intervening items. If this is at all likely, two levels are not sufficient. Singh *et al.* used a 2^3 design. The factors were age group, time lapse until memory tested, and the number of intervening commercials. They also ran a control experiment in which participants saw the test commercial only once. The levels for age were low, 20–35, and high, 62–83. The age in years and the sex of subjects were also recorded, and could be used as a concomitant variable. The levels for time lapse until recall was tested were low, soon after the video tape had been shown, and high, the next day. The levels for the number of intervening commercials were 1 and 4 for low and high respectively. More than 400 people agreed to take part in the study, which was ostensibly set up to elicit opinions about a news programme. Participants were not told the purpose of the study until they had completed the memory recall test. One hundred and twenty subjects were randomly chosen from the younger age group, and then randomly assigned to the four (2×2) time lapse by the number of intervening commercials combinations. Similarly, 120 of the older subjects were assigned to the four combinations. The remainder were assigned to the control groups which only saw the test commercial once, half being tested after a short time lapse and the others, the next day.

Three video tapes were used for the experiment. All included seven commercials within a news programme. The sequence for the low number of intervening commercials ran: $F1$, $F2$, $F3$, $F4$, X, $F5$, X where X was the test commercial and $F1, \ldots, F5$ were the intervening commercials. The sequence for the high number of intervening commercials was $F4$, X, $F1$, $F2$, $F3$, $F5$, X. Notice that the researchers set up the two tapes with the same commercial before the first X, and the same commercial before the second X. The control tape had the sequence $F4$, $F6$, $F1$, $F2$, $F3$, $F5$, X. The test commercial was promoting a vegetarian alternative to a hamburger and had a relatively high information content. The short time lapse group were asked to complete a news evaluation form and answer some questions about the test commercial soon after the tape finished. The long time lapse group were asked to do this on the day after seeing the tape. The recall score was 1 for remembering the brand name and 1 extra point for each of the eight product claims made in the commercial. The top score was 4.

Singh *et al.* reported the means for the different groups, and these are reproduced in Table 7.1. We will start by considering whether seeing commercial X twice improves recall. Set an indicator variable (x_4) at 0 for the single exposure control groups, and at 1 for the double exposure groups. There were 40 people in each of the 4 control groups and 30 people in each of the other 8 groups. A Minitab weighted regression of recall y on age (x_1), time lapse (x_3) and single or

Table 7.1 Mean recall scores for message spacing experiment. Reprinted with permission from Singh *et al.* (1994) in the Journal of Marketing Research published by the American Marketing Association

Age (x_1) 20–35 (−1) 62–83 (+1)	Intervening commercials (x_2) 1 (−1) 4 (+1) control group single exposure to X (C)	Time lapse until recall test (x_3) 5 minutes (−1) 1 day (+1)	Mean recall score (y)
−1	−1	−1	1.98
−1	−1	+1	1.17
−1	+1	−1	1.45
−1	+1	+1	1.38
+1	−1	−1	1.20
+1	−1	+1	0.32
+1	+1	−1	0.78
+1	+1	+1	0.83
−1	C	−1	1.37
−1	C	+1	0.85
+1	C	−1	0.52
+1	C	+1	0.24

double exposure is:

$$y = 0.745 - 0.360x_1 - 0.208x_3 + 0.394x_4$$
$$(0.097)\ (0.061)\quad (0.061)\quad (0.125)$$

where, as usual, the bracketed figures are the estimated standard deviations of the coefficients. The weights are proportional to the inverses of the variances of the responses, i.e. proportional to the sample sizes. There is evidence that the double exposure does improve recall (*t*-ratio is 3.15, $P = 0.014$) and even stronger evidence that the longer time lapse and higher age groups are associated with reduced recall. These conclusions are the same if an unweighted regression analysis is used. It is important to note that the increase due to double exposure is estimated as 0.394, since x_4 is 0 or 1, while the estimated decrease due to a longer time lapse is 0.416, because x_3 is coded as −1 or +1. We will now analyse the mean recall values for the 8 double exposure groups in the same way as we can analyse results from a single replicate of any 2^3 design (Table 7.2).

Starting with only the main effects of age, intervening commercials and time lapse, the fitted model is

$$y = -1.139 - 0.356x_1 - 0.029x_2 - 0.214x_3$$
$$(0.1082)\ (0.1082)\quad (0.108)\quad (0.108)$$

with $s = 0.306$ on 4 degrees of freedom. The sum of squares attributed to the regression is 1.387 44, out of a total of 1.761 89. Only the age effect is statistically significant beyond the 10% level, the *P*-value being 0.03. Compare this with a model that includes all the two-factor interactions.

$$y = 1.139 - 0.356x_1 - 0.029x_2 - 0.214x_3 + 0.051x_1 \times x_2 + 0.006x_1 \times x_3$$
$$+ 0.209x_2 \times x_3$$

Table 7.2 Design matrix for a 2^3 factorial design

x_1	x_2	x_3	$x_1 \times x_2$	$x_1 \times x_3$	$x_2 \times x_3$	$x_1 \times x_2 \times x_3$
−1	−1	−1	+1	+1	+1	−1
−1	−1	+1	+1	−1	−1	+1
−1	+1	−1	−1	+1	−1	+1
−1	+1	+1	−1	−1	+1	−1
+1	−1	−1	−1	−1	+1	+1
+1	−1	+1	−1	+1	−1	−1
+1	+1	−1	+1	−1	−1	−1
+1	+1	+1	+1	+1	+1	+1

with $s = 0.067$ on 1 degree of freedom. The sum of squares attributable to the regression is now 1.757 37. The estimated standard deviation of all the coefficients is 0.024. Notice that the coefficients of x_1, x_2 and x_3 have not changed. The reason for this is that any pair from $x_1, x_2, x_3, x_1 \times x_2, x_1 \times x_3, x_2 \times x_3$ and $x_1 \times x_2 \times x_3$ have zero correlation. In vector terms they are said to be orthogonal, and this term is preferred to uncorrelated because the components of x_1, x_2 and x_3 are carefully chosen rather than being naturally occurring values of some variable. The F-ratio for testing the hypothesis that the coefficients of all three two-factor interactions are zero is: $[(1.757\,37 - 1.387\,44)/3]/0.004\,51$ which equals 27.34. However, this would need to exceed $F_{3,1,0.10}$, which equals 53.6, to be statistically significant at even the 10% level. There is a shortage of degrees of freedom, but despite this the effects of age, time lapse and the interaction between the number of intervening commercials and time lapse are individually significant beyond the 10% level (the P-values are 0.042, 0.070 and 0.072 respectively). The estimated interaction effect is similar to that found in the earlier word recall experiments. The degrees of freedom for error can be increased by dropping the interactions $x_1 \times x_2$ and $x_1 \times x_3$ from the model. If this is done, s becomes 0.093 on 3 degrees of freedom, the standard deviations of the coefficients increase to 0.033, but the P-values all reduce to less than 0.01 due to the increase in degrees of freedom. The drawback is that pooling the sums of squares attributed to non-significant terms together, and treating this as the error sum of squares, will tend to create statistically significant results. However, there is some justification for doing so in this case since investigation of the interaction between the number of intervening commercials and the time lapse was one of the reasons for running the experiment. Another approach to the problem is to fit a model with all possible interactions, i.e. including $x_1 \times x_2 \times x_3$, and draw a normal plot of the seven estimates of the coefficients, i.e. ignoring the constant. The principle is that if all the population coefficients are zero then the seven estimates come from a normal distribution with mean 0. Clear deviations from a straight line are evidence of an effect, and Olguín and Fearn (1997) give a method for formally assessing significance.

Singh *et al.* avoided the problem by analysing data from individual subjects. With 240 subjects there were 222 degrees of freedom for error, after fitting all possible interactions. The value of s was 0.98 but this was the standard deviation of an individual. The point estimates of the effects were almost the same as those from the regression analysis of the means, but the age, time lapse and the

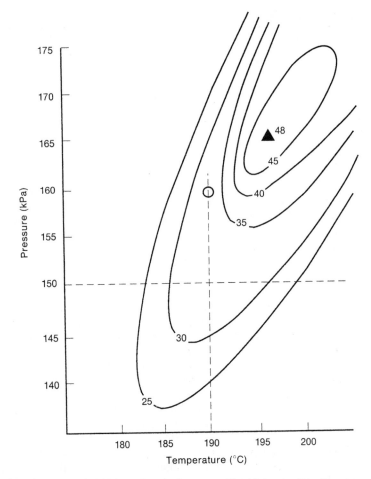

Figure 7.1   Contours of yield for a chemical process. The highest point with pressure fixed at 150 is obtained when the temperature is 190. Then the highest point with this temperature, marked by O, is obtained when the pressure is 160. The yield of 34 at this point is well below the maximum of 48

interaction between the number of intervening advertisements and the time lapse were highly statistically significant. In particular there was very strong evidence that people in the older group had less recall of the food commercial. However, extrapolation of this inference to all commercials, or memory generally, is based on judgement. The importance of the experiment to advertisers is that it suggests they should ensure repeat showings of their commercials are separated by several others rather than just one.

Singh *et al.* could have included subjects' sex and age in years in the model as concomitant variables. This might have reduced the standard deviation of the errors. For example, it would have been possible to try a model of the form:

$$Y_i = \beta_0 + \beta_1 x_{1i} + \cdots + \beta_7 x_{1i} \times x_{2i} \times x_{3i} + \beta_8 x_{5i} + \beta_9 x_{6i} + E_i$$

where x_{5i} is 0 or 1 for male or female and x_{6i} is (age-60), which assumes age only has an effect on memory after people have turned 60.

A limitation of 2^k designs is that they do not allow for a curved response. A solution is to include a mid-range level and use a 3^k design. However, the number of runs required soon becomes large even for moderate k. The researcher in the next example used a fractional 3^k design to circumvent this. Another useful device is to use central composite designs, and these will be discussed in the section on EVOP.

7.2.2 3^k design

Danaher (1997) investigated the effects of three factors on airline passengers' overall satisfaction with a flight. The objective was to identify the factor or factors most in need of improvement. The factors were the conduct of the cabin crew, the standard of meals and drinks, and flight comfort. Randomly selected passengers were asked to complete a questionnaire which contained nine questions about a hypothetical airline flight service. The levels for the factors were 'worse than expected', 'as expected' and 'better than expected' and a typical question was as follows.

> If the cabin crew were better than expected how satisfied or dissatisfied would you have been if also... the meals and drinks were worse than you expected, and the flight comfort better than you expected:

> Very dissatisfied   Very satisfied

> ☐   ☐   ☐   ☐   ☐   ☐   ☐

Notice that overall satisfaction was measured on a scale from 1 to 7.

Questionnaires

There were three versions of questionnaires. Let x_1, x_2 and x_3 be the cabin crew, meals and comfort factors and the values -1, 0, 1 represent worse, as, better than expected respectively. The questionnaire versions are given in Table 7.3. Taken together, there are 27 possible factor combinations, but each subject was only asked to answer questions based on nine of them on one version of

Table 7.3   Airline questionnaire versions (after Danaher, 1997)

	Version 1			Version 2			Version 3	
x_1	x_2	x_3	x_1	x_2	x_3	x_1	x_2	x_3
-1	-1	-1	-1	-1	0	-1	-1	1
-1	0	0	-1	0	1	-1	0	-1
-1	1	1	-1	1	-1	-1	1	0
0	-1	0	0	-1	1	0	-1	-1
0	0	1	0	0	-1	0	0	0
0	1	-1	0	1	0	0	1	1
1	-1	1	1	-1	-1	1	-1	0
1	0	-1	1	0	0	1	0	1
1	1	0	1	1	1	1	1	-1

the questionnaire, because asking someone to answer all 27 seemed unreason-able. There is a technical significance to the allocation of questions to the three versions, and this is taken up in Exercise 7.1. The researchers obtained 260 usable questionnaires from 47 flights, every sixth aisle passenger having been asked to complete one. There was a similar number of returns for each version.

Analysis

A full analysis is complicated by the fact that sets of nine questions are answered by the same person. It follows that if all 9×260 questions and responses are analysed using the standard multiple regression model the assumption about independence of the errors is not justified. This will affect significance levels and give more apparently statistically significant results than are warranted. However, it is not likely to affect the estimates of coefficients in the regression model, which were the main interest in this application. A repeated measures analysis (see Hand and Crowder, 1996, for example), would allow for the correct error structure. An advantage of the full analysis is that the effects of concomitant variables such as age, sex, frequency of travelling, and purpose of trip on overall satisfaction could be investigated. It would also be possible to allow for the particular flight on which the passengers completed the ques-tionnaire as a concomitant variable, although the questions were referring to hypothetical situations rather than the flight itself.

The analysis becomes much simpler, and is still sufficient for the prime purpose, if we average all the responses to each of the 27 questions, and then analyse these 27 means (y_i) by multiple regression. The 3^3 design allows us to fit a regression which includes all linear and quadratic terms and two linear factor interactions, that is a full quadratic surface, with 17 degrees of freedom left for error. See Exercise 7.2 for the model with 26 orthogonal components.

Let Y_i be the average overall satisfaction score with a factor combination (x_{1i}, x_{2i}, x_{3i}), for $i = 1, \ldots, 27$. Let x_{4i}, x_{5i} be indicator variables for the question-naire version number, coding x_4 as 1 for questionnaire 2, x_5 as 1 for questionnaire 3 and both 0 otherwise. The full quadratic surface, plus indicator variables is,

$$Y_i = \beta_0 + \beta_1 x_{1i} + \beta_2 x_{2i} + \beta_3 x_{3i} + \beta_4 x_{1i}^2 + \beta_5 x_{2i}^2 + \beta_6 x_{3i}^2$$
$$+ \beta_7 x_{1i} x_{2i} + \beta_8 x_{1i} x_{3i} + \beta_9 x_{2i} x_{3i} + \beta_{10} x_{4i} + \beta_{11} x_{5i} + E_i$$

The coefficients β_{10} and β_{11} represent any differences in the way in which the three groups of passengers completing the three versions of the questionnaire interpret the overall satisfaction scale. It is rather unlikely that the estimates of β_{10} and β_{11} will be significantly different from 0, and they can probably be dropped from the model once this check has been made. Danaher estimated the parameters in the equation as:

$$y = 4.61 + 1.01 x_1 + 0.72 x_2 + 0.88 x_3 - 0.42 x_1^2 - 0.52 x_2^2 - 0.58 x_3^2$$
$$+ 0.07 x_1 x_2 + 0.32 x_1 x_3 + 0.17 x_2 x_3$$

The quadratic model was a significant improvement over a model with the linear terms only. All but one of the individual coefficients were significantly different from zero, at the 5% level. The exception was the crew and meals interaction ($P = 0.07$) and it is usual to retain all two-factor interactions if squared terms are introduced. For example, the interpretation of the positive crew and comfort interaction is that an improvement in both will have more effect on increasing passenger satisfaction than the sum of the effects of the two improvements on their own. This suggests that a policy of improving the factors at the same time would be particularly beneficial. The negative coefficients of the squared terms indicate that either y peaks and then declines or that there is a diminishing returns effect as the factor values increase. Only the second explanation makes sense and it is necessary to check the location of the stationary points. The gradient of the function is

$$\begin{pmatrix} \dfrac{\partial y}{\partial x_1} \\[2mm] \dfrac{\partial y}{\partial x_2} \\[2mm] \dfrac{\partial y}{\partial x_3} \end{pmatrix} = \begin{pmatrix} 1.01 - 0.84x_1 + 0.07x_2 + 0.32x_3 \\ 0.72 + 0.07x_1 - 1.04x_2 + 0.17x_3 \\ 0.88 + 0.32x_1 + 0.17x_2 - 1.16x_3 \end{pmatrix}$$

At a stationary point the gradient is zero, and the solution of the three linear simultaneous equations is $(1.83, 1.05, 1.42)$. At this point y equals 6.54, which is a maximum so the surface is plausible for factor values in the range -1 to 1.

The main objective of the study was to find the direction of steepest ascent on the customer satisfaction surface, as a means of establishing the relative importance of improving crew, meals and comfort. The direction of steepest ascent is given by the value of the gradient at the present values of x_1, x_2 and x_3. This was estimated from a supplementary question. After they had completed a version of the questionnaire with the nine hypothetical scenarios, passengers were asked to rate the crew, meals and comfort for the flight they were on, using a scale from -1 to 1. The mean values for all passengers were $(0.32, 0.08, 0.13)$. Substituting into the expression for the gradient gives

$$(0.79, 0.68, 0.85)^T$$

as the direction of steepest ascent. The mathematical interpretation is that the quickest way to increase passenger satisfaction scores is to improve comfort, crew and meals in the ratio of $1 : 0.93 : 0.80$. All three factors should be improved, with the most emphasis on flight comfort, despite meals having the lowest current rating. In practice it might be much easier and cheaper to improve meals than change the other factors, but the analysis has not allowed for this possibility.

7.2.3   Fractional factorial design

A company prints two-coloured designs onto sheets of metal which are then made into tins. The performance of the printing machine can be checked by printing a test pattern of dots of 2 mm diameter. If the colour does not take to the metal properly the dots will be incomplete. If the second colour does not print exactly over the first the dots will become slightly oval. The

Table 7.4 Construction of a 2^{4-1} design. A lower-case letter in the treatment column means that the corresponding factor, upper-case same letter, is at the high level. Absence denotes the factor is at the low level. 1 denotes that all factors are at the low level. Factors A, B, C, D are represented by x_1, x_2, x_3, x_4 respectively.

x_1	x_2	x_3	x_4	Treatment	$x_1x_2x_3x_4$	$x_1x_2x_3$	x_1x_2	x_3x_4
−	−	−	−	1	+	−	+	+
−	−	−	+	d	−			
−	−	+	−	c	−			
−	−	+	+	cd	+	+	+	+
−	+	−	−	b	−			
−	+	−	+	bd	+	+	−	−
−	+	+	−	bc	+	−	−	−
−	+	+	+	bcd	−			
+	−	−	−	a	−			
+	−	−	+	ad	+	+	−	−
+	−	+	−	ac	+	−	−	−
+	−	+	+	acd	−			
+	+	−	−	ab	+	−	+	+
+	+	−	+	abd	−			
+	+	+	−	abc	−			
+	+	+	+	abcd	+	+	+	+

quality of the printing can be quantified by measuring the maximum and minimum diameters of the dots. A manager in the company wishes to carry out an experiment to investigate four factors which are thought to affect the quality of the printing, the speed at which the press is run, the pressure on the print roller, the type of ink and whether the metal sheets have been cleaned before printing. The speed (A) and pressure (B) can be set low or high. The ink (C) is one of two types and the cleaning treatment (D) is either applied or not. Therefore the factors can be represented by four variables $x_1, \ldots, x_4$ which take values of -1 or 1. Fifty sheets will be printed for each factor combination, and the maximum and minimum diameters of all dots in the test pattern on sheets 21-30 and 41-50 will be measured under a microscope. Four response variables will be considered: the mean and the variance of all the measurements of diameters and the mean and variance of the differences between the maximum and minimum diameter for each dot. A full factorial experiment would require 2^4, which is 16, runs. The four-factor interaction is estimated by the difference between the average response of the eight runs when $x_1x_2x_3x_4$ equals 1 and the average response of the eight runs when $x_1x_2x_3x_4$ equals -1. Either of these eight runs on its own is a half replicate (2^{4-1}) of the 2^4 design. It is no longer possible to estimate the four-factor interaction but it is probably reasonable to assume it is negligible. However, the number of other comparisons is also reduced. You can see from Table 7.4 that the estimate of the three-factor interaction $x_1x_2x_3$ will be indistinguishable from the estimate of the main effect of D, both equalling the difference between the average of the four responses when x_4 is $+1$ and the average of the four responses when x_4 is -1. Indistinguishable effects are known as aliases and it is easy to check that the alias of any effect can be found from the product of that effect with $ABCD$, subject to the rule that any letter squared equals 1. For example, the alias of the two-factor interaction AB is

$$AB(ABCD) = A^2 B^2 CD = CD$$

The two-factor interactions are aliased in the half-replicate design, but it would reduce the amount of work required. The manager might begin with a half-replicate design and then follow up with the other half-replicate design if, either the standard deviations of the estimates of the main effects are larger than anticipated, or there is evidence of an interaction and the manager wishes to identify the factors involved. This is an example of confounding because any systematic difference between the conditions under which the two half-replicate design experiments are carried out, such as warmer weather during the second, will be indistinguishable from the four-factor interaction. The term confounding is used rather than aliasing when the indistinguishable effect is not made up of the experimental factors.

If possible, the order of the runs should be randomised because it will tend to reduce the effect of any unknown source of systematic variation and it makes the assumption of independence of errors plausible. However, a random run order may not always be feasible in industrial experiments (e.g. Grove in the discussion of Montgomery, 1999, and Joiner and Campbell, 1976). In this case, changing the ink might be troublesome so it would be expedient to carry out the runs with one ink type consecutively. One drawback is that ambient temperature might increase during the course of the experiment. In principle, the temperature could be included as a concomitant variable, but this will not help if the experiment is carried out on two days and one is much colder than the other. Another limitation might be that all the ink is from the same container. The manager must decide whether ambient temperature and variation of ink between containers is likely to affect the process. Randomisation should not be relied upon to even out predictable variation. It may not, and even if it does it will give less precise results than using an appropriate experimental design.

In many processes there are more than three factors. Similar principles can be used to obtain smaller fractions of larger factorial designs.

Example 7.1
The cleanliness of carburettor assemblies was measured by the number of dirt particles in a sample of ten carburettors. There were six control variables, each with two levels: preclean wells in body (A), low (L) no, high (H) yes; prewash components (B), L no, H yes; flush time (C) L short, H long; double assembly of tubes and jets (D), L no, H yes; environment (E), L clean, H extra clean; speed of assembly (F), L slow, H normal. A full factorial design would require 2^6, which is 64 runs. A one-eighth fraction of this would have only 2^{6-3}, which is eight runs. This is sufficient to allow estimation of main effects, but as they are aliased with two-factor interactions the design is only suitable if the latter can be assumed negligible. The 2^{6-3} is said to have resolution III because a one-factor effect is aliased with a $III - 1$, equals two, factor effect. In general a design of resolution R is one in which no p-factor effect is aliased with any other effect containing less than $R - p$ factors, e.g. the 2^{4-1} design has resolution IV. Minitab and DEX can generate fractional two-factor factorial designs. Montgomery (1997) gives an appendix of fractional

Table 7.5   A 2^{6-3} design for the carburettor assembly experiment and the numbers of particles found in samples of ten assembled carburettors (y)

x_1	x_2	x_3	x_4	x_5	x_6	y
-1	-1	-1	$+1$	$+1$	$+1$	131
-1	-1	$+1$	$+1$	-1	-1	156
-1	$+1$	-1	-1	$+1$	-1	145
-1	$+1$	$+1$	-1	-1	$+1$	142
$+1$	-1	-1	-1	-1	$+1$	63
$+1$	-1	$+1$	-1	$+1$	-1	113
$+1$	$+1$	-1	$+1$	-1	-1	83
$+1$	$+1$	$+1$	$+1$	$+1$	$+1$	73

two-factor factorial designs for up to 11 factors, and Box *et al.* (1978) is another useful source. The 2^{6-3} design for the carburettor experiment is shown in Table 7.5. It is similar to the design matrix for the 2^3 design (Table 7.2), except that D, E and F have been identified with the interactions AB, AC and BC. These identities are known as the design generators: $D = AB$, $E = AC$ and $F = BC$. The full alias structure is given in Exercise 7.3, but the only essential fact is that the main effects are aliased with two-factor interactions. Since the variable is a count a Poisson regression is preferable to a standard regression analysis but the upshot is very similar whichever is used. A Poisson regression estimate of the mean μ of the distribution is

$$\hat{\mu} = 4.688 - 0.282x_1 - 0.028x_2 + 0.073x_3 - 0.022x_4 + 0.044x_5 - 0.089x_6$$

The deviance was 3.95 on 1 degree of freedom. The estimated standard deviation of the coefficients is 0.035 if the scale parameter is taken as 1, and multiplying by $\sqrt{3.95}$ to allow for over-dispersion increases this to 0.070. The t-ratio for the coefficient of x_1 is 4.05. Although this is high it is not statistically significant at the 10% level with 1 degree of freedom for error. If we fit a regression with x_1 as the only predictor variable, the deviance is 17.70 on 6 degrees of freedom. The t-ratio is $0.28/(0.035 \times (17.7/6)^{1/2}) = 4.7$ which is large when compared with t_6, $\Pr(t_6 > 4.7) = 0.002$. It is difficult to provide any precise P-value. It can be argued that as there are just six possible regressions on a single predictor variable, the P-value should be less than 0.002×6 with a one-sided alternative (see the Bonferonni inequality, Section 7.5). At the other extreme it can be argued that there are 2^6 possible regression analyses and that this is the one which is the most statistically significant. A sensible practical conclusion is that precleaning wells in the carburettor body probably reduces the number of particles but that the experiment was too small to estimate the effect very precisely. A replicate of the entire experiment should provide sound evidence for the effect.

7.2.4   Central composite design

This case is loosely based on work by Norman and Naveed (1990) who developed an expert system for the control of a rotary kiln for making cement, but the data are fictitious. The meal which is fed into the kiln is a mixture of limestone, clay, sand and iron ore. The main ingredient is the limestone and the management of the plant required the expert system to maintain the percentage of free lime in the cement product (y) between 1% and 2% whilst maximising the ratio of feed rate to fuel rate. Six control variables were identified: feed rate (x_1), rotation speed (x_2), proportion of fuel to oxidant (x_3), fuel rate (x_4), fan 1 speed (x_5) and fan 2 speed (x_6). Small changes about the standard operating conditions could be made to establish their effect on the percentage of free lime in the cement. The process was operated continuously and did not respond immediately to changes in the control variables, so two hours were allowed for the process to reach a steady state after each change. Each run lasted for four hours during which samples of cement were taken for analysis at five-minute intervals. The response (y) was the average of the 48 analyses, and plant engineers thought it might depend on quadratic effects and interactions of control variables as well as the linear terms. A 3^6 experiment would require 729 runs, and even a one-ninth replicate would need 3^{6-2} which equals 81 runs. If the control variables can be changed or adjusted over a continuous range a central composite design is a more attractive option. It is an extension of the two-factor factorial, or fractional factorial design, which allows for the estimation of quadratic effects. If there are k control variables, an additional $2k$ design points are needed at

$$(-\alpha, 0, \ldots, 0)$$
$$(+\alpha, 0, \ldots, 0)$$
$$\vdots$$
$$(0, 0, \ldots, +\alpha)$$

together with the point at the centre $(0, \ldots, 0)$ which is often replicated. The distance of the original design points from the centre can be found from Pythagoras' theorem and it is

$$\sqrt{1^2 + 1^2 + \cdots + 1^2} = \sqrt{k}$$

A reasonable value for α is $\sqrt{k}$. Montgomery (1997) gives a table of preferred values of α, and the number of runs at the centre point to give an orthogonal design, although as this can be quite large it is common to accept small correlations amongst the predictor variables. In this case there were six variables, and $\sqrt{k}$ would be 2.45, but the plant manager restricted α to 2.

A quarter replicate, 2^{6-2}, of a 2^6 factorial design was chosen with 13 extra points to make it into a central composite design. With only one run at the centre, there was a total of 29 runs. A 2^{6-2} design can be constructed by writing down a full factorial for four factors, and then aliasing E with the three-factor interaction ABC and F with the three-factor interaction BCD. You are asked to show that two-factor interactions are aliased in Exercise 7.4(a). The final design is shown in the first six columns of Table 7.6. Two concomitant variables were monitored, the 'burnability' (x_7) and water content (x_8) of the meal. These are given together with the response (y) in Table 7.6. The following regression model is a good fit to the data. The plant engineer thought the aliased interactions were likely to be almost entirely due

Table 7.6 Percentage free lime (percentage above 1% times 100) in a cement product

Feed rate x_1	Rotation speed x_2	Fuel/ oxidant x_3	Fuel rate x_4	Fan 1 speed x_5	Fan 2 speed x_6	Burn- ability x_7	Water content x_8	Free lime y
−1	−1	−1	−1	−1	−1	4	7	37
1	−1	−1	−1	1	−1	5	4	7
−1	1	−1	−1	1	1	7	7	131
1	1	−1	−1	−1	1	3	6	14
−1	−1	1	−1	1	1	11	−6	116
1	−1	1	−1	−1	1	1	4	91
−1	1	1	−1	−1	−1	2	−2	74
1	1	1	−1	1	−1	2	−24	48
−1	−1	−1	1	−1	1	3	16	102
1	−1	−1	1	1	1	3	−4	−12
−1	1	−1	1	1	−1	1	−18	−2
1	1	−1	1	−1	−1	0	12	−87
−1	−1	1	1	1	−1	10	5	150
1	−1	1	1	−1	−1	−2	10	−53
−1	1	1	1	−1	1	−4	−15	−12
1	1	1	1	1	1	−3	−3	−71
−2	0	0	0	0	0	4	−13	137
2	0	0	0	0	0	−4	−2	−16
0	−2	0	0	0	0	9	12	78
0	2	0	0	0	0	5	−15	15
0	0	−2	0	0	0	3	−3	49
0	0	2	0	0	0	5	1	88
0	0	0	−2	0	0	5	−1	136
0	0	0	2	0	0	2	−3	−5
0	0	0	0	−2	0	2	−22	−57
0	0	0	0	2	0	2	0	51
0	0	0	0	0	−2	−3	0	−23
0	0	0	0	0	2	4	−2	39
0	0	0	0	0	0	−4	11	63

to the factors included below.

$$y = 62.6 - 38.2x_1 - 6.98x_2 + 13.9x_3 - 29.4x_4 - 16.6x_5 + 11.0x_6$$
$$+ 2.53x_7 + 1.68x_8 - 21.1x_1x_4 - 18.3x_2x_4$$
$$- 12.0x_2x_6 - 8.14x_2^2 - 13.4x_5^2 - 13.9x_6^2$$

The objective is to keep y between 1% and 2% with x_1 as high as possible and x_4 as low as possible. In the coded units we set x_1 at 2, x_4 at −2 and try and adjust y to 50 with the other control variables, which are restricted to the range −2 up to 2. For example, if x_7 and x_8 are 0 the control equation is:

$$y = 129.4 + 29.6x_2 + 13.9x_3 + 16.6x_5 + 11x_6$$
$$- 12x_2x_6 - 8.2x_2^2 - 13.4x_5^2 - 13.9x_6^2$$

and it is quite easy to adjust y to 50 by, for example, setting x_5 and x_6 to −1.2, x_3 to −0.5 and x_2 to 0.

7.3   Taguchi methods

7.3.1   Introduction

Taguchi (e.g. 1980, 1986, 1987, 1988) has promoted the use of experimental design for devising products which are relatively unaffected by variation in component parts and changes in the environmental conditions under which they are used. This strategy is referred to as robust design. A fundamental principle of his approach is that there is a cost, usually modelled by a quadratic function, associated with any deviation from the target value. Most of the underlying theory on the design of experiments can be traced back to at least the 1930s, and the authoritative book edited by Davies (1954) was an early attempt to promote the methods within industry, but Taguchi's emphasis on robust design and novel applications has given his work great influence. Some of the statistical detail is open to criticism, particularly the reliance on, and analysis of, highly fractional factorial designs. These often confound main effects with two-factor interactions, and assuming the latter are negligible can lead to seriously misleading conclusions. Also, as we have seen in the analysis of the carburettor example, there is scope for some subjective pooling of effects to increase the degrees of freedom for error and this needs to be done advisedly rather than according to some standard scheme. Routine use of signal-to-noise ratios is also questionable and it is usually more helpful to treat the mean and standard deviation as separate responses. This is quite in keeping with Taguchi's ideas, as he advocates finding some control variables which primarily affect the mean level and others which also affect variability. The ideal is that the control variables can be chosen to minimise the variance, and then those control variables that only affect the mean can be used to adjust it to the required level.

7.3.2   Signal-to-noise ratios

Suppose T is the target value and y_i is the response. If there are n values of the response the average quadratic loss (L) is given by:

$$L = \sum (y_i - T)^2 / n = \hat{\sigma}^2 + \widehat{\text{bias}}^2$$

where

$$\hat{\sigma}^2 = \sum (y_i - \bar{y})^2 / n \quad \text{and} \quad \widehat{\text{bias}} = \bar{y} - T$$

A signal-to-noise ratio (SNR) is defined by

$$\text{SNR} = -10 \log_{10}(L)$$

Also see Exercise 7.5(a). Note that maximising SNR is equivalent to minimising L. In some cases T will be zero, for example the amount of impurity in a chemical, and the SNR becomes

$$\text{SNR} = -10 \log_{10} \left(\sum y_i^2 / n \right)$$

In yet other cases it will be desirable to maximise the response, for example the strength of hand brake cables. This is equivalent to minimising the reciprocal of

Table 7.7 Taguchi L_9 array (left) and identification for welding experiment (right)

Column number						
1	2	3	4	Current	Time	Steel
1	1	1	1	−1	−1	Mild
1	2	2	2	−1	0	Stainless
1	3	3	3	−1	1	Galvanised
2	1	2	3	0	−1	Stainless
2	2	3	1	0	0	Galvanised
2	3	1	2	0	1	Mild
3	1	3	2	1	−1	Galvanised
3	2	1	3	1	0	Mild
3	3	2	1	1	1	Stainless

y_i and the SNR becomes

$$\text{SNR} = -10\log_{10}\left(\sum(1/y_i^2)/n\right)$$

(See Exercise 7.5(b).) Two examples of Taguchi-style investigations, with conventional statistical analyses, follow.

Example 7.2
An experiment was carried out with the objective of making a spot-welding process robust against thickness of metal and welder's experience. A high average strength (y) with a small standard deviation was desirable. The control factors were current (x_1), time (x_2) and type of steel. A Taguchi L_9 orthogonal array (Table 7.7) was used for the control factors. Low, medium and high current were coded −1, 0 and 1 and associated with 1, 2 and 3 respectively of the first column. Short, middling and long weld times were associated with 1, 2 and 3 of the second column. The steel types, mild steel, stainless steel, and galvanised steel, corresponded to 1, 2 and 3 respectively in column three. The fourth column was left empty. The design for the control factors is known as the inner array. The noise factors, thin (−1) and thick (+1) steel, and experienced (−1) and apprenticed (+1) welder were assigned to the first two columns of an L_4 orthogonal array (shown below). This is known as the outer array. Apart from the coding of the levels, the L_4 orthogonal array is equivalent to the array of signs in a 2^2 factorial design.

Taguchi L_4			2^2 factorial design		
col_1	col_2	col_3	A	B	AB
1	1	2	−1	−1	+1
1	2	1	−1	+1	−1
2	1	1	+1	−1	−1
2	2	2	+1	+1	+1

You are asked to investigate the relationship between the Taguchi L_9 array and the 3^2 design in Exercise 7.6. Clarke and Kempson (1997) and Lochner and Matar (1990) discuss equivalences in more detail. The 'larger the better' SNR

Table 7.8 Results of welding experiment

			−1 −1	+1 −1	−1 +1	+1 +1	Welder material		
Current	Time	Steel		Strength (kN) y			Mean ȳ	SDs	SNR
−1	−1	M	4.6	5.4	5.6	5.8	5.35	0.53	14.46
−1	0	S	7.8	7.8	8.0	8.4	8.00	0.28	18.05
−1	1	G	3.6	2.9	2.8	1.9	2.80	0.70	8.23
0	−1	S	8.4	9.6	7.0	8.6	8.40	1.07	18.32
0	0	G	6.0	7.1	5.9	5.2	6.05	0.78	15.48
0	1	M	5.6	6.2	8.6	13.1	8.37	3.41	17.14
1	−1	G	6.6	7.8	7.6	6.2	7.05	0.77	16.84
1	0	M	6.2	6.4	6.6	13.6	8.20	3.60	17.06
1	1	S	12.0	12.8	14.6	14.8	13.55	1.37	22.54

$(-10\log_{10}(\sum(1/y_i^2)/n))$ seems appropriate, but in a re-analysis of the results Wahid and Metcalfe (1996) chose to use another heuristically reasonable response variable $(\bar{y} - 2s)$, which is an estimate of the lower 2.5% point of a normal distribution. Even if the distribution of weld strengths is not close to normal, $(\bar{y} - 2s)$ places more emphasis on reducing variability than SNR. This can be seen from a regression of SNR on $\bar{y}$ and s, which gives

$$\text{SNR} = 7.2 + 1.27\bar{y} - 0.24s$$

with an R^2-adjusted of 88%. In general, the lower the coefficient of variation the more closely will SNR depend mainly on $\bar{y}$, in which case maximising SNR becomes tantamount to maximising the mean. The results of the experiment are given in Table 7.8. Material variables were coded by two indicator variables:

	x_3	x_4
Mild steel	0	0
Stainless steel	1	0
Galvanised steel	0	1

The most convincing regression was

$$(\bar{y} - 2s) = 1.39 + 0.70x_1 - 0.38x_2 + 2.67x_1 \times x_2 + 5.89x_3 + 4.19x_4$$

$$(0.69) \quad (0.69) \quad (1.19) \quad\quad (1.37) \quad (1.81)$$

with an R^2-adjusted of 73%. There are only 3 degrees of freedom for error, and the distribution of $(\bar{y} - 2s)$ is not normal, so any conclusions are tentative. As a guide, the upper 5% point of t_3 is 2.35. There is reasonable evidence that stainless steel is the best material, and the indications are that the high current with a long time combination is best, while galvanised steel is better than mild steel, the t-ratios for the coefficients of both $x_1 \times x_2$ and x_4 being close to 2.35. The interaction and x_4 are correlated, and the coefficient of x_4 is reduced to 1.52 if the interaction is dropped from the model. However, it is reasonable to expect the interaction of time and current to be important in a welding operation.

Table 7.9   Crossed array for the semiconductor experiment (Montgomery, 1999, copyright Royal Statistical Society)

Controllable variables			Noise variables				Response average	Response standard deviation
x_1	x_2	x_3	$z_1 = -1,$ $z_2 = -1$	$z_1 = +1,$ $z_2 = -1$	$z_1 = -1,$ $z_2 = +1$	$z_1 = +1,$ $z_2 = +1$		
−1	−1	−1	118.9	65.7	95.3	92.4	93.08	21.77
+1	−1	−1	153.7	229.4	119.9	251.5	188.63	62.07
−1	+1	−1	196.7	170.9	234.2	166.6	192.10	31.06
+1	+1	−1	211.1	245.7	241.0	252.6	237.60	18.30
−1	−1	+1	145.2	132.2	167.1	137.9	145.60	15.29
+1	−1	+1	125.3	201.6	185.5	267.3	194.93	58.36
−1	+1	+1	283.0	251.1	263.4	190.4	246.98	39.94
+1	+1	+1	184.2	279.5	247.2	259.2	242.53	41.11

Example 7.3

Montgomery (1999) describes an experiment which was performed in a semi-conductor factory to determine how five factors influenced the transistor gain for a particular device. The specification for the gain was 200 ± 20. Three of the variables were relatively easy to control: the implant dose (x_1), drive-in time (x_2) and vacuum level (x_3). Two variables were difficult to control in routine manufacturing, although they could be controlled under experimental conditions, and were considered as noise factors: oxide thickness (z_1) and temperature (z_2). The experimenters used a 2^3 factorial design (three columns from a Taguchi L_8 orthogonal array) for the control variables and a 2^2 factorial design for the noise factors. The results are reproduced in Table 7.9. Montgomery (1999) suggested the following analysis in preference to the 'target-is-best' SNR of Exercise 7.5(a). The overall design is a full 2^5 factorial, and he fits the model

$$y = 192.68 + 23.24x_1 + 37.12x_2 + 14.83x_3 - 12.98x_1x_2$$
$$- 12.03x_1x_3 + 6.95w_1 + 25.48x_1w_1$$

with $s = 25.28$. The temperature, and the omitted interactions, have no significant effect on the gain (y). Now assume that under routine manufacturing conditions $w_1 \sim N(0, \sigma_w^2)$. Then the mean value of the gain is given by the regression equation with w_1 set equal to 0. The variance of the gain is estimated by

$$\widehat{\text{var}}(y) = (6.95 + 25.48x_1)^2\sigma_w^2 + \sigma^2$$

if it is assumed that w_1 is independent of x_1. Here σ^2 is the variance of the errors in the regression which represent unidentified variation. The purpose of the experiment was to find values of the control variables which minimise variability. The optimum value for x_1 is $-6.95/25.48$, which equals -0.27 and is within the range -1 to 1. Then the variance of y is estimated by s^2, which is 639. There is an infinite number of ways of adjusting the mean to 200 with the control variables constrained to the range -1 to 1. Unfortunately the minimum standard deviation of y, estimated as 25, is high in comparison with the specification.

7.4   EVOP

7.4.1   Case study

The yield of a liquid chemical used in the pharmaceutical industry is known to depend on the pressure and temperature inside the reactor. Both variables can be controlled quite accurately, and the specified settings have been 160 kPa and 190°C for as long as anyone can remember. It is safe to operate the process with changes in temperature and pressure up to at least 10%. A graduate student on an industrial placement proposed experimenting with small changes in these two variables. The plant would continue to produce satisfactory product and the experimental programme should identify optimum settings for pressure and temperature. If these turned out to be different from the current specification, substantial savings would be made. The site manager was interested in the possible benefits but remained sceptical because this was presumably how the specified values had been decided.

The student then suggested that previous experiments might have relied on a one-variable-at-a-time strategy, and explained that this might not have led to the best settings. A possible dependence of yield on pressure and temperature is represented by the contour diagram in Figure 7.1. This is an example of a response surface, and the general shape shown in the diagram is quite common in the chemical industry. If we fix pressure at 150 kPa and vary the temperature we will find a highest yield of 32 kg at 190°C. Now suppose we fix the temperature at 190°C and vary the pressure. The highest yield becomes 34 kg when the pressure is 160 kPa. This is a long way from the optimum of 48 kg. The pressure and temperature are said to interact – that is, the effect of changing one depends on the value of the other. This is just one possible scenario, and alternatives include the possibility that the present settings are already at the highest point or the response surface is of a quite different shape (multi-peaked or saddle-shaped, for example). The objective of an experimental programme is to infer the nature of the surface, as efficiently as possible, and to use this information to specify optimum operating conditions.

A simple design which makes changes in both variables is the 2^2 factorial design. Each factor appears as high (coded $+1$) or low (coded -1). This gives four runs, allowing a plane to be fitted with one degree of freedom for error. The student set up a small team which included the process engineer and an operator. They carried out such an experiment and the results are given in Table 7.10(a). The runs were carried out in a random order, which helps justify the assumption of random errors in the regression model. The fitted plane, where y is the yield, x_1 the pressure and x_2 the temperature, is

$$y = 44.50 + 1.05x_1 + 0.450x_2$$

with $s = 0.40$ and $R^2 = 97\%$.

The standard deviation of all the estimated coefficients is 0.20 but, with only one degree of freedom for error, the 90% confidence intervals are inevitably wide and include 0. As any movement would be away from the specified settings, the team thought it prudent to augment the experiment with another five runs arranged as a star design to give a composite design. The runs were carried out in a random order and the data are given in Table 7.10(b). The

Table 7.10 Yields from chemical process for (a) the first experiment, a 2^2 factorial, and (b) the follow-up star design. Together they make a composite design

(a)

Pressure (5 kPa from 160 kPa)	Temperature (5°C from 190°C)	Yield (kg)
1	1	46.2
−1	−1	43.2
1	−1	44.9
−1	1	43.7

(b)

0	−1.4	42.1
1.4	0	43.9
−1.4	0	41.9
0	0	43.6
0	1.4	45.1

plane was fitted to all nine data:

$$y = 43.84 + 0.88x_1 + 0.76x_2$$

with $s = 0.90$ and $R^2 = 69\%$.

The standard deviation of the coefficients of x_1 and x_2 has actually increased to 0.32 because the reduction due to additional data has been more than offset by the increase in s. The confidence intervals are, nevertheless, narrower because of the increase in degrees of freedom to six. In fact the 90% confidence intervals for the coefficients of x_1 and x_2 are 0.88 ± 0.62 and 0.76 ± 0.62 respectively.

It is possible to fit a quadratic surface to the data in the composite design, but this results in an increase in s to 1.17. As a result of all these analyses the team felt sufficiently confident to carry out a similar composite design based on a new centre point in the direction of steepest ascent of the plane. To find this direction, suppose x_1 is increased by 1 unit, and then find the increase in x_2 such that

$$\frac{\text{change in } y}{\text{change in vector } (x_1, x_2)}$$

is a maximum. If this value of x_2 is denoted by d, the objective is to maximise

$$g = \frac{0.884 + 0.758d}{\sqrt{1 + d^2}}$$

with respect to d. Elementary calculus leads to the result that

$$d = 0.758/0.884 = 0.86$$

An increase in pressure by 1 unit (5 kPa) did seem reasonable. It kept the process well within safe operating conditions, and the corresponding increase in temperature was rounded to 4°C. Another composite design experiment was carried out and the results are given in Table 7.11(a). A quadratic surface fitted to these data does have a smaller associated estimate of standard deviation

Table 7.11   Yields from chemical process for (a) the second composite design and (b) the 3^2 design

(a)

Pressure (5 kPa from 165 kPa)	Temperature (5°C from 194°C)	Yield (kg)
1.0	−1.0	46.7
−1.0	1.0	47.1
−1.4	0.0	45.9
1.4	0.0	43.7
0.0	−1.4	45.9
0.0	0.0	48.0
1.0	1.0	44.2
0.0	1.4	45.1
−1.0	−1.0	47.7

(b)

−1	1	47.8
1	1	44.8
−1	0	48.9
0	−1	49.7
1	0	47.1
0	1	47.1
0	0	48.2
1	−1	46.2
−1	−1	46.8

of the errors (1.16) than a plane (1.26). This surface is

$$y = 47.9 - 0.881x_1 - 0.533x_2 - 1.27x_1^2 - 0.475x_1x_2 - 0.909x_2^2$$

but the coefficients have not been determined very accurately, there are only three degrees of freedom for error and s is not strikingly smaller than the unconditional standard deviation of the yields, 1.50. The team did not wish to make any firm recommendations on the basis of this fit, and decided to carry out a further experiment centred on the same operating point. An alternative to the composite design, which still allows the fitting of interaction and quadratic terms and has a slight advantage that these are all uncorrelated, is the 3^2 design, in which each factor now appears at three levels, high, medium or low. The results from this design are given in Table 7.11(b). An analysis of all 18 results centred on 165 kPa and 194°C gives the equation

$$y = 48.805 - 0.889x_1 - 0.519x_2 - 1.533x_1^2 - 0.537x_1x_2 - 1.167x_2^2$$

with $s = 0.94$. The standard deviations of the six coefficients are 0.550, 0.252, 0.252, 0.389, 0.332 and 0.389 respectively, and the team was reasonably confident about the fitted model. A contour plot is shown in Figure 7.2 and the estimated optimum conditions are pressure and temperature corresponding to about −0.25 and −0.2 respectively. If y is partially differentiated with respect to x_1 and x_2, and the derivatives are set equal to zero, the exact maximum for the fitted surface at −0.26 and −0.16 will be obtained. The three-dimensional representation of the response surface (Figure 7.3) was produced with the

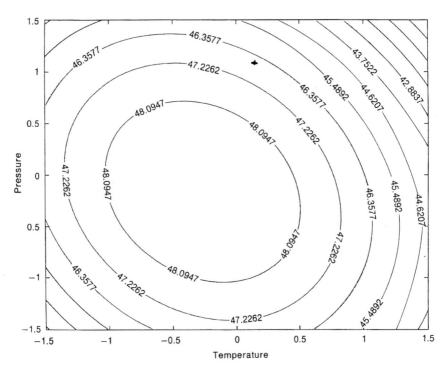

Figure 7.2   Contour plot for yields predicted from the quadratic regression surface fitted to all 18 data in Table 7.11

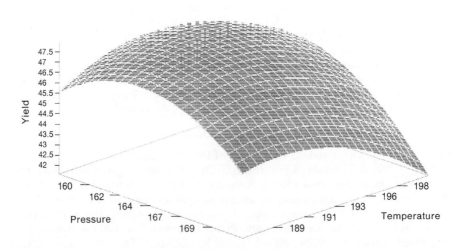

Figure 7.3   Dex plot of yield against temperature and pressure

experimental design package, DEX, which is particularly easy to use. The team recommended that the specified operating conditions be changed to a pressure of 163.7 kPa and a temperature of 193.2°C, that the yields be carefully monitored, and that occasional further experiments with changes in pressure of 2 kPa and temperature of 2°C be carried out. The 95% prediction limit for the yield at the estimated optimum is

$$48.96 \pm 2.34$$

7.4.2   Simplex algorithm and simplex designs

In the case study of Section 7.4.1 the process was well established and the team needed to be fairly sure that their recommendations for changed operating conditions would lead to an increased yield. With a new process, rather more speculative changes could be tried. The minimum number of points needed to fit a plane in 3D is three, and the simplex design for two control variables (the response being the third variable) is the vertices of an equilateral triangle. If there are three control variables the simplex design is the vertices of a tetrahedron. If the standard deviation of the random variation is low, simplex designs will lead to the optimum process conditions in fewer runs than factorial designs. Another variant is to use the modified simplex method of Nelder and Mead (1965), described in Appendix A5, instead of moving in the direction of steepest ascent.

7.4.3   EVOP and APC

There is no reason why EVOP has to involve operator interaction. The sequence of applications of any experimental design can be programmed. Automatic process control is implemented by linking the PC running the program to sensors and actuators on the process. The actuators are designed so that the process control variables are always set within a safe region of operation, and the overall controller can never become unstable. Luangpaiboon et al. (2000) investigated this strategy, and some of their results are described in the next section.

7.4.4   Genetic algorithm

Luangpaiboon et al. (2000) set up a computer simulation to compare the performance of various steepest ascent algorithms and a genetic algorithm (GA) for finding optimal operating conditions of chemical processes. Process yield was characterised by a response surface with noise added, and the study was restricted to two control variables. Process 1 had parabolic contours (Figure 7.4(a)), given by the equation:

$$y = 55 - 0.44x_1 - 0.26x_2 - 0.0076x_1^2 - 0.0027x_1x_2 - 0.0058x_2^2$$

The maximum is 62.85 when (x_1, x_2) equals $(-26.04, -16.35)$ and the safe region of operation is x_1 and x_2 between -40 and 40. In a physical context x_1 and x_2 might be changes in temperature and pressure from normal operating conditions. Process 2 was described by Rosenbrock's surface and had

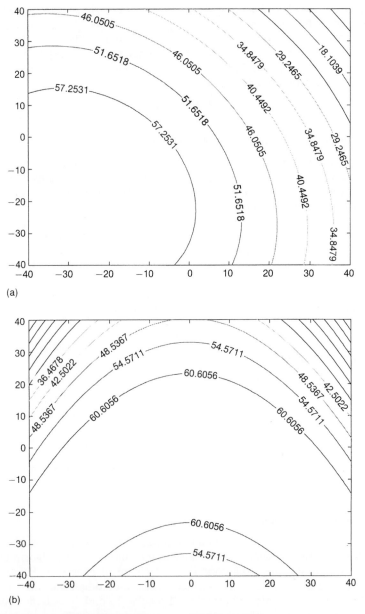

Figure 7.4   Contours of yield for (a) Process 1; (b) Process 2

banana-shaped contours (Figure 7.4(b)). The equation is

$$y = 66 + 0.032\{20 - [100((-x_2/17) - (-x_1/27)^2)^2 + (1 + x_1/27)^2]\}$$

and the maximum is 66.64 when (x_1, x_2) equals $(-27, -17)$. The GA is quite easily explained for a specific case. The values of x_1 and x_2 were coded using

16 binary bits, i.e.

$$x_1 = -40 + (\text{16-bit binary representation}/(2^{16} - 1)) \times 80$$

A Gray code was used rather than base 2 arithmetic (Exercise 7.7) but this is not essential. An initial set of 30 design points (x_1, x_2) was established by randomly assigning 0s or 1s to 30 sequences of 32 bits. In general, with k control variables using B-bit coding we would need m sequences of kB bits for a set of m design points. GAs are based on a genetic analogy and the literature reflects this. The set of 30 design points is referred to as the first 'generation' and is a 'population' of 30. The sequences of 32 bits are called 'chromosomes'. The yield of the process is measured at each design point. In the simulations the yield is calculated as the value of y from the defining equation with random variation added. The next step is to select 'parents' for the next generation. The probability of using a particular chromosome is proportional to the corresponding process yield, referred to as its 'fitness'. Sampling is with replacement. The first two chromosomes to be selected may be 'mated' by applying a 'crossover' operation. The probability of this occurring is some preset probability, P_c. If crossover is to be applied, a location along the chromosome is chosen at random and the bits beyond this location are swapped over to create two new chromosomes.

If crossover does not occur, the parents are copied to the new generations without alteration. The process is repeated until $m/2$ pairs of chromosomes have been considered, and hence the new generation maintains the population size at m. Finally, there is a 'mutation' operator to change, independently, the value at each position on the chromosome from one to zero or vice versa with some preset probability, P_m. The GA parameters P_c and P_m were chosen to be 0.9 and 0.01. These sequences continue for a preset number of generations. It is possible that 'twins' occur in a population, but this was not a practical problem with the parameter values chosen for this study. A common modification, known as 'elitism', is to pass the fittest from each generation to the next without change. In this sort of application at least, the performance of the GA is insensitive to the choice of parameter values (Luangpaiboon *et al.*, 1999, and Exercise 7.8).

On the parabolic surface, simplex algorithms seemed to be the most efficient, in terms of speed of convergence, when the level of error standard deviation was low. However, when the standard deviation of the errors was at higher levels, simplex algorithms could not be relied upon to locate the optimum. To overcome this limitation, some modifications such as simulated annealing, which admits a small probability of moving in the wrong direction, were considered, but these had only limited success. In particular, the number of runs required for convergence to the optimum was sometimes greater than from the algorithms based on factorial and triangular designs. The sequential use of factorial designs, as described in many statistical textbooks, was found to be the most efficient at the high level of noise. The genetic algorithm was the only strategy that found the maximum on the curved ridge surface, but the slope of the ridge was only slight. The GA was rather slow, five generations of a population of 30 being 150 runs, but this may not be a serious drawback in the context of automatic process control. The algorithms based on standard designs failed to find the maximum because they oscillated between one side of the ridge and the other. Operator intervention could be used to improve performance by moving the design point at right angles to the oscillations, and more sophisticated algorithms could do this automatically. Luangpaiboon *et al.* summarised their results as follows. If a 3D response surface can reasonably be assumed to be a paraboloid the choice between simplex algorithms and factorial designs depends on the level of noise. If there is no justification for any assumptions about the shape of the 3D response surface a GA may be preferable, especially if there could be several local maxima.

7.5 Completely randomised design

7.5.1 Introduction

In a study of personality and entrepreneurial leadership, Nicholson (1998) compared the personality profiles of a sample of chief executive officers (CEO) from UK independent companies with personality profiles of both a control group of managers and a normative sample from the USA. He used the NEO-PI-R personality inventory which has five dimensions: neuroticism, defined as propensity for emotional responsiveness; extraversion, defined as outgoing and active orientation; openness to experience, defined as creativity and adaptiveness; agreeableness, defined as nurturance and tender-mindedness; and conscientiousness, defined as a need for order and control. Each dimension has six facets, so each respondent has five dimension scores each being the sum of six facet scores. The sample of CEOs was obtained by writing to the CEO of the top 116 companies in a list of the UK's top performing independent companies in 1993–94, published by the *Independent* newspaper from a database maintained by Price Waterhouse. Forty-three CEOs returned fully completed questionnaires. The managerial control sample was drawn from managers attending professional development courses at London Business School or known to the research team through consulting assignments. Fifty-seven managers from the private sector, who did not have a corporate leadership role, agreed to complete the questionnaire. The NEO normative sample consisted of 500 respondents.

Any statistical analysis is based on an assumption that the respondents are random samples from the corresponding populations. Nicholson discusses possible sources of bias, but there is no compelling argument against the assumption. Consider the total extraversion scores, over six facets. It is possible to compare CEOs with managers, CEOs with the normative sample, and managers with the normative sample. This makes a total of three *t*-tests, and if they were independent, which they are not, the probability at least one test is significant at the 5% level is 0.14. One way around this is to test at a lower α level, and rely on the Bonferonni inequality. That is:

Pr(at least one out of N statistical tests significant at

$\alpha \times 100\%$ level|no differences in populations) $\leq N\alpha$

The inequality follows, by induction, from the elementary probability result that

$$\Pr(A \text{ or } B) \leq \Pr(A) + \Pr(B)$$

With large samples this may be adequate because any difference which is large enough to be of practical interest is likely to be highly statistically significant. Nicholson gives the means and standard deviations for all five dimensions, and their facets, for the three groups. He also gives a table of t-statistics for testing a hypothesis of equal population means, from two independent samples. The t-statistics are given for the leadership sample (leaders) versus the NEO normative sample (NEO) and for the leadership sample versus the management control sample (control). For extraversion, the former was 8.20 and the latter was 0.16. Given that a t-value with an absolute value greater than 2 is significant at the 5% level, it seems reasonable to conclude that managers, whether or not they are in a leadership role, tend to be extraverts. Any misgivings about this conclusion arise from the randomness assumption rather than precise P-values. The mean extraversion scores for leaders, control and NEO groups are 121, 120 and 108 respectively. There was no evidence of any difference in openness between any of the three groups. The neuroticism results had a negative sign: -2.46 and -1.47 for leaders versus NEO and control respectively. The mean neuroticism scores for leaders, control and NEO groups are 67, 74 and 75 respectively. The agreeableness results prompted a cartoon of an armadillo driving a steam roller at the beginning of the paper: -12.91 and -0.60 for leaders versus NEO and control respectively. The mean agreeableness scores for leaders, control and NEO groups are 105, 107 and 120 respectively. At least there is no evidence that leaders are less agreeable than the managers in the control group. The results for conscientiousness are not surprising: 6.77 and 2.20 for leaders versus NEO and control respectively. The mean conscientiousness scores for leaders, control and NEO groups are 132, 122 and 124 respectively. A multivariate analysis could have been applied, and it would be interesting to know what the correlations between the facets and dimensions are. However, it would be unlikely to change any conclusions about the means.

Nicholson had quite large samples. There was no doubt that there were significant differences, and the t-tests gave a clear indication of these. However, with small samples we need a more precise analysis.

7.5.2 One-way analysis of variance

The production manager of a company which manufactures filters for liquids, for use in the pharmaceutical and food industries, wishes to compare the burst strength of four types of membrane. The first (A) is the company's own standard membrane material, the second (B) is a new material the company has developed, and C and D are membrane materials from other manufacturers. The manager has tested five filter cartridges from ten different batches of each material. The mean burst strengths for each set of five cartridges are given in Table 7.12, and plotted in Figure 7.5. We will set up the following model in which Y_{ij} is the result for the batch j of material i (where 1, 2, 3 and 4 represent materials A, B, C and D respectively)

$$Y_{ij} = \mu + \alpha_i + E_{ij} \quad \text{for } i = 1, 2, 3, 4 \quad \text{and} \quad j = 1, ..., 10$$

where $\sum_{i=1}^{4} \alpha_i = 0$ and $E_{ij} \sim N(0, \sigma^2)$ and are independent.

Table 7.12 Burst strengths of filter membranes (kPa)

Type A	Type B	Type C	Type D
95.5	90.5	86.3	89.5
103.2	98.1	84.0	93.4
93.1	97.8	86.2	87.5
89.3	97.0	80.2	89.4
90.4	98.0	83.7	87.9
92.1	95.2	93.4	86.2
93.1	95.3	77.1	89.9
91.9	97.1	86.8	89.5
95.3	90.5	83.7	90.0
84.5	101.3	84.9	95.6

This model assumes that we have random samples of size 10 from four normal distributions A, B, C, D with means $\mu + \alpha_1$, $\mu + \alpha_2$, $\mu + \alpha_3$ and $\mu + \alpha_4$ respectively.

The quantity α_1, for example, is referred to as the *effect* of material A.

A regression model with indicator variables is equivalent (Exercise 7.9), but it becomes less convenient as the number of populations to be compared increases.

The null hypothesis is that there is no difference between materials.

$$H_0: \alpha_1 = \alpha_2 = \alpha_3 = \alpha_4 = 0$$

and the alternative is

$$H_1: \text{at least one of the } \alpha_i \neq 0$$

We will test the null hypothesis using an analysis of variance (ANOVA) procedure and one F-test. This avoids the imprecise overall significance level which arises if we perform the four possible two-sample t-tests. If we reject H_0 if any of the four tests is significant at the ε level then the overall significance level $<4\varepsilon$, by the Bonferonni inequality, but the lack of independence of the tests makes it difficult to be more precise.

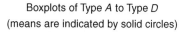

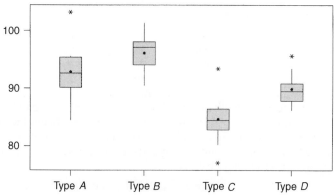

Figure 7.5 Burst strengths of filter membranes (kPa)

The model implies that the variance of the strengths for each material is the same value σ^2. We can then make four independent estimates of σ^2, one from the data for each material, and pool them to obtain a within-samples estimate (s_w^2) of this common population variance σ^2.

$$s_w^2 = \frac{s_A^2 + s_B^2 + s_C^2 + s_D^2}{4} = \frac{23.34 + 11.50 + 18.38 + 7.64}{4} = 15.22$$

If H_0 is true, another unbiased estimate of σ^2 is the between-samples estimate (s_b^2). The null hypothesis is that the samples are all from the same population so the variance of $\bar{Y}$, based on a sample of size 10, is $\sigma_{\bar{Y}}^2 = \sigma^2/10$. This is estimated by

$$s_{\bar{Y}}^2 = \frac{(92.84 - 90.86)^2 + (96.08 - 90.86)^2 + (84.63 - 90.86)^2 + (89.89 - 90.86)^2}{(4 - 1)}$$

$$= 23.64$$

Hence, the alternative estimate of σ^2, which is only valid if H_0 is true, is $s_b^2 = 10 \times 23.64 = 236.4$, with 3 degrees of freedom.

In contrast when H_0 is not true, $E[S_b^2] > \sigma^2$
If H_0 is true, we have two independent estimates of the same variance and $S_b^2/S_w^2 \sim F_{3,36}$

Note that if H_0 is not true we would expect large values of the calculated F-ratio. The critical region for testing H_0 against H_1 at the 1% level is values of F exceeding 4.38. In this case the calculated value of F is 15.54 so we have evidence to reject H_0 at the 1% level. We can obtain a P-value from Minitab, which is precise to four decimal places:

$$\Pr(S_b^2/S_w^2 > 15.54|H_0 \text{ true}) = 0.0000$$

The data provide strong evidence of a difference between materials. The Minitab command:

$$\text{Statistics} \triangleright \text{ANOVA} \triangleright \text{Oneway} \dots \text{Graphs}$$

provides an ANOVA table, which is explained below, and a useful summary. Both are shown in Figure 7.6.

Estimators of the parameters in the model

Consider the more general case of samples of size n from k populations. Remember the constraint that $\sum \alpha_i = 0$.

$$\bar{Y}_{i\cdot} = \sum_{j=1}^{n} Y_{ij}/n = \mu + \alpha_i + \bar{E}_{i\cdot}$$

$$\bar{Y}_{\cdot\cdot} = \sum_{i=1}^{k} \sum_{j=1}^{n} Y_{ij}/nk = \sum_{i=1}^{k} \bar{Y}_{i\cdot}/k = \sum_{i=1}^{k} (\mu + \alpha_i + \bar{E}_{i\cdot})/k = \mu + \bar{E}_{\cdot\cdot}$$

$$E[\bar{Y}_{i\cdot}] = \mu + \alpha_i \qquad E[\bar{Y}_{\cdot\cdot}] = \mu$$

```
MTB  > AOVOneway 'Type A' 'Type B' 'Type C' 'Type D';
SUBC>    GBoxplot.
```

One-way Analysis of Variance

Analysis of Variance

Source	DF	SS	MS	F	P
Factor	3	709.2	236.4	15.54	0.000
Error	36	547.8	15.2		
Total	39	1257.0			

Individual 95% CIs For Mean
Based on Pooled StDev

Level	N	Mean	StDev	
Type A	10	92.840	4.831	(----*----)
Type B	10	96.080	3.391	(----*----)
Type C	10	84.630	4.287	(----*----)
Type D	10	89.890	2.764	(----*----)

```
                                    -----+---------+---------+---------+
Pooled  StDev =    3.901           85.0      90.0      95.0      100.0
MTB > nooutfile
```

Figure 7.6   Minitab one-way ANOVA analysis of burst strengths of membranes

The least squares estimators $\hat{\mu}$, $\hat{\alpha}_i$ of μ and the α_i are obtained by minimising

$$\psi = \sum_{i=1}^{k} \sum_{j=1}^{n} (Y_{ij} - \mu - \alpha_i)^2$$

with respect to μ and $\alpha_1, \ldots, \alpha_k$. They are

$$\hat{\mu} = \bar{Y}_{..} \quad \text{and} \quad \hat{\alpha}_i = \bar{Y}_{i.} - \bar{Y}_{..}$$

For any value of i,

$$\sum_{j=1}^{n} (Y_{ij} - \bar{Y}_{i.})^2 / (n-1)$$

is an unbiased estimator of σ^2. Hence

$$S_w^2 = \sum_{i=1}^{k} \sum_{j=1}^{n} (Y_{ij} - \bar{Y}_{i.})^2 / k(n-1)$$

is also unbiased for σ^2 and will henceforth be written as S^2 for the estimator and s^2 for the estimate. In the previous example $\hat{\mu} = 90.86$, $\hat{\alpha}_1 = 1.98$, $\hat{\alpha}_2 = 5.22$, $\hat{\alpha}_3 = -6.23$, $\hat{\alpha}_4 = -0.97$ and $s^2 = 15.22$.

Partitioning of the sum of squares

It is convenient to define S_{yy} as a shorthand for the sum of squared deviations of y_{ij} from the overall mean. This can be partitioned as follows.

$$
S_{yy} = \sum_{i=1}^{k} \sum_{j=1}^{n} (y_{ij} - \bar{y}_{..})^2
$$

$$
= \sum_{i=1}^{k} \sum_{j=1}^{n} (y_{ij} - \bar{y}_{i.} + \bar{y}_{i.} - \bar{y}_{..})^2
$$

$$
= \sum_{i=1}^{k} \sum_{j=1}^{n} (y_{ij} - \bar{y}_{i.})^2 + \sum_{i=1}^{k} \sum_{j=1}^{n} (\bar{y}_{i.} - \bar{y}_{..})^2 + 2 \sum_{i=1}^{k} \sum_{j=1}^{n} (y_{ij} - \bar{y}_{i.})(\bar{y}_{i.} - \bar{y}_{..})
$$

The last of these three terms is zero since

$$
\sum_{i=1}^{k} \sum_{j=1}^{n} (y_{ij} - \bar{y}_{i.})(\bar{y}_{i.} - \bar{y}_{..}) = \sum_{i=1}^{k} (\bar{y}_{i.} - \bar{y}_{..}) \sum_{j=1}^{n} (y_{ij} - \bar{y}_{i.}) = 0
$$

The final result is

$$
S_{yy} = \sum \sum (y_{ij} - \bar{y}_{i.})^2 + n \sum_{i=1}^{k} (\bar{y}_{i.} - \bar{y}_{..})^2
$$

The first term on the right is referred to as the *within-samples* (corrected) *sum of squares* and the second as the *between-samples* (corrected) *sum of squares*.

Expected value of the between-samples sum of squares

Consider

$$
\sum_{i=1}^{k} (\bar{Y}_{i.} - \bar{Y}_{..})^2 = \sum_{i=1}^{k} (\alpha_i + \bar{E}_{i.} - \bar{E}_{..})^2
$$

$$
= \sum_{i=1}^{k} (\alpha_i^2 + (\bar{E}_{i.} - \bar{E}_{..})^2 + 2\alpha_i(\bar{E}_{i.} - \bar{E}_{..}))
$$

Taking expectation gives

$$
E\left[\sum_{i=1}^{k} (\bar{Y}_{i.} - \bar{Y}_{..})^2 \right] = \sum_{i=1}^{k} \alpha_i^2 + (k-1)\sigma^2/n
$$

The expected value of the between-samples sum of squares is the product of this expression with n.

Calculations for ANOVA

The calculations described in the preceding two subsections can be summarised in an analysis of variance (ANOVA) table. There are k samples each of size n.

Source of variation	(Corrected) sum of squares	d.f.	Mean square	Expected value of the mean square
Between samples	$n\sum\limits_{i=1}^{k}(\bar{y}_{i.} - \bar{y}_{..})^2$	$k-1$	$\dfrac{CSS}{k-1} = s_b^2$	$\sigma^2 + n\sum\limits_{i}\alpha_i^2/(k-1)$
Within samples	$\sum\limits_{i=1}^{k}\sum\limits_{j=1}^{n}(y_{ij} - \bar{y}_{i.})^2$	$k(n-1)$	$\dfrac{CSS}{k(n-1)} = s_w^2 \ (=s^2)$	σ^2
Total	$\sum\limits_{i=1}^{k}\sum\limits_{j=1}^{n}(y_{ij} - \bar{y}_{..})^2$	$nk-1$		

The model is a special case of the general multiple regression model, and the residuals are the differences between observations and their fitted values. In this context:

$$y_{ij} - (\hat{\mu} + \hat{\alpha}_i) = y_{ij} - \bar{y}_{i.}$$

The within-samples sum of squares is just the sum of squared residuals, and is often referred to as 'residual' or 'error'.

Difference between fixed effects and random effects

We refer to the α_i as fixed effects because they refer to specific different types of material. Suppose, instead, that A, B, C and D had referred to different months of manufacture of the same type of material. It would be appropriate to treat the months as a random sample of all possible months. The month effects would then be described as random effects and, writing σ_α^2 for the variance between months,

$$n\left\{E\left[\sum_{i=1}^{k}(\bar{Y}_{i.} - \bar{Y}_{..})^2\right]\right\} \Big/ (k-1) = \sigma^2 + n\sigma_\alpha^2$$

This is the components of variance analysis of Section 6.2. The designation of effects as fixed or random has important consequences for hypothesis testing in more complicated designs, because it determines which components of variability are taken as the error.

Multiple comparisons

If H_0 is rejected it remains to find where the differences lie. The simplest way is to use *least significant differences* (LSD). This technique involves making all possible t-tests, so the probability of declaring a statistically significant difference if there is none will be higher than the nominal level. Such a procedure was avoided at the start of the analysis, but we do now have a significant F-test.

The standard deviation of the difference of any two group means, for the filter material example, is estimated by

$$\sqrt{[s_w^2(1/n + 1/n)]} = \sqrt{15.22(1/10 + 1/10)} = 1.74$$

The relevant percentage points of the *t*-distribution for nominal 5% and 1% significance levels are

$$t_{36,0.025} = 2.028 \quad \text{and} \quad t_{36,0.005} = 2.719 \text{ respectively}$$

Therefore the least difference in means to be significant at a nominal 5% level is $2.028 \times 1.74 = 3.53$. The least significant difference at the nominal 1% level is $2.719 \times 1.74 = 4.73$. These can be written concisely as:

$$\text{LSD}(5\%) = 3.53 \quad \text{and} \quad \text{LSD}(1\%) = 4.73$$

For the filter material data

$$\bar{y}_A = 92.840 \quad \bar{y}_B = 96.080 \quad \bar{y}_C = 84.630 \quad \bar{y}_D = 89.89$$

There is evidence that material *B* is stronger than materials *C* and *D*, and that *A* and *D* are stronger than *C*. There is no substantial evidence of any difference between materials *A* and *B*. Remember that confidence intervals for individual means can overlap although the difference in the means is statistically significant (Exercise 7.10).

7.6   Randomised block designs

Twelve middle managers in a large company, based in the UK, asked the manager responsible for professional development (PDM) to pay for them to attend a course on leadership skills, which included seminars and outdoor activities, over a long weekend in the Lake District. PDM was sceptical about the value of this expensive course and concerned about setting a precedent of sending people on it. However, PDM compromised by agreeing to set up an experiment to evaluate the course. It would be compared with both a one-day course of seminars on leadership skills, without any associated outdoor activities, and no formal training on leadership. PDM enlisted the support of the human resources manager (HRM). For each of the 12 managers, HRM found two others who were similar in terms of age, education and level of responsibility within the company. Each person in a group of three similar managers was assigned at random to one of the weekend course (*W*), the day course (*D*), or no formal training (*C*). The randomisation was subject to a restriction that four of the original 12 managers would attend *W*, four would attend *D* and four would be in the control group *C*. All 36 managers agreed to take part in the experiment with this random allocation. Their leadership skills during the six months following the weekend course would be assessed by a psychologist (PHG) from a local university. PHG would ask all the staff who have day-to-day dealings with the manager to complete a questionnaire during a short interview. PHG would also interview each manager and award him or her a score, between 0 and 100, on the basis of the interview and questionnaire results. The scores would be held anonymously, by PHG, but managers would be given details of their own performances if they wished. PDM and HRM would only be given the results of the analysis.

If there is just one observation on each of *k* treatments in each of *b* blocks, and blocks are considered to be fixed effects, a *two-way classification fixed effects model* can be used. This is:

$$Y_{ij} = \mu + \alpha_i + \beta_j + E_{ij}$$

where

$$\sum_{i=1}^{k} \alpha_i = 0 \quad \sum_{j=1}^{b} \beta_j = 0$$

the $E_{ij} \sim N(0, \sigma^2)$ and are independent of the α_i, β_j and each other, μ is the overall mean, α_i, for $i = 1, \ldots, k$, are the treatment effects, and β_j, for $j = 1, \ldots, b$ are the block effects.

The hypothesis of interest is

$$H_0: \alpha_1 = \cdots = \alpha_k = 0$$

The least squares estimates of the parameters in the model are

$$\hat{\mu} = \bar{y}_{..} \quad \hat{\alpha}_i = \bar{y}_{i.} - \bar{y}_{..} \quad \hat{\beta}_j = \bar{y}_{.j} - \bar{y}_{..}$$

The general form of the ANOVA table is:

Source of variation	(Corrected) sum of squares	d.f.	Mean square	Expected value of the mean square
Treatments	$b \sum_{i=1}^{k} (\bar{y}_{i.} - \bar{y}_{..})^2$	$k - 1$	CSS divided	$\sigma^2 + b \sum_{i=1}^{k} \alpha_i^2 / (k-1)$
Blocks	$k \sum_{j=1}^{b} (\bar{y}_{.j} - \bar{y}_{..})^2$	$b - 1$	by the degrees of freedom	$\sigma^2 + k \sum_{j=1}^{b} \beta_j^2 / (b-1)$
Residual	$\sum_{i=1}^{k} \sum_{j=1}^{b} (y_{ij} - \bar{y}_{i.} - \bar{y}_{.j} - \bar{y}_{..})^2$	$(k-1)(b-1)$		σ^2
Total	$\sum_{i=1}^{k} \sum_{j=1}^{b} (y_{ij} - \bar{y}_{..})^2$	$kb - 1$		

If block effects are treated as random, the expected value of the blocks mean square becomes

$$\sigma^2 + k\sigma_\beta^2$$

The data are given in Table 7.13, and the results of the Minitab Two-way Analysis of Variance are shown in Figure 7.7. The three columns of 12 data, shown in Table 7.13, were stacked into a single column called 'score'. A column of 12 1s, 12 2s and 12 3s was set up using

Calc ▷ Make Patterned Data

and called 'mode'. A column of 1 ... 12, repeated three times, was also set up with Make Patterned Data and called 'group'. The command for the analysis is

Stat ▷ ANOVA ▷ Two-Way ... Graphs
Response: score

Table 7.13   Leadership scores for managers

Group	Weekend	Day	Control
1	70	63	65
2	59	72	60
3	52	56	61
4	68	59	73
5	74	79	58
6	82	73	67
7	49	58	44
8	73	82	68
9	57	53	46
10	61	72	65
11	73	79	59
12	64	73	66

```
MTB  > Twoway 'score' 'group' 'mode';
SUBC>  Means 'group' 'mode'.

Two-way  Analysis of Variance

Analysis  of Variance for score
Source    DF      SS       MS       F       P
group      11    2092.3    190.2    4.70    0.001
mode        2     317.7    158.9    3.93    0.035
Error      22     889.6     40.4
Total      35    3299.6
                    Individual 95% CI
group     Mean    --------+---------+---------+---------+---
  1       66.0                    (-------*-------)
  2       63.7                  (-------*-------)
  3       56.3          (-------*-------)
  4       66.7                   (-------*-------)
  5       70.3                     (-------*-------)
  6       74.0                       (-------*-------)
  7       50.3    (-------*-------)
  8       74.3                       (------*-------)
  9       52.0    (-------*-------)
 10       66.0                    (-------*-------)
 11       70.3                     (------*-------)
 12       67.7                   (-------*-------)
                    --------+---------+---------+---------+---
                        50.0      60.0      70.0      80.0

                    Individual 95% CI
mode      Mean    --------+---------+---------+---------+---
  1       65.2                  (---------*---------)
  2       68.3                      (---------*---------)
  3       61.0    (---------*---------)
                    --------+---------+---------+---------+---
                        60.0      64.0      68.0      72.0

MTB  > nooutfile
```

Figure 7.7   Minitab two-way ANOVA analysis of leadership courses

Row factor: group   ✓ Display means
Column factor: mode ✓ Display means

There is evidence against a hypothesis that W, D and C are identical. The means are:

$$\bar{y}_W = 65.17 \quad \bar{y}_D = 68.25 \quad \bar{y}_C = 61.00$$

The LSD (5%) is $2.07 \times \sqrt{[40.4 \times (1/12 + 1/12)]}$ which equals 5.37. There is evidence that managers who attend the day course do better than those who are given no formal training. No other differences are statistically significant at the 5% level. Since the day course was considerably cheaper than the weekend course, PDM decided to use the day course in future.

7.7   Difference between RBD and CRD with two factors

The 2^2 and 3^2 experiments described at the beginning of Section 7.4.1 are examples of completely randomised designs (CRD) with two factors. In general there are two factors at r and c levels respectively, so the experiment consists of rc runs carried out in a random order. CRD with two factors can be analysed by two-way ANOVA, in a similar way to the RBD, but there is a crucial difference in the interpretation of the results as the following example shows.

Anderson and McLean (1974) discuss a project to compare different designs of prosthetic cardiac valves. Four valve types were investigated in a mechanical apparatus which had been constructed to simulate the human circulatory system. Tests were carried out at six pulse rates, from 20 to 220 beats per minute. The response was maximum flow gradient (measured in millimetres of mercury).

A completely randomised design would need six valves of each type. There are 24 valve by pulse by rate combinations, and the allocation of valves of each type to the runs for each pulse rate, and the order of testing all combinations, should be randomised. A criticism of this as an experimental programme is the need for six valves of each type. Prototypes of precision equipment are expensive to produce, and we might be tempted to treat valves as blocks and set up a randomised blocks design. We could then manage with one valve of each type. But we would have no information about the variability of valves of a given type. If we do find any significant differences we cannot tell from the experiment whether it is an exceptionally good valve of its type or whether its type is the best design.

A possible compromise, and the one adopted, was to design an experiment around two valves of each type. Each valve of each type was tested with all six pulse rates. This needed 48 runs, but the time taken for a run is negligible compared with that needed to make the prototype valves. The order of the runs was randomised, and the apparatus was reassembled after each run, even if the same valve was to be used. It would have saved time to try all pulse rates in sequence on the same valve, but this would leave the possibility that an apparently poor performance of a valve was due to a poor assembly of the apparatus. A regression-based analysis is given in Metcalfe (1994).

7.8 Summary

1. A 2^n factorial design allows the researcher to investigate the linear effects and interactions of n control variables, the factors, on some response variable. Although the control variables are usually continuous they are restricted to two levels, 'high' and 'low'. If n is large the number of runs for a full factorial design soon becomes prohibitive. It is possible to generate fractions, involving 2^{n-m} runs, that can provide information on the linear effects and the low-order interactions which could be of practical importance. A star design can be used to follow-up a factorial or fractional factorial design. It allows quadratic effects to be investigated. The overall design, factorial plus star, is known as a composite design. It is only practical for continuous control variables.
2. In general the factors can be at several levels and, for example, a completely randomised design with two factors at r and c levels respectively involves rc runs. The runs should be carried out in a random order.
3. Taguchi-style experiments test the different control variable combinations at different levels of noise variables. In the experiment the noise variables can be set at specific levels but in normal operation they would vary randomly. One objective is to find values of the control variables that will be insensitive to noise (robust).

Exercises

7.1 Two columns of a design are said to be orthogonal if they are uncorrelated. If the sum of the row entries of at least one of two columns is 0 then they are orthogonal if the sum of the products of their row entries is 0.

(i) Verify that any two from the seven columns in Table 7.2 are orthogonal.
(ii) Verify that if a column of 1s is added at the beginning of Table 7.2, before the x_1 column, the regression model with all interactions can be written

$$y = XB$$

where $y = (y_1, \ldots, y_8)^T$, $B = (\beta_0, \beta_1, \ldots, \beta_7)^T$, $X = (1, x_1, \ldots, x_7)$, $\mathbf{1}$ is a column of 1s and x_j has entries x_{ji} for $i = 1, \ldots, 8$.
(iii) Two vectors a and b are said to be orthogonal if they are at right angles to each other. A necessary and sufficient condition for a and b to be orthogonal is that their scalar product

$$a \cdot b = \sum a_i b_i = 0$$

Consider $1, x_1, x_2$ as vectors and verify that they are orthogonal in a vector sense.
(iv) Suppose low and high are coded 1 and 2 respectively so that x_1 and x_2 become

$$x_1 = (1 \quad 1 \quad 1 \quad 1 \quad 2 \quad 2 \quad 2 \quad 2)^T$$
$$x_2 = (1 \quad 1 \quad 2 \quad 2 \quad 1 \quad 1 \quad 2 \quad 2)^T$$

Verify that x_1 and x_2 are uncorrelated, but not orthogonal in the vector sense. They are, however, still referred to as orthogonal in a statistical context. Are x_1 and x_1^2, and x_1 and $x_1 \times x_2$, uncorrelated?

7.2 (a) The three questionnaire versions are combined into one array of 27 rows.

Verify that there are: 6 single-factor terms, $x_1, \ldots, x_3^2$; 12 two-factor terms, $x_1 x_2, \ldots, x_1^2 x_2, \ldots, x_2^2 x_3^2$; and 8 three-factor terms, $x_1 x_2 x, \ldots,$ $x_1^2 x_2^2 x_3, x_1^2 x_2^2 x_3^2$. Check that any two from the 26 terms are orthogonal.

(b) The three versions are each a one-third replicate 3^{3-1} of a 3^3 design. Verify, for any version, that the linear and quadratic terms are all orthogonal to each other, but are correlated with two-factor interactions. Notice that the two-factor interactions are not aliased with any of the single-factor terms.

7.3 The aliases of A in a 2^{6-3} design with generators $D = AB$, $E = AC$ and $F = BC$ are found by multiplying A into each of DAB, EAC, FBC and products of 2 or 3 from these 3, subject to a letter squared being 1.

(i) Check that this is equivalent to multiplying A into: ABD, ACE, BCF, $BCDE$, $ACDF$, $ABEF$, DEF.

(ii) Hence verify that the full list of aliases for A is: BD, CE, CDF, BEF, $ABCF$, $ADEF$, $ABCDE$.

7.4 (a) The cement kiln experiment had design generators $E = ABC$ and $F = BCD$. Verify that the main effects are aliased with three-factor interactions and that the two-factor interactions are aliased as:

$$AB = CE, \quad AC = BE, \quad AD = EF, \quad AE = BC = DF,$$

$$AF = DE, \quad BD = CF, \quad BF = CD$$

(b) Investigate the correlations between $x_1, \ldots, x_6$, their squares $x_1^2, \ldots, x_6^2$ and two-factor interactions $x_1, x_2, \ldots, x_5, x_6$.

7.5 (a) A 'target-is-best' SNR is sometimes defined by

$$\text{SNR} = 10 \log_{10}(\bar{y}/s^2)$$

Show that this is linearly related to

$$-10 \log_{10} \left(\sum (y_i - T)^2 / n \right)$$

(b) The 'larger-the-better' SNR is defined by

$$\text{SNR} = -10 \log_{10} \left[\left(\sum 1/y_i^2 \right)/n \right]$$

Explain, by reference to the smaller-the-better SNR, why this may be appropriate if the objective is to achieve a high mean and small variance. [Hint: use the fact that minimising y is equivalent to maximising $1/y$.]

7.6 Recode the 1, 2, 3 in the L_9 Taguchi array as $-1, 0, 1$ and compare it with a 3^2 design. [Hint: calculate squares and cross-products of the columns in the two designs.]

7.7 Base two arithmetic and Gray coding of the integers 0 to 7 are given below.

Integer	Base 2	Gray
0	000	000
1	001	001
2	010	011
3	011	010
4	100	110
5	101	111
6	110	101
7	111	100

Why might a Gray coding be an advantage in a *GA*?

7.8 A design point (x_1, x_2) is defined by two 32-bit chromosomes. If the probability of mutation is 0.01, what is the probability at least one bit changes during the mutation operation of a *GA*?

7.9 Analyse the membrane burst strengths in Table 7.12 by using a regression model with indicator variables for materials *B*, *C*, *D* relative to material *A*.

7.10 (i) Explain why 95% confidence intervals for the mean of population *A* and the mean of population *B* can overlap when the two sample means are significantly different at the 5% level. [Hint: the variances of independent variables are additive, the standard deviations are not!]

(ii) Explain why two sample means are significantly different, beyond the 5% level, if the 95% confidence intervals for the individual means do not overlap.

8

Forecasting

8.1 Introduction

All businesses need to estimate the future demand for their services or products for both short-term planning and long-term strategic decisions. Manufacturers of pharmaceuticals, many of which have short shelf lives, need to forecast demand up to a few weeks ahead for production scheduling. The manufacturer must also attempt to predict what will happen to demand in a few years' time. Some things may be known in advance, such as a patent coming to the end of its term, but other things are very uncertain. An improved treatment may be found or, despite careful trials, the drug may be found to have unexpected side effects. Another example is an airport which is being considered for expansion and for which demand needs to be forecast for many years ahead.

The best way of making forecasts is to find some other variable which is known to affect, or at least be associated with, future values of the variable which is to be forecast. A leading world-wide marine coatings company uses statistics available in the public domain to determine the number, type, and sizes of ships to be built over the next three years. For example, *World Shipyard Monitor* contains brief details of orders in about 320 shipyards. The coatings company has set up an internal data base of individual ships, to which a regression analysis is applied with the aim of predicting areas to be painted from a ship's size and type. Then the number of ships, typical areas to be painted, dry film thickness, spread rates, number of coats and loss rates are used to determine the likely paint demand for the coming years. In 1997 the total area to be coated at newbuilding work, which excludes repairs to ships at sea, corresponded to about 80 million litres of paint. The company monitors its market share carefully and uses the forecast for planning production and for setting prices.

Another useful strategy is to find some other variable which is associated with the variable to be forecast and easier to predict. For example, gas suppliers in the UK have to place orders for gas from the offshore fields one day ahead. Variation about the average for the time of year depends on temperature and, to some extent, the wind speed. Weather forecasts for the next day are now quite accurate and well worth incorporating into the forecasting procedure.

However, even if some associated variables can be found, most forecasts rely either entirely or mainly on past values of the variable to be forecast. A sequence of values of a variable over time is known as a time series. The time series of sales of Australian wines, given in TableWSozwine, were compiled by the Australian Bureau of Statistics and are reproduced on the book website with the Bureau's permission. The wine growers need forecasts so that they can set their prices, plan whether to extend their vineyards, choosing in doing so which grape varieties to cultivate, and decide whether to produce fortified wines from some of their harvest.

Whilst time series generally have a rather random appearance it is often possible to discern some systematic cyclical patterns and perhaps an overall trend. For example, in the UK sales of cartons of ice cream from supermarkets are higher during the warmer months, although there is always some demand for ice cream. Demand for domestic gas is correlated with temperature and is highest in winter. Such systematic variation within the year is known as a seasonal component, and in the case of sales which are related to temperature it may be reasonably modelled by a harmonic function (sine curve) with a period of one year. However, many business time series show marked monthly effects which do not vary smoothly throughout the year. The demand for package holidays is highest during school vacations and traditional holiday periods. Until recently, sales of cars in the UK were artificially high at the end of August when the next year's registration plates were issued. In these cases independent monthly indices can be calculated.

Time series can be collected with any time interval between observations. For some purposes hourly demand for gas needs to be forecast. There is marked systematic variation throughout the 24-hour day, with a greatly reduced demand throughout the night. Clothing retailers may monitor weekly sales and order stock each week.

Seasonal patterns can often be assumed to persist in the future, although there may be changes. An example is the new legislation concerning vehicle registration in the UK which has resulted in a more even demand for cars. In contrast, trends are unlikely to persist for long periods. Sales cannot increase indefinitely, and an increasing trend may be associated with the boom period of some business cycle. Unfortunately, even if business cycles provide a description of past market behaviour they are notoriously hard to predict. For short-term predictions, we often assume that the present trend will continue.

There are several strategies for identifying a seasonal variation and trend. If seasonal variation and trend are removed from a time series we can then investigate whether there is any further relationship that can be used for making predictions. If this month's sales are well above the average for the time of year, it may be that next month's sales will also be relatively good. Other uses of time-series analysis include automatic process control and simulation studies.

In some applications, such as queueing processes, the state variable, which is the number of people in the queue, is discrete. It may also be more appropriate to treat time as continuous than as a sequence of steps. Models for such processes are described in Section 8.7.

8.2   Index numbers

8.2.1   Introduction

The business development manager in a statistical consultancy has records of the amount of money turned over each year since its inception 11 years ago. Turnover has increased, but so have prices in general. The manager wishes to adjust turnover for these general price increases before analysing the company's performance. In the UK the Office for National Statistics (ONS) maintains the retail price index (RPI). January 1987 is currently the base period and corresponds to 100. The June 1999 value is 165.6. The interpretation of this is that a salary of £2000 per month in January 1987 would be equivalent to a salary of

$$2000 \times 165.6/100 = 3312$$

now. Conversely, a salary of £2500 a month now would be equivalent to

$$2500 \times 100/165.6 = 1510$$

then. Yearly values for 1981 and the period 1994–97 are: 74.8, 144.1, 149.1, 152.7 and 157.5 respectively. Monthly records are available from the Incomes Data Services (IDS) website (www.incomesdata.co.uk/index.html) (Exercise 8.1). Some pensions and UK government investments such as the Index Linked National Savings Certificates, are linked to the RPI. Most countries have their own version of a retail price index. There is also a wide range of more specialist indices which are often used in contracts to allow for inflation. Such allowances are particularly important in the construction industry because contracts can extend over several years and the choice of index can make a substantial difference to the out-turn cost.

Example 8.1
Beal (1995) argues that the construction output index (COPI) does not reflect adequately the mix of work typically undertaken by water companies in the UK. For example, mechanical and electrical plant (M&E) are more significant for water company projects than they are for construction or road-building projects. This motivated the Water Research Centre (WRC) to develop the construction cost index (CCI) for the water industry, based on stylised projects, e.g. a water pumping station consists of 120 cubic metres of concrete, 1.02 tonnes of reinforcement and three 6 kW pumps. Since its inception in 1988/89 and the date of the article, 1993/94, the CCI has stayed between the COPI, which dropped due to a recession in the construction industry, and the retail price index (RPI) which increased steadily. This feature could be explained by inclusion of a substantial M&E component. The CCI will not necessarily stay between the COPI and the RPI, but it is likely to be less volatile than the COPI.

An article in *Water News* (1995) attributed to London Economics, the company that worked in association with WRC on the CCI, suggested that companies might modify the CCI to form an objective company-specific capital cost index which would be based on their own mix of work and nationally agreed indices for specific items. For example, a major sewer contract for Northumbria Water would be: 50% labour, 17.5% plant, 7% aggregates, 7% bricks, 8% cement, 3.5% fuel, 3.5% timber and 3.5% reinforcing steel. The indices for these items are published monthly by the Department of the Environment in *The Monthly Bulletin of Construction Indices*. A national all civil engineering costs index is the EC Harris/NCE index which is published regularly in the *New Civil Engineer*.

8.2.2   Construction of an index

A price index is the ratio of the cost of a basket of goods now to its cost in some base year. In the Laspeyre formulation the basket of goods is based on typical purchases in the base year. The Paasche index makes the comparison on the basis of typical purchases in the current year. A sales manager set up an index for the cost of a car for a sales representative, several years ago. In the base year cars were replaced every three years. Tyres were used to represent parts that were expected to wear out. Two clutches were chosen to represent typical yearly expenditure on other parts. The data for the base year, 1994, and the year 1998 are:

Item	Base year (1994)		Current year (1998)	
	Quantity q_{i0}	Price/unit p_{i0}	Quantity q_{it}	Price/unit p_{it}
Car	0.33	9000	0.5	10 000
Petrol (litres)	2000	0.40	1500	0.80
Servicing (hours)	40	20	20	30
Tyres	3	40	2	60
Clutches	2	100	1	180

The formula for the Laspeyre index (LI_t) is

$$LI_t = \frac{\sum_{i=1}^{m} q_{i0}\, p_{it}}{\sum_{i=1}^{m} q_{i0}\, p_{i0}}$$

In this case the Laspeyre index for 1998 relative to a base year of 1994 is

$$\frac{0.33 \times 10\,000 + 2000 \times 0.80 + 40 \times 30 + 3 \times 60 + 2 \times 180}{0.33 \times 9000 + 2000 \times 0.40 + 40 \times 20 + 3 \times 40 + 2 \times 100} = 1.358$$

The formula for the Paasche index (PI_t) is

$$PI_t = \frac{\sum_{i=1}^{m} q_{it}\, p_{it}}{\sum q_{it}\, p_{i0}} \times 100\%$$

In this case the Paasche index for 1998 relative to 1994 is

$$\frac{0.5 \times 10\,000 + 1500 \times 0.80 + 20 \times 30 + 2 \times 60 + 1 \times 180}{0.5 \times 9000 + 1500 \times 0.40 + 20 \times 20 + 2 \times 40 + 1 \times 100} = 1.250$$

The Paasche index will usually be lower because people substitute cheaper alternatives for items which become particularly expensive. In the above example it was found more economic to replace cars every two years and hence reduce petrol and maintenance costs due to more efficient and reliable designs. A disadvantage of the Paasche index is that the content of the basket of goods has to be estimated every year. A compromise is to use a Laspeyre index and bring the basket of goods up to date every few years. You are asked to consider two indices which fall between the Laspeyre and Paasche indices in Exercise 8.2.

8.3 Time series

8.3.1 Model definition

We consider a continuous state variable measured at discrete time points, e.g. litres of wine sold per month. The discrete time points may arise because the variable has been summed over the sampling interval, as is the case for sales, or may be a sample from an underlying continuous signal. Two examples of sampling a continuous signal are temperature measurements from a chemical process, and end-of-week share prices for some company. We will assume the time interval between observations is constant, although this is not exact for calendar months. The differing numbers of days in months will have a more noticeable effect on an aggregated variable such as sales, but it will be incorporated into the monthly effect. An alternative would be to take sales/day as the state variable. Let y_t represent the value of the state variable at time t. The sequence of measurements

$$\{y_t\} \quad \text{for } t = 1, 2, \ldots, n$$

is known as a time series. We assume the observed time series is a realisation of some underlying random process

$$\{Y_t\} \quad \text{for } t = \ldots, 1, 2, \ldots$$

The mean of this random process is defined by

$$\mu_t = \mathrm{E}[Y_t]$$

where the expectation is taken across the ensemble; that is, it is the average of all possible time series that might have been generated by the random process (Figure 8.1). In principle it can vary at every time point, but this generality would not be helpful. Usually, we only have one time series and we need to assume some structure such as a linear trend or seasonal variation in the mean if we are to make useful progress. The first step in the analysis is to plot the data and identify and remove any trend or seasonal patterns, with the objective of obtaining a series which can plausibly be thought of as a realisation from a stationary random process. We can then fit probability models to

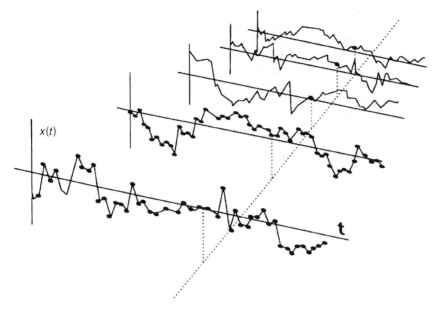

$x(t)$

t

Figure 8.1   Ensemble averaging

the stationary series, and use these to predict future values of the stationary series. The trend and seasonal effects are then replaced to give forecasts.

8.3.2   Moments of a random process, stationarity and ergodicity

A random process is stationary in the mean if the mean does not change over time, i.e.

$$E[Y_t] = \mu \quad \text{for all } t$$

It is second-order stationary if the covariance structure depends only on the time lag between variables (k), and not on absolute time (t). That is

$$E[(Y_t - \mu)(Y_{t+k} - \mu)] = \gamma(k)$$

where the autocovariance function (acvf), $\gamma(k)$, does not depend on t. The lag k can be negative, but since $\gamma(-k)$ equals $\gamma(k)$ the acvf is symmetric about 0 and there is no need to consider explicitly negative lags. The variance of the process is given by $\gamma(0)$. The autocorrelation function (acf) is defined by

$$\rho(k) = \gamma(k)/\gamma(0)$$

and like any correlations these lie between -1 and 1 (see Exercise 8.4). A plot of $\rho(k)$ against k is called the correlogram. The process is stationary if all its moments, including for example skewness, are independent of time. Apart from some rather specialist applications, it is unusual to distinguish between second-order stationarity and stationarity. In particular, if the process is Gaussian, that is $\{Y_t\}$ is multivariate normal, then second-order stationarity implies stationarity because a normal distribution is completely specified by its mean and variance–covariance matrix.

A stationary random process is ergodic in the mean if the time average over one realisation equals the ensemble average μ. It is said to be ergodic, with no restrictions, if all time averages equal the corresponding ensemble averages. For sales or other series, in real time, we nearly always assume an underlying ergodic process. Examples of random processes which need not be ergodic arise in the chemical industry. A multinational company manufactures polypropylene on five different sites. All the sites use an identical design of process plant which is operated continuously. One of the important process variables is the viscosity of the polymer before it is extruded. This viscosity is measured every hour, and we will assume that the process is stationary. If the process is ergodic, statistics from any of the sites will tend towards the same population values as the length of the time series increases: that is, there is no variability between sites. The process is not ergodic if there are systematic differences between sites, for example the means might be slightly different. The expected value of the process would then be the mean of the five site means.

In the remainder of the chapter we will assume that all time series are realisations of ergodic processes.

8.3.3   Identifying trends and seasonal effects

Three methods for identifying seasonal effects and a trend will be described for monthly data, although the same principles can be used for any other period and sampling interval. Two variants of the first method are given.

Centred moving average (CMA variant 1)

I think the simplest approach is to estimate seasonal effects, deseasonalise the data and then estimate the trend. The seasonal effects can be chosen to be multiplicative or additive. If the seasonal variation is fairly constant, despite changes in the mean level, additive effects are appropriate, but if the seasonal variation is approximately proportional to the mean then multiplicative effects should be used. The method will be described for multiplicative effects. The modification for additive effects is given in (v) below. The implied model for a linear trend and multiplicative seasonal effects, with $\{X_t\}$ stationary is:

$$Y_t = (\beta_0 + \beta_1 t + X_t) \times \text{seasonal index}$$

(i)   Calculate a centred 12-month moving average:

$$Sm(y_t) = (\tfrac{1}{2} y_{t-6} + y_{t-5} + \cdots + y_{t+5} + \tfrac{1}{2} y_{t+6})/12$$

Thus, if the data run from January 1980 to July 1995, as they do in Table-WSozwine, the first 12-month average will be centred on July 1980 and the last will be centred on January 1995. The reasoning behind the formula given is that the average

$$(y_{t-6} + y_{t-5} + \cdots + y_{t+5})/12$$

corresponds to the average value at time $t - \tfrac{1}{2}$, and the average

$$(y_{t-5} + y_{t-4} + \cdots + y_{t+6})/12$$

corresponds to time $t + \frac{1}{2}$. The formula given is the average of these two averages and corresponds to time t.

(ii) Calculate monthly ratios for t equal 7 up to $n - 6$:

$$y_t / Sm(y_t)$$

(iii) If there are 15 years of data there will now be 14 ratios for each month. Average these 14 ratios to obtain an unadjusted index for each month.

(iv) Average the 12 unadjusted indices, and divide each by this average to obtain an adjusted index for each month. The average of these adjusted indices is exactly 1, so if they are applied to a series of constant values the total will be unchanged.

(v) The data are deseasonalised by dividing by the appropriate, adjusted, indices, e.g. the deseasonalised January 1980 datum is the original datum divided by the January index. The method described gives multiplicative seasonal indices. An alternative is to assume additive seasonal effects: calculate differences $y_t - Sm(y_t)$; calculate average differences for each month and adjust these so they average 0; and deseasonalise by subtracting the additive effects.

(vi) Plot the deseasonalised data, and consider fitting a trend. Trend curves can be fitted by least squares, but the standard statistical tests should not be relied on because the errors are likely to be correlated. We should be wary of extrapolating trends, although it may be reasonable to argue that a linear trend changes relatively slowly when making short-term forecasts. In such cases it might be preferable to fit a local trend to later values instead of an overall trend. Polynomials such as,

$$y_t = \beta_0 + \beta_1 t + \beta_2 t^2$$

are sometimes fitted as a trend curve, but apart from the special case of a straight line, are rarely suitable for extrapolation (see Exercise 8.7). The modified exponential and logistic curves tend towards finite upper limits, and might be suitable for sales of an innovative product, e.g. CD players when they were first marketed. The equation for the modified exponential curve is

$$y_t = a - b r^t \quad \text{where } 0 < r < 1$$

If y_t is replaced by $\ln y_t$ this is known as the Gompertz curve. The equation for the logistic curve is

$$y_t = a / (1 + b e^{-ct})$$

The least squares equations for the parameters in these two curves have to be solved iteratively. Minitab fits the logistic model and an exponential growth model (Exercise 8.5).

The estimated trend can be removed by subtracting the trend value from the deseasonalised data. Forecasts can be made by extrapolating trend curves into the future and then applying seasonal effects. However, there is often no physical basis for choosing a trend curve and several plausible curves may lead to widely different forecasts. In such cases we could present a range of scenarios.

Centred moving average (Minitab Decomposition)

Either a multiplicative model

$$Y_t = (\beta_0 + \beta_1 t) \times \text{seasonal} + X_t$$

or an additive model

$$Y_t = \beta_0 + \beta_1 t + \text{seasonal} + X_t$$

can be fitted using the Minitab Time Series Decomposition routine which involves the following steps (Minitab Inc., 1997).

(i)   MINITAB fits a trend line to the data, using least squares regression.
(ii) The trend is removed by either dividing the data by the trend component (multiplicative model) or subtracting the trend component from the data (additive model).
(iii) Then, the detrended data are smoothed using a centred moving average with a length equal to the length of the seasonal cycle. When the seasonal cycle length is an even number, a two-step moving average is required to synchronise the moving average correctly.
(iv) Once the moving average is obtained, it is either divided into (multiplicative model) or subtracted from (additive model) the detrended data to obtain what are often referred to as raw seasonals.
(v)   Within each seasonal period, the median value of the raw seasonals is found. The medians are also adjusted so that their mean is one (multiplicative model) or their sum is zero (additive model). These adjusted medians constitute the seasonal indices.
(vi) The stationary series $\{x_t\}$ is obtained by subtracting the product of the trend value with the appropriate seasonal index from $\{y_t\}$.

Example 8.2

The monthly seasonal indices for the Australian dry white wine series from January 1980 until December 1994 using CMA variant 1 and Minitab Decomposition are given in Table 8.1. There is no substantial difference between the two sets of indices. The trend fitted with CMA variant 1 is:

$$\text{deseasonalised sales} = 2818 + 4.62t$$

where t is the number of months from the beginning of the record, January 1980 being taken as one. The standard deviation for the coefficient of t is 0.47. The standard deviation of the residuals is 329 and the lag 12 auto-correlation of the residuals is 0.14. The trend given by the Minitab Decomposition is:

$$\text{deseasonalised sales} = 2754 + 5.38t$$

The standard deviation of the residuals is 334 and the lag 12 autocorrelation is 0.20. The residuals have smaller correlations at other lags, so it is a simplifying approximation to assume that $\{X_t\}$ is a sequence of independent variables with mean 0. If future values of X_t are assumed to be equal to 0, predictions are made in the same way with both models. In general, predictions would be slightly different because the predicted value of X_t would be added before multiplying

Table 8.1   Seasonal indices for the Australian dry white wine series: January 1980 until December 1994

	CMA variant 1	Minitab Decomposition
January	0.7187	0.7243
February	0.8555	0.8175
March	0.9529	0.9557
April	0.8843	0.8988
May	0.9128	0.8947
June	0.8830	0.8625
July	1.0250	1.0242
August	1.1260	1.1564
September	0.9685	0.9980
October	1.0352	1.0261
November	1.2604	1.2699
December	1.3777	1.3718

by the seasonal index with CMA variant 1, and afterwards using the Minitab Decomposition. The prediction for January 1995 using CMA variant 1 is

$$(2818 + 4.62 \times 181) \times 0.7187 = 2626$$

and using Minitab Decomposition it is

$$(2754 + 5.38 \times 181) \times 0.7243 = 2700$$

The predictions and actual sales for January 1995 until July 1995 are given below.

	CMA variant 1	Minitab Decomposition	Actual sales
January	2626	2700	2367
February	3130	3052	3819
March	3491	3573	4067
April	3244	3365	4022
May	3352	3354	3937
June	3247	3238	4365
July	3774	3851	4290

There are several criteria that can be used to assess the performance of different methods for making predictions. These include the mean error, the mean of the absolute values of the percentage errors (MAPE), the mean of the absolute values of the errors (MAE) and the mean squared error (MSE). For the CMA variant 1 predictions these are -4003, 16.4%, 646 and 476 752 respectively. For the Minitab Decomposition the corresponding statistics of the errors are -3734, 16.2%, 629 and 453 943. In fact the sales were uncharacteristically high at the beginning of 1995, as can be seen in Figure 8.2(a), and no forecasting method based on past values of the time series only would predict that. In practice there is usually little difference between CMA variant 1 and the Minitab Decomposition, which is described in more detail in Makridakis *et al.* (1998). Fitting the trend first could lead to a slight bias if there is not an exact number of years and the record is short (Exercise 8.6). Minitab provides

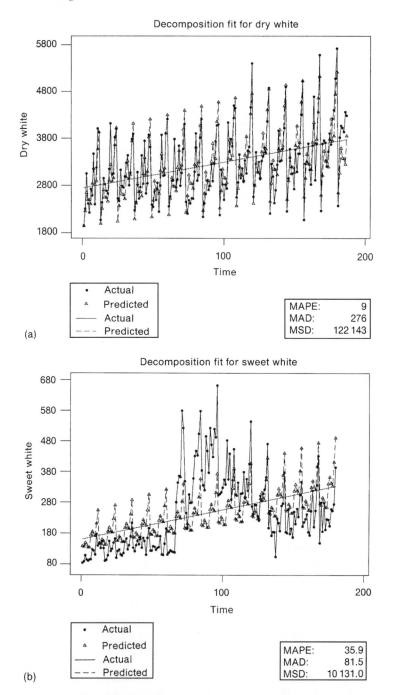

Figure 8.2   Minitab graphical analysis of sales of Australian wine: (a) dry white wine; (b) sweet white wine; (c) components of sweet white wine time series; (d) seasonal analysis of sweet white wine time series

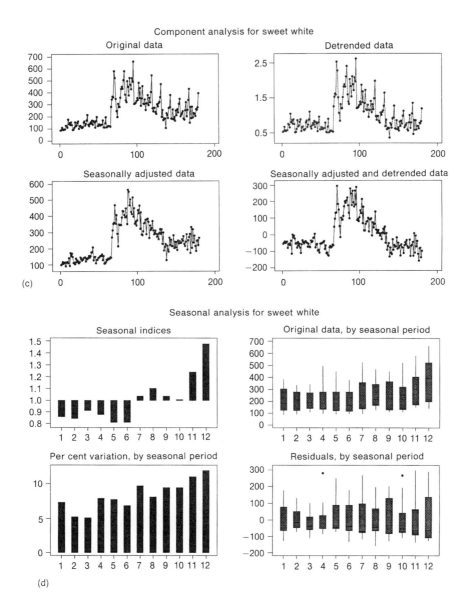

Component analysis for sweet white

(c)

Seasonal analysis for sweet white

(d)

a very useful set of graphs with the Decomposition routine. These are shown for the sweet white wine time series in Figure 8.2(b), (c) and (d).

Standardisation method

We begin by fitting and removing any trend, after which we estimate a mean and standard deviation for each month. It requires many years of data for the 24

Table 8.2   Monthly means and standard deviations for the Australian dry white wine series with the trend removed: January 1980 until December 1994

	Mean	Standard deviation
January	−931	299
February	−473	259
March	−135	313
April	−382	321
May	−319	334
June	−372	342
July	78	308
August	342	530
September	−72	342
October	138	254
November	856	263
December	1270	456

estimates to have acceptable precision, but a compromise is to smooth them. The procedure involves the following steps.

(i)   Calculate the means for each calendar year and plot them. If appropriate, fit a trend curve and then subtract the trend from the data.
(ii) For each month, calculate the sample mean and standard deviation.
(iii) The data are deseasonalised by subtracting the monthly means and dividing by the monthly standard deviations.

Example 8.3
The yearly mean monthly sales for the first 15 years of the Australian dry white wine series are: 2853.9, 3065.2, 3078.7, 3007.8, 2985.2, 3116.3, 2963.8, 3076.6, 3351.5, 3441.9, 3425.0, 3330.0, 3380.3, 3598.2 and 3929.2. If time (t) is measured in months from the beginning of the record with January 1980 as 1, then the yearly means correspond to times 6.5, 18.5, ..., 174.5. A regression analysis gives the trend line

$$y = 2812.5 + 4.726t$$

and the detrended time series is

$$w_t = y_t - (2812.5 + 4.726t)$$

The mean and standard deviation of the 15 w_t values for each month are given in Table 8.2. The deseasonalised and detrended series $\{x_t\}$ is found from w_t by subtracting the appropriate monthly mean and dividing by the monthly standard deviation. For example, w_{12} and w_{169} are 1055 and −1346 respectively and are a December and a January sales figure respectively. Therefore

$$x_{12} = \frac{1055 - 1270}{456} = -0.471$$

$$x_{169} = \frac{-1346 - (-931)}{299} = -1.388$$

The mean and standard deviation of the supposed stationary time series $\{x_t\}$ are approximately 0 and 1 respectively.

Harmonic cycles

The centred moving average method estimates 12 indices, and the standard-isation method estimates 24 parameters. If it is reasonable to suppose that monthly variation is part of a harmonic cycle, we can fit the three-parameter model

$$Y_t = \beta_0 + \beta_1 \cos(2\pi t/12) + \beta_2 \sin(2\pi t/12) + X_t$$

where X_t represents random variation with zero mean, and the frequency $2\pi/12$ corresponds to 1 cycle per year. This can be done with a standard multiple regression routine. The predictor variables are calculated as the cosine and sine of $(2\pi t/12)$ as t goes from 1 to n and the residuals are the deseasonalised time series. A trend can be incorporated by adding, for example, $\beta_3 t$. A more complex seasonal pattern could be allowed for by adding more harmonics. Two cycles per year has a frequency of $4\pi/12$ and we would add $\beta_3 \cos(4\pi t/12)$ and $\beta_4 \sin(4\pi t/12)$ to the model. However, we should be cautious about adding extra harmonics simply because at least one of the pair of coefficients appears to be marginally significant. The X_t are likely to be positively correlated, and the standard deviations of coefficients, and hence *P*-values, given by a standard regression routine, will be smaller than they should be. The slight lack of efficiency of the standard regression, compared with a weighted least squares, is unimportant if we have a reasonable length series.

The method does not allow for any seasonal variation in standard deviation, but if the original series has a constant coefficient of variation throughout the year, the series of logarithms will have a constant standard deviation. Another possible advantage of working with logarithms is that forecast and simulated data will always be positive.

Example 8.4
The data in Table 8.3 are the total number of passengers per month (y_t) using Newcastle International Airport between January 1996, when t equals 1, and December 1998, when t equals 36. The figures exclude passengers in transit.

Table 8.3 Total number of passengers per month using Newcastle International Airport between January 1996, $t = 1$, and December 1998, $t = 36$.

Month	t	Passengers	t	Passengers	t	Passengers
January	1	126 980	13	130 794	25	141 833
February	2	133 272	14	130 998	26	144 218
March	3	156 607	15	156 550	27	165 617
April	4	156 364	16	159 392	28	172 858
May	5	237 614	17	265 958	29	301 279
June	6	277 394	18	292 592	30	332 805
July	7	275 984	19	298 382	31	355 889
August	8	283 705	20	307 514	32	353 641
September	9	277 503	21	295 271	33	340 509
October	10	244 126	22	266 315	34	314 022
November	11	152 051	23	172 001	35	173 708
December	12	134 314	24	143 520	36	153 479

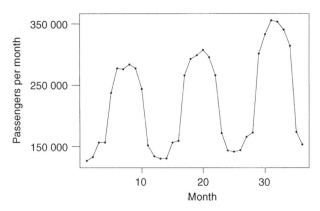

The time series is plotted in Figure 8.3. The regression

$$y_t = 195\,098 - 73\,573\cos(2\pi t/12) - 66\,702\sin(2\pi t/12) + 1503.8t$$

has an R^2 of 91.5%. The residuals from the regression are considered to be a time series $\{x_t\}$ from a stationary random process. The mean and standard deviation of $\{x_t\}$ are 0 and 22 760. The mean and standard deviation of $\{y_t\}$ are 222 918 and 78 181 respectively.

8.3.4   Estimation of the correlogram

Assume a time series $\{x_t\}$, $t = 1, \ldots, n$, is a realisation of a stationary stochastic process $\{X_t\}$, any trend or seasonal pattern in the mean or standard deviation of the original series having been removed. We can now concentrate on estimating the correlation structure, which will be helpful for making predictions when, for example, consecutive values of $\{X_t\}$ tend to be close together. The mean (μ) is estimated by

$$\bar{x} = \sum_{t=1}^{n} x_t/n$$

and the acvf ($\gamma(k)$) is estimated for lags $k = 0, \ldots, L$ by

$$c(k) = \sum_{t=1}^{n-k} (x_t - \bar{x})(x_{t+k} - \bar{x})/n$$

The divisor n is commonly used because it guarantees the sample acf will be between -1 and 1. The slight bias,

$$E[c(k)] \simeq \gamma(k)(1 - k/n)$$

is negligible in applications. The acf ($\rho(k)$) is estimated by

$$r(k) = c(k)/c(0)$$

and a plot of $r(k)$ against k is known as the sample correlogram. If $\{X_t\}$ is just a sequence of independent, identically distributed, random variables, which is conveniently referred to as discrete white noise (DWN), then

$$E[r(k)] \simeq -1/n$$

$$\text{var}(r(k)) \simeq 1/n$$

It is helpful to add to the correlogram a pair of lines, corresponding to 5% significance, parallel to the lag axis at distances

$$-1/n \pm 2/\sqrt{n}$$

from it. If the process is DWN we expect an average of 1 in 20 $r(k)$ to lie outside these limits. Minitab also provides the ratios of $r(k)$ to their estimated standard deviations

$$\sqrt{\widehat{\text{var}(r(k))}} = \left\{ \left[1 + 2 \sum_{j=1}^{k-1} (r(j))^2 \right]^{1/2} \middle/ n \right\}^{1/2}$$

and the Ljung–Box Q (LBQ) statistic

$$Q(k) = n(n+2) \sum_{j=1}^{k} (r(j))^2 \middle/ (n-j)$$

If $\{X_t\}$ is DWN then $Q(k)$ has an approximate chi-square distribution with $k-1$ degrees of freedom. Large values of $Q(k)$ are evidence against the hypothesis that $\{x_t\}$ is a realisation of DWN, but if the time series is long we may have evidence of non-zero autocorrelations which are of no practical significance. If $\{X_t\}$ are autocorrelated the estimators $r(k)$ will be correlated, e.g. if $\rho(3)$ is under-estimated it is likely that $\rho(2)$ and $\rho(1)$ will be as well.

Example 8.5
The correlogram for the deseasonalised and detrended dry white wine series is shown in Figure 8.4(a). The $r(12)$ value of 0.20 is statistically significant, which suggests some slight seasonality remains. However, it will not have any appreciable effect on forecasts. The LBQ statistic is significant, but the correlations are small and of little practical significance in a forecasting application. The correlogram for the sweet white wine in Figure 8.4(b) is much more interesting. A relatively simple ARMA model accounts for the correlation structure and can be used to substantially improve forecasts.

8.3.5 ARMA models

Autoregressive moving average (ARMA) models represent stationary random processes by linear combinations of DWN. Here, $\{E_t\}$ will denote a sequence of DWN with a mean of zero and a variation σ_E^2. It follows from the definition of DWN that all autocovariances of $\{E_t\}$ are zero.

Moving average processes

A process $\{X_t\}$ defined by

$$X_t = \mu + \beta_0 E_t + \beta_1 E_{t-1} + \cdots + \beta_q E_{t-q}$$

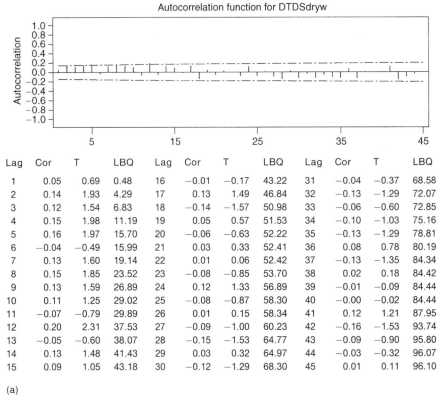

Figure 8.4 data table:

Lag	Cor	T	LBQ	Lag	Cor	T	LBQ	Lag	Cor	T	LBQ
1	0.05	0.69	0.48	16	−0.01	−0.17	43.22	31	−0.04	−0.37	68.58
2	0.14	1.93	4.29	17	0.13	1.49	46.84	32	−0.13	−1.29	72.07
3	0.12	1.54	6.83	18	−0.14	−1.57	50.98	33	−0.06	−0.60	72.85
4	0.15	1.98	11.19	19	0.05	0.57	51.53	34	−0.10	−1.03	75.16
5	0.16	1.97	15.70	20	−0.06	−0.63	52.22	35	−0.13	−1.29	78.81
6	−0.04	−0.49	15.99	21	0.03	0.33	52.41	36	0.08	0.78	80.19
7	0.13	1.60	19.14	22	0.01	0.06	52.42	37	−0.13	−1.35	84.34
8	0.15	1.85	23.52	23	−0.08	−0.85	53.70	38	0.02	0.18	84.42
9	0.13	1.59	26.89	24	0.12	1.33	56.89	39	−0.01	−0.09	84.44
10	0.11	1.25	29.02	25	−0.08	−0.87	58.30	40	−0.00	−0.02	84.44
11	−0.07	−0.79	29.89	26	0.01	0.15	58.34	41	0.12	1.21	87.95
12	0.20	2.31	37.53	27	−0.09	−1.00	60.23	42	−0.16	−1.53	93.74
13	−0.05	−0.60	38.07	28	−0.15	−1.53	64.77	43	−0.09	−0.90	95.80
14	0.13	1.48	41.43	29	0.03	0.32	64.97	44	−0.03	−0.32	96.07
15	0.09	1.05	43.18	30	−0.12	−1.29	68.30	45	0.01	0.11	96.10

(a)

Figure 8.4 Autocorrelation functions for (a) the dry white wine series; (b) the sweet white wine series, after the trend and seasonal effects have been removed

is a *moving average process* of order q (MA(q)), but be aware that some books, and Minitab, define the MA process with minus signs for the β_k. Since $E[E_t] = 0$ for any time, $E[X_t] = \mu$. The autocovariance is given by

$$\gamma(k) = E[(X_t - \mu)(X_{t+k} - \mu)]$$
$$= E[(E_t + \cdots + \beta_q E_{t-q})(E_{t+k} + \cdots + \beta_q E_{t+k-q})]$$
$$= \sigma_E^2 \sum_{i=0}^{q-k} \beta_i \beta_{i+k} \quad \text{for } k \leq q$$

The parameter β_0 is usually set at 1. If k exceeds q, then $\gamma(k) = 0$. The process is stationary for any finite value of q, but see Exercise 8.10 for an example of the invertibility condition.

Example 8.6
An altimeter in an aircraft estimates height every second. Errors $\{E_t\}$ are independent and normally distributed with a mean of 0 and a standard

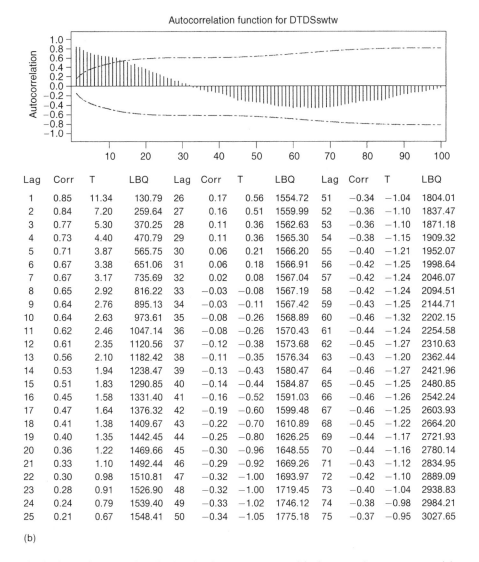

Autocorrelation function for DTDSswtw

Lag	Corr	T	LBQ	Lag	Corr	T	LBQ	Lag	Corr	T	LBQ
1	0.85	11.34	130.79	26	0.17	0.56	1554.72	51	−0.34	−1.04	1804.01
2	0.84	7.20	259.64	27	0.16	0.51	1559.99	52	−0.36	−1.10	1837.47
3	0.77	5.30	370.25	28	0.11	0.36	1562.63	53	−0.36	−1.10	1871.18
4	0.73	4.40	470.79	29	0.11	0.36	1565.30	54	−0.38	−1.15	1909.32
5	0.71	3.87	565.75	30	0.06	0.21	1566.20	55	−0.40	−1.21	1952.07
6	0.67	3.38	651.06	31	0.06	0.18	1566.91	56	−0.42	−1.25	1998.64
7	0.67	3.17	735.69	32	0.02	0.08	1567.04	57	−0.42	−1.24	2046.07
8	0.65	2.92	816.22	33	−0.03	−0.08	1567.19	58	−0.42	−1.24	2094.51
9	0.64	2.76	895.13	34	−0.03	−0.11	1567.42	59	−0.43	−1.25	2144.71
10	0.64	2.63	973.61	35	−0.08	−0.26	1568.89	60	−0.46	−1.32	2202.15
11	0.62	2.46	1047.14	36	−0.08	−0.26	1570.43	61	−0.44	−1.24	2254.58
12	0.61	2.35	1120.56	37	−0.12	−0.38	1573.68	62	−0.45	−1.27	2310.63
13	0.56	2.10	1182.42	38	−0.11	−0.35	1576.34	63	−0.43	−1.20	2362.44
14	0.53	1.94	1238.47	39	−0.13	−0.43	1580.47	64	−0.46	−1.27	2421.96
15	0.51	1.83	1290.85	40	−0.14	−0.44	1584.87	65	−0.45	−1.25	2480.85
16	0.45	1.58	1331.40	41	−0.16	−0.52	1591.03	66	−0.46	−1.26	2542.24
17	0.47	1.64	1376.32	42	−0.19	−0.60	1599.48	67	−0.46	−1.25	2603.93
18	0.41	1.38	1409.67	43	−0.22	−0.70	1610.89	68	−0.45	−1.22	2664.20
19	0.40	1.35	1442.45	44	−0.25	−0.80	1626.25	69	−0.44	−1.17	2721.93
20	0.36	1.22	1469.66	45	−0.30	−0.96	1648.55	70	−0.44	−1.16	2780.14
21	0.33	1.10	1492.44	46	−0.29	−0.92	1669.26	71	−0.43	−1.12	2834.95
22	0.30	0.98	1510.81	47	−0.32	−1.00	1693.97	72	−0.42	−1.10	2889.09
23	0.28	0.91	1526.90	48	−0.32	−1.00	1719.45	73	−0.40	−1.04	2938.83
24	0.24	0.79	1539.40	49	−0.33	−1.02	1746.12	74	−0.38	−0.98	2984.21
25	0.21	0.67	1548.41	50	−0.34	−1.05	1775.18	75	−0.37	−0.95	3027.65

(b)

deviation of 10 m. The display in the passenger cabin is a moving average, with equal weights, of the current and last three estimates. Suppose the aircraft is cruising at a height μ. The display is

$$X_t = \mu + 0.25E_t + 0.25E_{t-1} + 0.25E_{t-2} + 0.25E_{t-3}$$

The mean of X_t is μ, and the variance is $25\,\text{m}^2$. The autocorrelations at lags 1, 2 and 3 are 0.75, 0.50 and 0.25 respectively. The autocorrelations at higher lags are all 0. A realisation of the process, and the corresponding sample correlogram are shown in Figure 8.5(a) and (b).

Now suppose we have a time series $\{x_t\}$ from a stationary process. If the correlogram, after lag q, appears to be consistent with sampling variation,

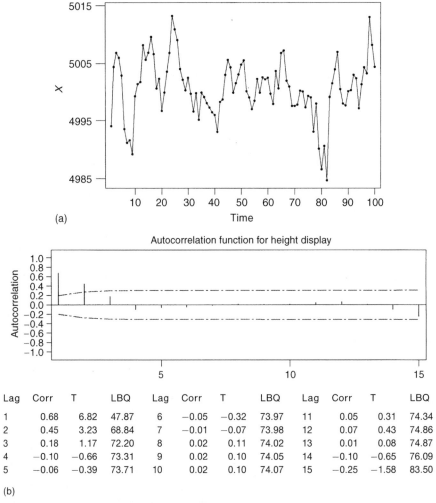

Figure 8.5 Altimeter display in passenger cabin of aircraft: (a) time series plot; (b) auto-correlation function calculated from time series

about $-1/n$, an MA(q) model would be suitable. It cannot be fitted by multiple regression because we do not know $E_{t-1}, \ldots, E_{t-q}$. The method of moments (MOM) estimates given by equating the sample and theoretical correlograms up to lag q are inefficient, although they do provide good starting values for the following iterative least squares estimation (LSE) procedure, described for q equal to 2.

(i)   Estimate μ by $\bar{x}$.
(ii) Guess, or use MOM estimates for, β_1 and β_2. Denote these estimates by $\hat{\beta}_1$ and $\hat{\beta}_2$, and write e_t for estimates of the values taken by E_t.

(iii) Take e_1 and e_2 equal to zero. Calculate

$$e_3 = x_3 - \bar{x} - \hat{\beta}_1 e_2 - \hat{\beta}_2 e_1$$

and hence

$$e_4 = x_4 - \bar{x} - \hat{\beta}_1 e_3 - \hat{\beta}_2 e_2$$

and so on down to

$$\vdots$$

$$e_n = x_n - \bar{x} - \hat{\beta}_1 e_{n-1} - \hat{\beta}_2 e_{n-2}$$

Then calculate the error sum of squares

$$\psi = \sum e_t^2$$

(iv) Use a numerical optimisation routine to minimise ψ with respect to $\hat{\beta}_1$ and $\hat{\beta}_2$. The LSE are the values of $\hat{\beta}_1$ and $\hat{\beta}_2$ at this minimum.
(v) Estimate σ_E^2 by $\psi/(n-3)$.
(vi) Use a histogram of the e_t to choose a plausible distribution for the E_t.

A refinement would be to treat e_1 and e_2 as additional parameters to be estimated.

Example 8.6 (continued)
The Minitab estimates of the parameters of an $MA(3)$ distribution, made from the realisation shown in Figure 8.5(a), are:

$MA1 - 0.9331$
$MA2 - 0.8953$
$MA3 - 0.9257$
Mean 5000.57
$MS = 7.678$ on 96 degrees of freedom

These are based on a definition of an $MA(q)$ process with $\beta_0 = 1$, and negative signs before $\beta_1, \ldots, \beta_k$. The model defined in this example has been estimated as

$$\hat{x}_t = 5000.57 + 0.25E_t + 0.23E_{t-1} + 0.22E_{t-2} + 0.23E_{t-3}$$

with a standard deviation of errors equal to

$$\sqrt{7.678}/0.25 = 11.08$$

Autoregressive processes

A process $\{X_t\}$ is autoregressive of order p (AR(p)) if:

$$(X_t - \mu) = \alpha_1(X_{t-1} - \mu) + \cdots + \alpha_p(X_{t-p} - \mu) + E_t$$

The mean μ is estimated by $\bar{x}$. It will make the algebra simpler, and nothing will be lost, if we assume the process has a mean of 0. We will start by deriving the acvf for an important special case. This is the AR(1) process, which is also known as a Markov process because it has the Markov property (Section 8.7.1). The AR(1) process is defined by

$$X_t = \alpha X_{t-1} + E_t$$

Repeated substitution gives

$$X_t = \alpha(\alpha X_{t-2} + E_{t-1}) + E_t = \alpha^2(X_{t-3} + E_{t-2}) + \alpha E_{t-1} + E_t$$

and so on until

$$X_t = E_t + \alpha E_{t-1} + \alpha^2 E_{t-2} + \cdots + \alpha^t E_0$$

The variance of X_t is given by

$$\mathrm{var}(X_t) = \sigma_E^2(1 + \alpha^2 + \alpha^4 + \cdots + \alpha^{2t})$$

If we use the standard result for the sum of a geometric progression (Appendix 1), we obtain

$$\sigma_X^2 = \sigma_E^2 \frac{1 - \alpha^{2(t+1)}}{1 - \alpha^2}$$

If α is greater than, or equal to, 1 the variance will increase over time and the process is not stationary. If α is less than 1, α^{2t} will rapidly tend to zero as t increases. It is usual to assume t is large enough for α^{2t} to be negligible, and then the process is stationary. The autocovariance at lag k is,

$$\gamma(k) = E[X_t X_{t+k}]$$
$$= E[(E_t + \alpha E_{t-1} + \alpha^2 E_{t-2} + ...)(E_{t+k} + \alpha E_{t+k-1} + \alpha^2 E_{t+k-2} + ...)]$$
$$= E\left[\left(\sum \alpha^i E_{t-i}\right)\left(\sum \alpha^j E_{t+k-j}\right)\right]$$

where the summations are formally from $i = 0$ to $i = \infty$. Since the E_t is DWN the only non-zero expectations are when $t - i = t + k - j$, and

$$\gamma(k) = \sigma_E^2 \sum \alpha^i \alpha^{k+i}$$
$$= \alpha^k \sigma_E^2 \sum \alpha^{2i}$$
$$= \alpha^k \sigma_E^2/(1 - \alpha^2)$$

It follows that the acf is

$$\rho(k) = \alpha^k \quad \text{for } |\alpha| < 1, \; k = 0, \pm 1, \ldots$$

If α is positive the correlogram decays exponentially. This can be quite realistic for many time series. If α is negative the correlogram oscillates between positive and negative when decaying. This could be appropriate if there is overcorrection of some process after each sample is taken. Realisations and the acfs of AR(1) processes with α equal to 0.9 and -0.9 are shown in Figure 8.6(a) and (b).

A more convenient way of deriving the acf for higher-order AR processes is to assume they are stationary, and then set up, and solve, the Yule–Walker equations. The conditions for stationarity can be deduced from their solutions. We will demonstrate the technique by deriving the acf, and variance, of an AR(2) process defined by

$$X_t = \alpha_1 X_{t-1} + \alpha_2 X_{t-2} + E_t$$

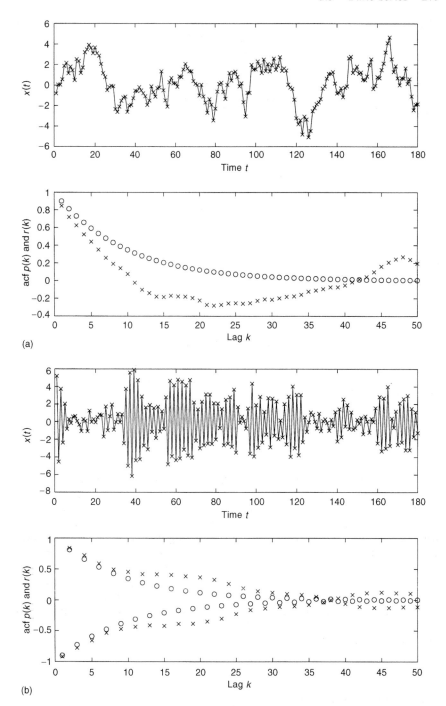

Figure 8.6   Realisation and theoretical (○) and sample (×) acfs of an AR(1) process with (a) $\alpha = 0.9$; (b) $\alpha = -0.9$

Multiply both sides of the defining equations by X_{t-k}, take expectation, and divide by σ_X^2, to obtain a linear difference equation.

$$\rho(k) = \alpha_1 \, \rho(k-1) + \alpha_2 \, \rho(k-2) \quad \text{for } 0 < k$$

This result relies on k being positive, in which case E_t is independent of the earlier variable X_{t-k}. A linear difference equation can be solved by trying a solution of the form

$$\rho(k) = \theta^k$$

For the AR(2) process we only have to solve the quadratic equation

$$\theta^2 - \alpha_1 \theta - \alpha_2 = 0$$

The standard solution of this equation is

$$\theta = \left(\alpha_1 \pm \sqrt{\alpha_1^2 + 4\alpha_2} \right) / 2$$

Since $\rho(k) = \theta^k$, the process is stationary if, and only if, $|\theta| < 1$. This is satisfied if:

$$\alpha_1 + \alpha_2 < 1 \quad \alpha_1 - \alpha_2 > -1 \quad \alpha_2 > -1$$

See Exercise 8.8 (after Chatfield, 1989). The general solution is

$$\rho(k) = A_1 \theta_1^k + A_2 \theta_2^k$$

where θ_1 and θ_2 are the two roots, which may be complex, of the equation and A_1, A_2 are arbitrary constants. We now need two conditions to determine the constants. The first is the fact that $\rho(0) = 1$. The second is obtained from the Yule–Walker equation with $k = 1$, and the equality of $\rho(k)$ and $\rho(-k)$,

$$\rho(1) = \alpha_1 \rho(0) + \alpha_2 \rho(-1)$$
$$= \alpha_1 + \alpha_2 \rho(1)$$

If θ_1 and θ_2 are complex the acf is a damped harmonic. A realisation from an AR(2) process with α_1 and α_2 equal to 1 and -0.5 respectively is shown in Figure 8.7, together with its acf. The variance of $\{X_t\}$ can be found by taking the variance of both sides of the defining equation and using the expression for $\rho(k)$ (see Exercise 8.9). It is not possible to give $\rho(k)$ explicitly in terms of the parameters of the model if p exceeds 4. This is a consequence of the renowned proof, by Galois (1802–23), that it is not possible to construct a formula for solving a general quintic equation, or higher-order polynomial equation, in terms of its coefficients. Although it is harder to find the acf for an AR process than for an MA process, fitting the models is relatively easy. The LSE can be found by multiple regression, which has the advantage that we can add other predictor variables.

ARMA models

A process $\{X_t\}$ of the form:

$$X_t = \alpha_1 X_{t-1} + \cdots + \alpha_p X_{t-p} + E_t + \beta_1 E_{t-1} + \cdots + \beta_q E_{t-q}$$

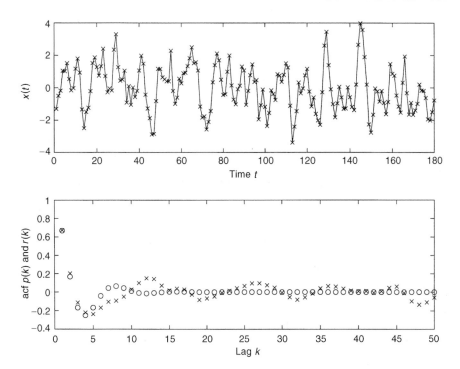

Figure 8.7   Realisation and theoretical ($\circ$) and sample ($\times$) acfs of an AR(2) process with $\alpha_1 = 1$ and $\alpha_2 = -0.5$

is ARMA(p, q) with a zero mean. The rationale for introducing it is that it may be possible to model a series with fewer parameters than with an AR or MA on its own. It is often written more concisely by introducing the backward shift operator B, defined by:

$$BX_t = X_{t-1}$$

$$B^2 X_t = B(BX_t) = BX_{t-1} = X_{t-2}$$

and generally

$$B^m X_t = X_{t-m}$$

Then

$$\phi(B)X_t = \theta(B)E_t$$

where $\phi(B)$ and $\theta(B)$ are polynomials in B of order p and q respectively. Formal use of B reduces the algebra involved in derivations. For example, for an AR(1) process

$$(1 - \alpha B)X_t = E_t$$

$$X_t = (1 - \alpha B)^{-1} E_t$$

which is expanded as a Taylor series

$$= (1 + \alpha B + (\alpha B)^2 + \cdots)E_t$$
$$= E_t + \alpha E_{t-1} + \alpha^2 E_{t-2} + \cdots$$

Model order

ARMA models are simple approximations to complex physical processes. We need to choose a suitable order for the process, but should remember that there is no correct answer. The residuals, after fitting the chosen ARMA model, should appear to be uncorrelated and have a minimum, or near minimum, standard deviation amongst the models tried. The chosen model should also be as simple as possible (see Exercise 8.11). An exponential decay of the acf suggests an AR(1) model and a damped harmonic indicates that an AR(2) might be suitable. An apparent cut-off after lag q is consistent with an $MA(q)$ process. The sample partial autocorrelation function (pacf) is a plot of α_k, when fitting an AR(k) model, against k. If an AR(p) process is suitable the pacf should be close to 0 after lag p.

Example 8.7
The correlogram for the sweet white wine in Figure 8.4(b) has a typical shape of an AR(2) model. The pacf shown in Figure 8.8 also indicates an AR(2) model is suitable. The estimates of the parameters in the model, from the Minitab ARMA routine, are:

	Coef.	St. Dev.
AR 1	0.4721	0.0673
AR 2	0.4498	0.0676
Constant	−1.063	3.619
Mean	−13.61	46.31

Residual $MS = 2311$ on 177 degrees of freedom.
    If the model is written as

$$(X_t - \mu) = \alpha_1(X_{t-1} - \mu) + \alpha_2(X_{t-2} - \mu) + E_t$$

then $\hat{\mu}$, $\hat{\alpha}_1$ and $\hat{\alpha}_2$ are −13.61, 0.4721 and 0.4498 respectively. The acf of the residuals (Figure 8.9(a)) is consistent with $\{E_t\}$ being DWN, and the estimate of the variance of the $\{E_t\}$ is 2311. Hence σ_E is estimated as 48.1. Since this is substantially less than the standard deviation of $\{X_t\}$, which is 100.9, use of the model will substantially reduce the width of prediction intervals. A histogram of the residuals is shown in Figure 8.9(b).
    The model can be rewritten as

$$X_t = \mu(1 - \alpha_1 - \alpha_2) + \alpha_1 X_{t-1} + \alpha_2 X_{t-2} + E_t$$

and the Minitab constant, −1.063, is $\hat{\mu}(1 - \hat{\alpha}_1 - \hat{\alpha}_2)$. The model can also be fitted using a standard multiple regression routine. The predictor

Partial autocorrelation function for DTDSswtw

Lag	PAC	T	Lag	PAC	T	Lag	PAC	T
1	0.85	11.34	16	−0.16	−2.14	31	0.04	0.55
2	0.43	5.74	17	0.15	2.00	32	−0.07	−0.92
3	0.03	0.39	18	−0.12	−1.65	33	−0.07	−0.90
4	−0.00	−0.01	19	−0.04	−0.60	34	−0.01	−0.09
5	0.09	1.26	20	−0.03	−0.45	35	−0.03	−0.37
6	0.00	0.04	21	−0.06	−0.85	36	0.05	0.73
7	0.09	1.23	22	−0.10	−1.35	37	−0.02	−0.32
8	0.07	0.88	23	0.07	0.89	38	0.10	1.36
9	0.04	0.54	24	−0.05	−0.71	39	−0.01	−0.09
10	0.06	0.83	25	−0.07	−0.94	40	−0.00	−0.06
11	−0.01	−0.09	26	−0.04	−0.53	41	−0.07	−0.89
12	0.01	0.15	27	0.04	0.51	42	−0.05	−0.71
13	−0.13	−1.76	28	−0.08	−1.09	43	−0.06	−0.75
14	−0.08	−1.02	29	0.05	0.64	44	−0.09	−1.19
15	0.05	0.65	30	−0.05	−0.71	45	−0.07	−0.94

Figure 8.8   pacf for sweet white wine series (trend and seasonal effects having been removed)

variables are obtained by moving the column $\{x_t\}$ down by 1 and 2 cells respectively to give the lagged sequences $\{x_{t-1}\}$ and $\{x_{t-2}\}$. The estimated model is:

$$x_t = -0.462 + 0.471x_{t-1} + 0.449x_{t-2}$$

Making predictions

The Minitab Decomposition for the sweet white wine series gives

$$y_t = 160.50 + 0.956\,93t$$

with seasonal indices of 0.857, 0.841, 0.911, 0.876, 0.809, 0.808, 1.038, 1.101, 1.038, 1.006, 1.239, 1.476 for January, ..., December. The AR(2) model fitted to the stationary series $\{x_t\}$ is

$$x_t = -1.063 + 0.4721x_{t-1} + 0.4498x_{t-2} + E_t$$

The last two values of the stationary series were: $x_{179} = -136.08$ and $x_{180} = -97.22$ respectively. Predictions for the next four months are made by

Autocorrelation function for RESI1

Lag	Corr	T	LBQ	Lag	Corr	T	LBQ	Lag	Corr	T	LBQ
1	−0.02	−0.21	0.04	26	−0.03	−0.36	26.43	51	0.04	0.46	57.23
2	−0.02	−0.28	0.13	27	0.03	0.36	26.62	52	−0.05	−0.53	57.83
3	−0.03	−0.46	0.34	28	−0.09	−1.06	28.32	53	0.05	0.59	58.59
4	−0.09	−1.24	1.94	29	0.09	1.02	29.92	54	0.02	0.17	58.66
5	0.02	0.27	2.02	30	−0.05	−0.59	30.46	55	−0.06	−0.68	59.69
6	−0.08	−1.03	3.15	31	0.09	1.07	32.28	56	−0.09	−0.98	61.88
7	0.01	0.10	3.16	32	0.05	0.64	32.95	57	−0.04	−0.38	62.21
8	−0.01	−0.12	3.18	33	−0.09	−1.10	34.94	58	0.05	0.49	62.78
9	−0.00	−0.05	3.18	34	0.02	0.25	35.05	59	0.03	0.28	62.97
10	0.07	0.98	4.25	35	−0.09	−1.05	36.93	60	−0.13	−1.36	67.42
11	0.05	0.64	4.71	36	−0.01	−0.06	36.94	61	−0.01	−0.09	67.45
12	0.19	2.47	11.67	37	−0.11	−1.22	39.50	62	−0.02	−0.25	67.61
13	−0.02	−0.24	11.73	38	0.06	0.67	40.29	63	0.10	1.08	70.53
14	0.00	0.03	11.74	39	−0.02	−0.18	40.34	64	−0.07	−0.76	72.01
15	0.04	0.54	12.09	40	0.07	0.76	41.39	65	−0.01	−0.08	72.03
16	−0.15	−1.87	16.44	41	0.06	0.70	42.28	66	−0.02	−0.23	72.16
17	0.17	2.13	22.38	42	0.01	0.15	42.32	67	−0.07	−0.72	73.53
18	−0.07	−0.86	23.40	43	0.07	0.81	43.56	68	−0.03	−0.31	73.79
19	0.05	0.64	23.98	44	−0.04	−0.42	43.89	69	0.03	0.36	74.13
20	0.07	0.78	24.85	45	−0.22	−2.51	55.97	70	−0.03	−0.28	74.35
21	−0.02	−0.22	24.92	46	0.04	0.41	56.31	71	−0.03	−0.30	74.60
22	−0.04	−0.48	25.25	47	−0.04	−0.40	56.65	72	−0.09	−0.95	77.13
23	0.04	0.54	25.67	48	−0.00	−0.03	56.65	73	−0.05	−0.56	78.04
24	0.04	0.51	26.05	49	0.00	0.04	56.65	74	0.01	0.09	78.06
25	−0.03	−0.35	26.23	50	−0.02	−0.23	56.76	75	0.03	0.33	78.37

(a)

Figure 8.9 (a) acf of residuals; (b) histogram of residuals, after fitting an AR(2) model to the sweet white wine series (trend and seasonal effects having been removed)

assuming $E_{t+1}, ..., E_{t+4}$ are equal to their expected values of 0. Then

$$\hat{x}_{181} = -1.063 + 0.4721 \times (-97.22) + 0.4498 \times (-136.08) = -108.17$$

$$\hat{x}_{182} = -1.063 + 0.4721 \times (-108.17) + 0.4498 \times (-97.22) = -95.86$$

$$\hat{x}_{183} = -1.063 + 0.4721 \times (-95.86) + 0.4498 \times (-108.17) = -94.97$$

$$\hat{x}_{184} = -1.063 + 0.4721 \times (-94.97) + 0.4498 \times (-95.86) = -89.02$$

Descriptive statistics

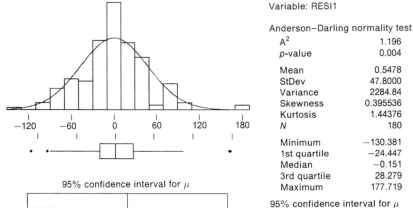

Variable: RESI1

Anderson–Darling normality test

A^2	1.196
p-value	0.004
Mean	0.5478
StDev	47.8000
Variance	2284.84
Skewness	0.395536
Kurtosis	1.44376
N	180
Minimum	−130.381
1st quartile	−24.447
Median	−0.151
3rd quartile	28.279
Maximum	177.719

95% confidence interval for μ

−6.483	7.578

95% confidence interval for σ

43.320	53.322

95% confidence interval for median

−6.493	6.266

(b)

With the Minitab Decomposition these are added to the seasonally adjusted trend. So

$$\hat{y}_{181} = (160.50 + 0.956\,93 \times 181) \times 0.857 + (-108.17) = 178$$

$$\hat{y}_{182} = (160.50 + 0.956\,93 \times 182) \times 0.841 + (-95.86) = 186$$

$$\hat{y}_{183} = (160.50 + 0.956\,93 \times 183) \times 0.911 + (-94.97) = 211$$

$$\hat{y}_{184} = (160.50 + 0.956\,93 \times 184) \times 0.876 + (-89.02) = 206$$

If the centred moving average CMA variant 1 had been used the predictions for the stationary series would be added to the trend, and their sum would be multiplied by the seasonal indices.

Simulating future values

Wine producers might ask for a set of 1000 time series, simulated over the next two years, so that they can gauge the chances of either failing to meet demand or being left with wine which they cannot sell. The simulation is carried out in a similar way to making predictions with the difference that $\{E_t\}$ are replaced by a sequence of independent random numbers rather than their expected value of 0. The histogram of residuals in Figure 8.9(b) is not particularly well modelled by a normal distribution. A normal distribution has a kurtosis, defined as κ in Appendix 2, of 3. Excess kurtosis, which is also commonly referred to as kurtosis, is $\kappa - 3$. The residuals have an excess kurtosis of 1.44. A versatile approach is to fit back-to-back Weibull distributions. The Weibull distribution has the cdf, in which a and b are shape and scale parameters,

$$F(x) = 1 - e^{-(x/b)^a} \quad 0 < x$$

A fitting procedure is described in Exercise 8.12. Figure 8.10(a) and (b) shows the Minitab Weibull probability plots for (a) the positive residuals and (b) the absolute values of the negative residuals. Since 90 out of the 180 residuals are positive the generation of a random number proceeds as follows. Let U_i and U_{i+1} be random numbers from $U[0, 1]$.

If $U_i < 0.5$

$$E_i = 35.6866[-\ln(1 - U_{i+1})]^{1/0.998\,601}$$

If $U_i \geq 0.5$

$$E_i = -35.3435[-\ln(1 - U_{i+1})]^{1/1.058\,10}$$

8.3.6 Integrated and seasonal ARMA processes

Another way of fitting polynomial trends is to *difference* the data. The difference operator is written as ∇, and defined by

$$\nabla Y_t = Y_t - Y_{t-1}$$

Notice that $\nabla = 1 - B$. First-order *differencing* will remove a linear trend, but the structure of the differenced process is different from that obtained by fitting a line and subtracting the trend (see Exercise 8.13). Second-order differencing will remove a quadratic trend:

$$\nabla^2 Y_t = \nabla(\nabla Y_t) = \nabla(Y_t - Y_{t-1}) = Y_t - 2Y_{t-1} + Y_{t-2}$$

If the dth difference of $\{Y_t\}$ is ARMA(p, q) we say that Y_t is ARIMA(p, d, q). The 'I' stands for 'integrated' (summing in the discrete time context), which is the inverse operator to differencing (see Exercise 8.14). The stationary ARMA(p, q) model fitted to the differenced data has to be summed to provide a model for the original time series. A seasonal pattern in the mean, of period s, can also be removed by differencing. A general seasonal ARIMA model, sometimes referred to as a SARIMA model, ARIMA pdq PDQ s is of the form

$$\phi(B)\phi_s(B^s)(1 - B)^d(1 - B^s)^D Y_t = \theta(B)\theta_s(B^s)E_t$$

where $\phi(B)$ and $\theta(B)$ are polynomials in B of degrees p and q respectively and $\phi_s(B^s)$ and $\theta_s(B^s)$ are polynomials in B^s of degrees P and Q respectively, where s is the seasonal period. The operator B^s is defined by

$$B^s Y_t = Y_{t-s}$$

D is almost always 0 or 1 and for monthly variables s is 12. Then

$$(1 - B^{12})Y_t = Y_t - Y_{t-12}$$

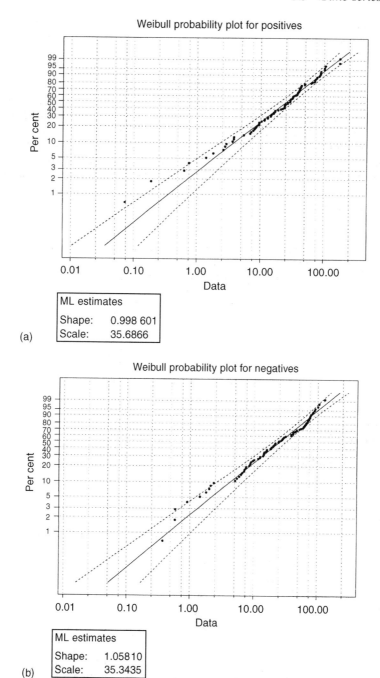

Figure 8.10   Weibull probability plots for residuals after fitting the AR(2) model to the sweet white wine series (trend and seasonal effects having been removed): (a) positive residuals; (b) absolute values of negative residuals

i.e. the difference between a monthly variable and its value in the same month of the previous year. If d equals 1, $(1 - B)(1 - B^{12})Y_t$ are the differences of the sequence $\{(Y_t - Y_{t-12})\}$, i.e.

$$\{(Y_t - Y_{t-1}) - (Y_{t-12} - Y_{t-13})\}$$

However, $(1 - B^{12})$ removes a linear trend as well as seasonal effects, so using $(1 - B)$ is redundant. Notice that, whatever the parameter values, a SARIMA model expresses Y_t as a linear combination of past values of Y, E_t and past errors. Reasonable choices of P and Q can be made from the correlogram of differenced values by looking at lags which are multiples of s. In particular, suppose we have a systematic monthly pattern with DWN, $\{E_t\}$, added. Differencing, with $D = 12$, will give a sequence $\dots (E_t - E_{t-12}), (E_{t+1} - E_{t-11}) \dots$ and Q will be 1.

Example 8.8
Given what we already know about the sweet white wine series, and taking account of the preceding comments, an ARIMA $(2, 0, 0)(0, 1, 1)12$ model should be suitable. Minitab gives the following results.

	Coef.	St. Dev.
AR 1	0.4837	0.0706
AR 2	0.4251	0.0708
SMA 12	0.7416	0.0555
Constant	0.850	1.081

Differencing 0 regular, 1 seasonal of order 12.
Number of observations: Original series 180, after differencing 168
Residuals $MS = 2773$ $DF = 164$

The fitted model is

$$(1 - 0.4837B - 0.4251B^2)(1 - B^{12})y_t = (1 - 0.7416B^{12})e_t$$

which can be rewritten as

$$y_t = y_{t-12} + 0.4837(y_{t-1} - y_{t-13}) + 0.4251(y_{t-2} - y_{t-14}) + e_t - 0.7416e_{t-12}$$

To predict sales one step ahead from the end of the series, i.e. for $t = 181$, we need $y_{180}, y_{179}, y_{169}, y_{168}, y_{167}$ and the residual at time $t = 169$, as an estimate of e_{169}. Minitab will automatically produce predictions, and for four steps ahead they are:

		95% limits	
t	Forecast	Lower	Upper
181	168	65	271
182	214	99	329
183	200	67	333
184	225	81	369

The 95% prediction limits are slightly wider than $\pm 2 \times \sqrt{2773}$, which is ± 105, because of uncertainty in the parameter estimates. They are sensitive to an assumption that the errors are normally distributed, and in this case should be even wider because the errors have a distribution with heavier tails than a normal distribution. Including differencing of order 1, without or with an MA1 term, results in higher mean square values, 2817 on 163 degrees of freedom and 2832 on 162 degrees of freedom, and slightly different forecasts. For example, the SARIMA $(2, 1, 1)(0, 1, 1)12$ predictions are: 172, 196, 197, 195 (see Exercise 8.15).

Using SARIMA models for time-series analysis, especially short-term forecasts, is very convenient with Minitab. However, these models have the disadvantages that: moving average terms are introduced by differencing, they are less straightforward to interpret than methods which explicitly estimate trends and seasonal effects, and they are not suitable for long simulations because differencing leads to instability (see Exercise 8.16). SARIMA models also implicitly assume additive seasonal effects, but this limitation can be circumvented by analysing some transformation of the original series $\{W_t\}$. The most common form of transform is the Box–Cox transform (Box and Cox, 1964):

$$Y_t = \begin{cases} [(W_t - L)^\lambda - 1]/\lambda & \text{for } \lambda \neq 0 \\ \ln(W_t - L) & \text{for } \lambda = 0 \end{cases}$$

The subtraction of 1 and division by λ makes the transform continuous with respect to λ, although this is only an advantage if λ is to be estimated by maximum likelihood.

If the data are transformed, the inverse transform of the prediction for y_t will not be an unbiased prediction for the mean value of w_t. For instance if $y_t = \ln w_t$, then $\hat{E}[W] = \exp(\hat{y}_t + s^2/2)$ where s is the estimated standard deviation of the $\{E_t\}$ in the model. Another common transform is $y_t = \sqrt{w_t}$, in which case $\hat{E}[W] = \hat{y}_t^2 + s^2$ (Exercise 8.17). These adjustments only apply to point estimates of means and are not used in simulations.

The parameters of the Box–Cox transform are often chosen so that $\{Y_t\}$ has a near normal distribution, but it does not follow that the E_t in an ARMA model need be near normal. In general $\{Y_t\}$ may be a linear combination of several $\{E_t\}$ and, as a consequence of the central limit theorem, $\{Y_t\}$ will be nearer to normal than the $\{E_t\}$.

8.3.7 EWMA for forecasting

The exponentially weighted moving average (EWMA) at time t is of the form

$$\tilde{x}_t = (1 - \theta)(x_t + \theta x_{t-1} + \theta^2 x_{t-2} + \cdots)$$

where θ is a smoothing constant. It is easy to verify that this is equivalent to

$$\tilde{x}_t = (1 - \theta)x_t + \theta\tilde{x}_{t-1}$$

by substituting for $\tilde{x}_{t-1}$. The EWMA at time t can be used as a forecast for time $t + 1$, i.e.

$$\hat{x}_{t+1} = \tilde{x}_t$$

The calculation of the EWMA can conveniently be written in terms of forecasts

$$\hat{x}_{t+1} = (1 - \theta)x_t + \theta\hat{x}_t$$

Table 8.4 Amount of money spent by tourists in Northumbria

Year	Spend (£ million)	RPI	£ million @ 1989 prices	SMOO1
1989	343	115.5	343	330.231
1990	306	126.8	279	309.059
1991	344	133.9	297	304.076
1992	373	138.5	311	306.937
1993	444	140.7	364	330.519
1994	455	144.1	365	344.768
1995	571	149.1	442	384.950
1996	434	152.7	328	361.415
1997	585	157.5	429	389.345

This is an optimum forecasting procedure if $\{X_t\}$ is ARIMA(0, 1, 1), and the following argument justifies this claim. If X_t is ARIMA(0, 1, 1), sometimes abbreviated to IMA(1), then

$$X_{t+1} = X_t + E_{t+1} + \theta E_t$$

Since $E[E_{t+1}] = 0$, the best estimate of x_{t+1} given x_t is

$$\hat{x}_{t+1} = x_t + \theta e_t$$

where e_t is the value taken by E_t. If we have an optimum forecasting procedure then

$$e_t = \hat{x}_t - x_t$$

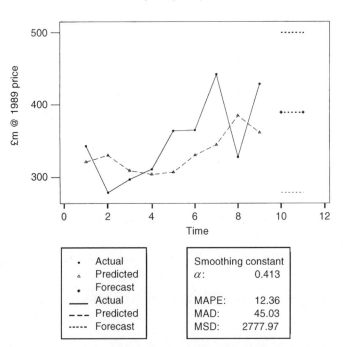

•	Actual	Smoothing constant	
▲	Predicted	α:	0.413
•	Forecast		
——	Actual	MAPE:	12.36
- - -	Predicted	MAD:	45.03
·····	Forecast	MSD:	2777.97

Figure 8.11 EWMA of tourist spending in Northumbria, 1989–97, at 1989 prices. The smoothing constant is obtained by minimising the sum of squared errors

Substituting gives

$$\hat{x}_{t+1} = x_t + \theta(\hat{x}_t - x_t)$$
$$= (1 - \theta)x_t + \theta\hat{x}_t$$

Example 8.9
The data in Table 8.4 show the expenditure by tourists in Northumbria from 1989 to 1997. In the fourth column the figures have been adjusted to 1989 prices. The Minitab single exponential smoothing routine provided the graph shown in Figure 8.11. The last calculation was

$$\tilde{x}_9 = 0.413x_9 + 0.587\tilde{x}_8$$

with $\tilde{x}_8$ and x_9 equal to 361.4 and 429 respectively. Hence $\tilde{x}_9$, and $\hat{x}_{10}$, equals 389.3. The Minitab smoothing constant corresponds to $(1 - \theta)$, so the smaller it is the greater the smoothing. The EWMA works best when there are no obvious trend or seasonal components (Exercise 8.18).

8.4   Multivariate time series

8.4.1   Inclusion of other series as predictor variables

Example 8.10
Ranne (1999) modelled inflation Y_t in year t as a linear function of inflation in the preceding two years and the oil price X in year t and the preceding year.

$$Y_t = m + \alpha_1(Y_{t-1} - m) + \alpha_2(Y_{t-2} - m) + b_0X_t + b_1X_{t-1} + E_t$$

This can be fitted by standard multiple regression of Y_t on Y_{t-1}, Y_{t-2}, X_t and X_{t-1}. The constant term is an estimate of $m(1 - \alpha_1 - \alpha_2)$, and since α_1 and α_2 are estimated directly an estimate of m can be deduced. This was used as part of an investment simulation model for a Finnish pension company. The model includes price movements of all the investment categories that make up the portfolio. The working capital is a crucial response variable and a set of 1000 scenarios is shown in Figure 8.12.

Example 8.11
The data in Table 8.5 reflect passenger throughput at Newcastle International Airport together with tourist spending in the region and inward investment at 1989 prices. A regression of passengers (y) on time (t) provides strong evidence of an increasing trend over this period. The fitted equation is

$$y_t = 1\,178\,455 + 156\,409t$$

$$(79\,226)\quad(11\,681)$$

with $s = 122\,514$. The lag 1 and lag 2 autocorrelations of the residuals are 0.26 and -0.46, and with such a short series both can reasonably be attributed to sampling error. A regression of tourist expenditure (x) on time (t) provides

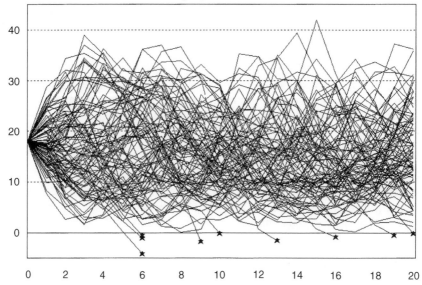

Figure 8.12   Set of scenarios for working capital of pension portfolio (Ranne, 1999)

some evidence of an increasing trend

$$x_t = 267 + 13.9t$$

$$(37) \quad (5.65)$$

If the trend is removed and the detrended series w_t included in the regression of y on t the result is

$$y_t = 615\,954 + 132\,579t + 1965w_t$$

$$(263\,713) \quad (18\,919) \quad (926)$$

Table 8.5   Passenger throughput at Newcastle International Airport, tourist spending and inward investment at 1989 prices

Year	t	Passengers	Tourist spending	Inward investment	inwdinv_1
1988	1	1 428 011	*	*	*
1989	2	1 544 707	343	*	*
1990	3	1 529 309	279	27 915	*
1991	4	1 616 056	297	53 422	27 915
1992	5	1 965 556	311	14 314	53 422
1993	6	2 135 141	364	26 384	14 314
1994	7	2 437 362	365	27 556	26 384
1995	8	2 569 937	442	42 811	27 556
1996	9	2 457 397	328	24 788	42 811
1997	10	2 652 613	429	16 461	24 788
1998	11	2 949 922	*	*	16 461

with $s = 107\,253$. However, this is not directly comparable with the first regression in this example because the number of years has been reduced due to the missing values. If the inward investment (v_t) lagged by one year is included, the fitted regression model is:

$$y_t = 443\,363 + 117\,942t + 2507w_t + 2.74v_{t-1}$$

$$(586\,779) \quad (40\,627) \quad (1773) \quad (5.54)$$

There are only three degrees of freedom for error and no substantial conclusions can be drawn, beyond the fact that the coefficients do at least have the expected sign.

If the trend continues, a 95% prediction interval for the year 2006, using the first regression, is between 3.7 and 4.6 million passengers per year. This would justify a major extension to the airport building. There are plans for further industrial and recreational development in the region and this lends support to the assumption that the trend will continue. It is also argued that there is runway congestion at another northern airport and that a substantial number of passengers, about 700 000 per year, would find Newcastle more convenient if the services were improved. The only discouragement is the Department of Transport forecast that a significant number of passengers will transfer from air to rail for short cross-Channel journeys. However, the further a passenger originates from London the less likely they are to use the service. It is estimated that about 20 000 passengers per year will be lost by the airport to Channel rail services. All these factors having been taken into consideration, a decision is made to go ahead with the airport extension.

8.4.2 Multivariate Gaussian processes

Australian wine growers need to understand how the wine market moves as a whole, rather than the behaviour of a single sector alone. For example, they may need to consider the question of whether low sales of red wine in one year are likely to be compensated for by high sales of white that year, or whether low sales of white in one year might be followed by high sales of fortified wine in the next. Simulations, and prediction intervals for total sales, should take account of the correlations between sales.

There is a high correlation of 0.63 between rosé and fortified wine but this is probably mainly accounted for by the fact that there has been a downward trend in sales for both these wine types. It is much more useful to look at correlations between the residuals of the series, having first removed any trend and seasonal effects, and having fitted ARMA models when appropriate. I removed the trend and seasonal effects from all six series using the Minitab decomposition. The stationary sweet white wine series is well modelled by an AR(2) process and the stationary rosé series is well modelled as an MA(1) process

$$X_t = 0.394 + E_t + 0.307E_{t-1}$$

with a standard deviation of errors of 16.8. The term pre-whitened is sometimes used for the residuals of a time series after removing trend and seasonal effects and fitting an ARMA model. The lower half of the correlation matrix between the stationary series, e.g. DTDSfort for the fortified wine series after removing the trend and seasonal effects, and the pre-whitened sweet white wine (PWswtw) and rosé (PWrose) series, is given below.

	DTDSfort	DTDSdryw	PWswtw	DTDSred	PWrose
DTDSdryw	0.258				
PWswtw	0.191	0.235			
DTDSred	0.239	0.515	0.226		
PWrose	0.258	0.297	0.046	0.313	
DTDSspkl	0.336	0.391	0.325	0.479	0.276

All the correlations, except that between PWrose and PWswtw wine, are statistically significant at the 1% level or beyond. It is possible that the, apparently, uncorrelated series are correlated if one is shifted in time (lagged) with respect to the other. Logically, two DWN series can only be correlated at one particular lag, if they are correlated at all. The cross-covariance function (ccvf) between two stationary time series $\{X_t\}$, $\{Y_t\}$ is a function of the lag (k) and is defined as:

$$\gamma_{xy}(k) = E[(X_t - \mu_X)(Y_{t+k} - \mu_Y)]$$

and estimated by:

$$c_{xy}(k) = \sum_{t=1}^{n-k}(x_t - \bar{x})(y_{t+k} - \bar{y})/n$$

The ccvf is not symmetric about $k = 0$, and it follows from the definition that

$$c_{yx}(-k) = c_{xy}(k)$$

Cross-correlation is defined as the cross-covariance divided by the product of the standard deviations of $\{X_t\}$ and $\{Y_t\}$. The sample estimates are calculated with the divisor n, i.e.

$$r_{xy}(k) = c_{xy}(k)/\sqrt{(c_{xx}(0)c_{yy}(0))}$$

Example 8.12
Two, very short fictitious, time series are given below. The numbers have been chosen so that there is an obvious correlation between x_t and y_{t-1}.

$$x_t \quad 1, \quad 4, \quad 1, \quad 1, \quad 1, \quad 6, \quad 1, \quad 1, \quad 1, \quad 7$$
$$y_t \quad 5, \quad 1, \quad 1, \quad 1, \quad 8, \quad 1, \quad 1, \quad 1, \quad 6, \quad 1$$
$$c_{xy}(-1) = 0.919 \quad c_{xy}(0) = -0.393 \quad \text{and} \quad c_{xy}(1) = -0.166$$

Note the high value of $c_{xy}(-1)$.

Example 8.13
The Pinkham Medicine Company sales (y_t, dollars), given in TableWSPinkham, are quite convincingly modelled by an AR(2) process:

$$(y_t - 1734.6) = 1.3977(y_{t-1} - 1734.6) - 0.4918(y_{t-2} - 1734.6)$$

The standard deviation of the residuals is 204.7, and this is much less than the standard deviation of $\{y_t\}$, 632.7. It is surprising that an empirical model which takes no explicit account of the worldwide Depression in the 1930s is so effective. An AR(2) process with these coefficients is quite close to the boundary of stability, and it may be modelling the behaviour of a volatile economy rather than the specific fortunes of the Pinkham Medicine Company. It would be interesting to repeat the analysis after adjusting sales by a price index or the Dow Jones index. The residuals from the AR(2) model are pre-whitened sales figures. The cross-correlation between the residuals at time t and the advertising expenditure at time $t + k$ is

$$r(-2) = -0.175$$

$$r(-1) = -0.171$$

$$r(0) = 0.135$$

$$r(1) = 0.371$$

$$r(2) = 0.235$$

The largest correlation, in absolute magnitude, is the positive correlation of 0.371 between pre-whitened sales in year t and advertising expenditure in year $t + 1$. An explanation is that the advertising budget was partly, at least, a proportion of the previous year's sales. The negative correlation between pre-whitened sales in year t and advertising one or two years earlier suggests that the increases in advertising expenditure may not have been effective.

Returning to the Australian wine growers, the cross-correlation between the PWrose and PWswtw is:

$$r_{xy}(-2) = -0.001$$

$$r_{xy}(-1) = -0.151$$

$$r_{xy}(0) = 0.043$$

$$r_{xy}(1) = -0.026$$

$$r_{xy}(2) = -0.055$$

None of the cross-correlations is striking, although the slight negative correlation between PWswtw in one year and PWrose the next is just statistically significant at the 5% level. We will ignore it in the remainder of this section but a method for incorporating it in a model is given in Section 8.4.4.

The covariances between the series are given below.

	DTDSfort	DTDSdryw	PWswtw	DTDSred	PWrose	DTDSspkl
DTDSfort	97 402.38					
DTDSdryw	27 034.95	112 414.16				
PWswtw	2847.21	3761.32	2284.84			
DTDSred	14 182.44	32 748.85	2049.59	36 029.78		
PWrose	1349.18	1665.93	36.72	994.62	280.67	
DTDSspkl	38 488.57	48 063.10	5694.35	33 317.03	1696.83	134 474.53

It is straightforward to simulate correlated DWN sequences of normal variables. If V is the required covariance matrix there will be a matrix Q such that

$$V = Q^2$$

Let $Z_t = (Z_{1,t}, ..., Z_{m,t})^T$ be a vector of independent standard normal random numbers. The correlated $E_{j,t}$ are related to the $Z_{i,t}$ by

$$E_t = QZ_t$$

where $E_t = (E_{1,t}, ..., E_{m,t})$. The distribution of E_t is multivariate normal because any linear combination of normal random variables is also normally distributed. No similar convenient property holds for any other form of distribution, since it is a consequence of the central limit theorem that the linear combination will be closer to a normal distribution. If we are willing to approximate the DWN wine sequences by normal distributions this method can be used with:

$$Q = \begin{pmatrix} 304.7 & & & & & \\ 35.9 & 322.0 & & & & \\ 5.4 & 7.0 & 45.2 & & & \\ 19.3 & 57.9 & 4.1 & 171.2 & & \\ 3.1 & 3.5 & -0.9 & 3.0 & 15.5 & \\ 53.4 & 63.6 & 11.9 & 54.5 & 3.1 & 352.8 \end{pmatrix}$$

Once we have realisations of the DWN sequences, we can calculate realisations of the AR(2) process for the sweet white wine and the MA(1) process for the rosé. It is convenient to set the first terms of the AR(2) process equal to the mean value. Finally, we include the trend and seasonal effects to obtain a realisation of wine sales.

If we are not prepared to assume the DWN series are from normal distributions we can use the methods of Section 8.4.3.

8.4.3 Gibbs sampler

The Gibbs sampler, described in Section 2.7.2, is a very versatile method that can be used to generate DWN sequences, from any shaped distribution, that are inter-correlated at time t. For the wine example we have to fit six regressions, one for each wine type on the other five types. The fitting procedure includes choosing suitable distributions for the residuals. The first fitted regression is

$$\text{DTDSfort} = 4.82 + 0.0907 \times \text{DTDSdryw} + 0.5627 \times \text{PWswtw}$$

$$+ 0.025\text{DTDSred} + 2.98\text{PWrose} + 0.186\text{DTDSspkl}$$

Eighty-four of the 180 residuals are positive and a Weibull distribution with shape and scale parameters 1.090 74 and 229.626, respectively, is a good fit to the empirical distribution. A Weibull distribution with shape parameter 1.1354 and scale parameter 203.792 provides a good fit for the absolute values of the 96 negative residuals. The sixth regression is,

$$\text{DTDSspkl} = -3.51 + 0.2028 \times \text{DTDSfort} + 0.1280 \times \text{DTDSdryw} + 1.4678$$
$$\times \text{PWswtw} + 0.5890 \times \text{DTDSred} + 2.032 \times \text{PWrose}$$

Back-to-back Weibull distributions again give a good fit for the residuals. The intervening regressions are similar. The simulation starts by generating a random DTDSfort variate from the first regression. If the predictor variables are set at zero this is a random number from the back-to-back Weibull distributions fitted to the residuals. The next step is to use the second regression to generate a DTDSdryw variate with the DTDSfort variate and a random number from the distribution fitted to the residuals. The third, fourth, fifth and sixth regressions are used in a similar way. For example, the sixth regression will be used to generate a DTDSspkl variate from the DTDSfort, ..., PWrose variates, and a random number from the back-to-back Weibull distributions fitted to the residuals. The result of these calculations will be a first multivariate datum $(\text{DTDSfort}, \ldots, \text{DTDSspkl})_1$. The process is repeated, except that the predictor variables for the first regression will be taken from $(\text{DTDSfort}, \ldots, \text{DTDSspkl})_1$. The predictor variables for the second regression will be the $(\text{DTDSfort})_2$ and $(\text{PWswtw})_1, \ldots, (\text{DTDSspkl})_1$, and so on until the sixth regression is used with predictor variables $(\text{DTDSfort})_2, \ldots, (\text{PWrose})_2$. The process continues and, for example, every 50th multivariate datum might be used for the simulation.

8.4.4   Conditional distributions

A useful practical method of simulating from a multivariate distribution relies on a factorisation into conditional distributions. For example:

$$f(x_1, x_2, x_3, x_4) = f(x_1) f(x_2 \mid x_1) f(x_3 \mid x_1, x_2) f(x_4 \mid x_1, x_2, x_3)$$

The conditional distributions can be estimated by the regressions: x_2 on x_1; x_3 on x_1 and x_2 ; and x_4 on x_1, x_2 and x_3. The distribution of the errors in the regressions will usually have to be estimated from the residuals. The technique is quite versatile and the variance of the errors could depend on the conditioning variables if this gives an improved fit.

This method could be used to incorporate the lag 1 cross-correlation between PWswtw and PWrose. The sequence of steps for a simulation of length n would be as follows:

1. Generate a sequence of n PWswtw values as DWN from a suitable distribution.
2. Generate PWrose at time t, conditional on PWswtw at time $t - 1$.
3. Generate DTDSfort at time t, conditional on PWrose and PWswtw at time t.
4. Generate DTDSdryw at time t, conditional on DTDSfort, PWrose and PWswtw at time t.
5. Generate DTDSred at time t, conditional on DTDSdryw, DTDSfort, PWrose and PWswtw at time t.
6. Generate DTDSspkl at time t, conditional on DTDSred, DTDSdryw, DTDSfort, PWrose and PWswtw at time t.

The conditional distributions would be estimated as regressions with suitable distributions of residuals.

8.5   Dynamic linear models

8.5.1   Definition of the dynamic linear model

The dynamic linear model (DLM) has many engineering applications, including Kalman filtering. The book by Pole *et al.* (1994) is an excellent introduction, and includes the Bayesian Analysis of Time Series (BATS) software. An example concerns a construction company which has a house-building division. The sales manager uses the following model to allow for the influence of a general level of sales in the sector and the company's own pricing policy on its sales.

$$\text{sales}_t = \text{level}_t + \beta_t \text{price}_t + E_t$$

$$\text{level}_t = \text{level}_{t-1} + \Delta\text{level}_t$$

$$\beta_t = \beta_{t-1} + \Delta\beta_t$$

The first equation is a linear regression with one predictor variable, and the modification that the slope and intercept can change over time. The E_t, Δlevel_t and $\Delta\beta_t$ are random error terms which are assumed to be independent over time. The E_t are assumed to be independent of the changes to the level and the slope β. This is an example of the known variance model described by Pole *et al.*:

observation equation: $Y_t = F_t^T \theta_t + v_t$
system equation: $\theta_t = G\theta_{t-1} + w_t$
error distributions: $v_t \sim N(0, V_t)$; $w_t \sim N(0, W_t)$

In these equations: Y_t is the observed dependent variable at time t; F_t is a vector of values of the predictor variables at time t; and θ_t is a vector of unknown time varying parameters, often known as system states, at time t. The errors, v_t and w_t, are assumed not to be correlated over time, and v_t is independent of w_t. The usual development assumes the errors are normally distributed. In the following, D_t represents knowledge at time t. The prior information is:

$$\theta_{t+1} \mid D_t \sim N(a_{t+1}, R_{t+1})$$

In the example, Y_t is sales, F_t is the price, and θ_t includes the slope and intercept.

Let the distribution of θ at time $t-1$, based on all our information at that time, be

$$\theta_{t-1} \mid D_{t-1} \sim N(m_{t-1}, C_{t-1})$$

Then our prior distribution for θ at time t is

$$\theta_t \mid D_{t-1} \sim N(a_t, R_t)$$

where $a_t = Gm_{t-1}$ and $R_t = GC_{t-1}G^T + W_t$.

We will derive the posterior distribution for θ_t given y_t, after first finding the one-step-ahead forecast distribution for Y_t. Since the posterior distribution for θ_t given D_{t-1} and y_t is the updated distribution for θ_t given our information at time t, D_t is the combination of D_{t-1} and y_t.

8.5.2   One-step-ahead forecast

$$E[Y_t \mid D_{t-1}] = E[F_t^T \theta_t + v_t \mid D_{t-1}]$$
$$= F_t^T E[\theta_t \mid D_{t-1}]$$
$$= F_t^T a_t$$
$$\text{var}[Y_t \mid D_{t-1}] = \text{var}[F_t^T \theta_t + v_t \mid D_{t-1}]$$
$$= F_t^T R_t F_t + V_t$$

8.5.3   Posterior distribution

$\theta_t \mid D_t$ is equivalent to $\theta_t \mid D_{t-1}, y_t$, and using Bayes' theorem

$$f(\theta_t \mid D_{t-1}, y_t) \propto f(y_t \mid \theta_t) f(\theta_t \mid D_{t-1})$$
$$\propto \exp[-\tfrac{1}{2}(y_t - F_t^T \theta_t)^2 V_t^{-1}] \times \exp[-\tfrac{1}{2}(\theta_t - a_t)^T R_t^{-1}(\theta_t - a_t)]$$
$$\propto \exp[-\tfrac{1}{2}(\theta_t - m_t)^T C_t^{-1}(\theta_t - m_t)]$$

where m_t and C_t are given by the following algorithm, in which e_t is the error in the one-step-ahead forecast:

$$e_t = y_t - F_t^T a_t$$
$$Q_t = F_t^T R_t F_t + V_t$$
$$A_t = R_t F_t / Q_t$$
$$m_t = a_t + A_t e_t$$
$$C_t = R_t - A_t A_t^T Q_t$$

8.5.4   Intervention

Our prior distribution for θ at time t would be

$$\theta_t \mid D_{t-1} \sim N(a_t, R_t)$$

but we now have some external information I_t about θ_t such as a change in taxation law relating to house purchase, or a new competitor in the market. We therefore adjust our prior distribution to

$$\theta_t \mid D_{t-1}, I_t \sim N(a_t^*, R_t^*)$$

Then a_t^* and R_t^* replace a_t and R_t in the algorithm. We should also increase G for at least one step:

$$G^* = (Z_t^{-1} U_t)^T G$$

where Z_t and U_t are obtained from R_t and R_t^* through the relationships

$$R_t^* = U_t^T U_t \quad R_t = Z_t^T Z_t$$

8.6   Non-linear time series

8.6.1   Bilinear models

A first-order bilinear model is of the form

$$X_t = (\alpha + \beta E_t)X_{t-1} + E_t$$

or variants of this such as βE_{t-1} replacing βE_t, where E_t is DWN. They can be used to model series that show sudden bursts of large amplitude oscillations at irregular intervals, and you are asked to investigate one of these in Exercise 8.19. ARMA models, with normal errors, have the moment property, i.e. moments of all orders exist. If the coefficients are subject to random perturbations, high-order moments will no longer be finite. However, a lack of moments does not prevent a stochastic process from being stationary. Stationarity only requires that the multivariate distribution of any subset of variables $(X_1, X_2, \ldots, X_p)$ is the same as that of $(X_{1+k}, X_{2+k}, \ldots, X_{p+k})$. Either form of the first-order bilinear model is stationary if

$$\alpha^2 + \beta^2 \sigma_E^2 < 1$$

in which case the first two moments, at least, are finite.

Tong (1990) gives a comprehensive account of non-linear models, and their intriguing properties, from a statistical perspective. The least squares estimators of α and β are the values which minimise:

$$\sum_{t=2}^{n} E_t^2 = \sum_{t=2}^{n} [(X_t - \alpha X_{t-1})/(1 + \beta X_{t-1})]^2$$

with

$$\beta^2 < (1 - \alpha^2)/\sigma_E^2$$

Example 8.14
Inward investment in North-East England between the first quarter of 1990 and the third quarter of 1998 is shown in Table 8.6. The figures are in thousands of pounds sterling and have not been adjusted by the RPI. The mean and standard deviation are 8525 and 9176 respectively. The whole sum for projects is given on the date of the first payment, which accounts for the very uneven distribution, and a negative correlation at lag 1 might also be expected. The autocorrelation at lag 1 is −0.11. The natural logarithms of the data are well fitted by a normal distribution and have a slightly larger, in absolute terms, lag 1 correlation of −0.15. If a bilinear model is fitted to the standardised series, i.e. the mean has been subtracted and the difference has been divided by the standard deviation, the estimates of α and β are −0.125 and −0.04 respectively. The residuals have a mean of 0.015, a standard deviation (divisor 34−1) of 1.006 and a minimum value of −0.734. After adding 0.8 an exponential distribution with a mean of 0.8417 is a good fit. Although the bilinear model provides an adequate description of the time series it is not a statistically significant improvement on an assumption of an independent sequence.

Table 8.6   Inward investment in North-East England (thousands of pounds sterling)

1990	Q1	4719	1994	Q3	3131
1990	Q2	6755	1994	Q4	7846
1990	Q3	4618	1995	Q1	19 839
1990	Q4	12 525	1995	Q2	8538
1991	Q1	6748	1995	Q3	3681
1991	Q2	12 633	1995	Q4	4255
1991	Q3	41 622	1996	Q1	38 791
1991	Q4	1619	1996	Q2	2860
1992	Q1	6059	1996	Q3	5197
1992	Q2	2026	1996	Q4	4583
1992	Q3	6275	1997	Q1	20 132
1992	Q4	3675	1997	Q2	4319
1993	Q1	5189	1997	Q3	2797
1993	Q2	3534	1997	Q4	9919
1993	Q3	4624	1998	Q1	5412
1993	Q4	6978	1998	Q2	4139
1994	Q1	17 005	1998	Q3	2756
1994	Q2	3563			

8.6.2   Catastrophe theory

Suppose you have started a small business selling T-shirts, with portraits of management gurus, and associated aphorisms, printed on them. You hope to sell these in business schools, and wish to model students' intentions to buy (y). You assume intention to buy will depend on the person's enthusiasm (x) for a particular design of T-shirt, but realise that someone might be discouraged from making a purchase for fear of being considered socially inept. Let w represent perceived social pressure against purchase. The first person you ask thinks the T-shirts would be very fashionable and his intention to buy would increase with enthusiasm for the design (Figure 8.13(a)). However, the second person is, perhaps, more discerning and thinks it would be embarrassing to be seen in one. If social pressures against purchase are high, the postulated 'threshold effect' is that an increase in enthusiasm will have little effect on intention to buy until a certain threshold (U) is reached, when social pressures are overcome and intention to buy jumps to a new high level (Figure 8.13(b)). The 'delay effect' is the converse, and intention to buy remains high, despite decreasing enthusiasm until a lower threshold (L) is reached when intention to buy falls to a low level (Figure 8.13(b)).

The dependence of purchase behaviour on enthusiasm and perceived social pressures can be modelled by a cusp catastrophe (Thom, 1972). Let x range from -1, corresponding to no enthusiasm, to $+1$, corresponding to great enthusiasm. Let w range from 0, representing no social pressures against purchase, to 1, representing high social pressure against purchase. If y is the logit of the probability of purchase, the surface

$$y^3 = x + kwy$$

in which k is a parameter which sets the positions of L and U, is a simple example of a cusp catastrophe (Figure 8.14, after Burghes and Wood, 1980). For a given value of w, x is a cubic function of y, and the thresholds L and U correspond to the value of x at the stationary points, $\pm\sqrt{4(kw)^3/27}$ (Exercise 8.20).

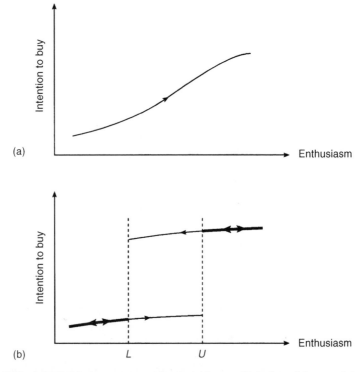

Figure 8.13   Intention to buy versus enthusiasm for product when: (a) no social pressure against purchase; (b) high social pressure against purchase

The parameter k could be estimated from a survey in which people are asked for their perceived value of w, and their corresponding assessments of L and U. It would be possible to investigate the effects of advertising on perceptions of w and corresponding values of k. You are asked to consider a modified surface, which gives a more realistic model,

$$x = (3y^3 - 3kwy)/(3 + kw)$$

in Exercise 8.20.

Poston and Stewart (1978) give an accessible account of the theory of mathematical catastrophes, a part of singularity theory, together with some fascinating examples. The general theme of modelling and predicting sudden changes has many potential applications in a management context (see, for example, Zeeman *et al.*, 1976).

8.6.3   Brownian motion and the Black–Scholes model

A stochastic process $W(t)$, defined over continuous time and a continuous state space, is Brownian motion if increments, $W(t + \tau) - W(t)$, are independent and have a Gaussian (normal) distribution with mean $\mu\tau$ and variance $\sigma^2\tau$. The parameters μ and σ^2 are known as the drift and diffusion coefficients

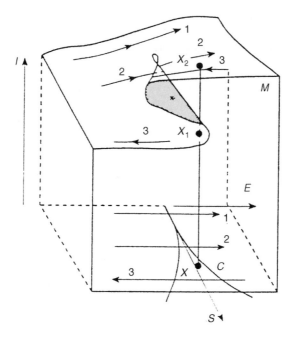

Figure 8.14   Cusp catastrophe (after Burghes and Wood, 1980)

respectively. An alternative name for this model is a Wiener process, after the American mathematician Norbert Wiener (1894–1964) who is also known for his work on cybernetics (Wiener, 1950). The process is Markovian because the probability distribution for $W(t+\tau)$, conditional on $W(t)$, does not depend on the history of the process before time t. Share prices, $S(t)$, are commonly modelled as a geometric Brownian motion:

$$S(t) = S_0\, e^{\sigma W(t) + (\mu - \sigma^2/2)t}$$

where $W(t)$ is a Brownian motion with drift 0 and diffusion 1, μ is the expected price change expressed as a force of interest (Appendix 1), and σ is the volatility of stock price changes. The parameters μ and σ can be estimated from a time series of logarithms of daily prices of the share. The gradient of the regression line for the logarithm of daily share price on day number is an estimate of μ, and the standard deviation of the residuals provides an estimate of σ. This formula is easier to interpret if it is written as

$$S_t = S_0\, e^{\mu t}\left(e^{\sigma W(t) - \sigma^2 t/2}\right)$$

Since

$$\sigma W(t) \sim N(0, \sigma^2 t)$$

$e^{\sigma W(t)}$ has a mean of $e^{0 + \sigma^2 t/2}$, and it follows that $e^{\sigma W(t) - \sigma^2 t/2}$ has a mean of 1.

A European call option is the right to buy a share at some specified time in the future, T, at a set price X. The value of the call option at a time t depends on: the share price $S(t)$, its volatility σ, the time to go $(T - t)$, and the risk-free interest rate r. The final value of the option is the amount by which the share price $S(T)$ exceeds X. The option is worthless if $S(T)$ is less than X. The Black–Scholes formula (Black and Scholes, 1973) for the value at time t is:

$$c(S(t), T - t; X, r, \sigma) = S(t)\Phi(d_1) - X\,e^{-r(T-t)}\Phi(d_2)$$

where

$$d_1 = [\ln(S_t/X) + (r + \sigma^2/2)(T - t)]/\left(\sigma\sqrt{(T - t)}\right)$$

and

$$d_2 = d_1 - \left(\sigma\sqrt{(T - t)}\right)$$

The derivation of this formula rests on the assumptions that: the risk-free interest rate (r) is constant, the share price is a geometric Brownian motion with $r \leq \mu$, there are no dealing costs, and no dividends are paid over the life of the contract (Salopek, 1997). Notice that it only makes sense to consider shares for which $r \leq \mu$, since $\mu < r$ would imply a risk-free investment with a higher expected yield. It is surprising that the formula does not depend on μ, but setting σ equal to 0 provides an explanation. The formula becomes,

$$c = S(t) - X\,e^{-r(T-t)}$$

and the value of the option on day t is the difference between the price of the share on day t and the discounted value of X at time T. The formula is useful for comparing an option to buy in the future with purchase now, rather than for choosing which of several shares to invest in.

A proof of the formula uses results from stochastic calculus (e.g. Lamberton and Lapeyre (1996), Salopek (1997)) and Guttorp (1995, and Exercise 8.23) explains the physical models behind stochastic calculus results. It is often convenient to resort to computer simulation. A Brownian motion with drift 0 and diffusion 1 can be approximated by a random walk along a line with time increment Δt and space increment $\sqrt{\Delta t}$. At each time step the process is equally likely to move right or left by $\sqrt{\Delta t}$.

8.6.4   Chaotic dynamics

A logistic function is often used to model the proportion (y) of a population who buy an innovative product over time. That is

$$\dot{y} = ay(1 - y/b) \quad \text{for } 0 < y < b$$

If $\dot{y}$ is replaced by $y_{t+1} - y_t$ a discrete form of the equation is obtained:

$$y_{t+1} = ky_t(1 - y_t/c)$$

and with c equal to 1 this becomes the logistic mapping

$$y_{t+1} = ky_t(1 - y_t)$$

It has a fixed point if k lies between 0 and 3, but as k increases from 3 towards 4 the mapping becomes increasingly odd. It becomes chaotic when k exceeds 3.58 (Stewart, 1997). You are asked to investigate this in Exercise 8.21. A two-state example from Goodwin (1990) is given in Exercise 8.22.

8.7   Markov chains and processes

8.7.1   Irreducible Markov chains

A man serves tea, coffee, cakes and biscuits from a caravan in a lay-by on the main road between Newcastle and Carlisle. He has noticed that his regular customers, who call in once a day, frequently change their preferences between coffee and tea. Sixty per cent of customers who have coffee on one day have tea the next, with the other 40% choosing coffee again. If a customer had tea yesterday, the probabilities for tea and coffee today are assessed as 0.2 and 0.8, respectively, from corresponding percentages. Two features of this example are that there is a discrete state space, coffee and tea, and that there are changes in state, known as transitions, occurring at a set of discrete times, which here are days. It is convenient to treat staying in the same state as a transition from that state to itself. The Markov property, which was implicit in the description of customers' behaviour, is that the probabilities of being in the various states after the next transition depend only on the state occupied now. This property, which is often assumed for models of random processes, is named after the Russian mathematician Andrei Andreevich Markov (1856–1922), who made a systematic study of such sequences.

Let $p_1^{(t)}$ and $p_2^{(t)}$ represent the probabilities that a customer chooses coffee and tea, respectively, at time t. Then the state space is the set $\{1, 2\}$ and transition probabilities are written as p_{ij} where i and j can be 1 or 2. For example, p_{12} is the probability a randomly selected customer who has coffee one day has tea the next, which is 0.6. The components of the probability vector $\boldsymbol{p}^{(t)}$ at time t are the probabilities for each of the possible states and must therefore add to 1. In this case

$$\boldsymbol{p}^{(t)} = (p_1^{(t)}, p_2^{(t)})$$

The transition probabilities can be gathered into a transition matrix, $\boldsymbol{M}$, which has the properties that all elements are between 0 and 1 and that all rows add to 1. The probability that a customer has coffee at time $t+1$ can be expressed in terms of the probabilities for states at time t and the transition probabilities

$$p_1^{(t+1)} = p_1^{(t)} p_{11} + p_2^{(t)} p_{21}$$

This and the other such probabilities are included in the matrix relationship

$$\boldsymbol{p}^{(t+1)} = \boldsymbol{p}^{(t)} \boldsymbol{M} \tag{8.1}$$

Example 8.15
A randomly selected customer had tea today. What is the probability that the person has tea in four days' time?

$$\boldsymbol{p}^{(0)} = (0, 1)$$

$$\boldsymbol{M} = \begin{pmatrix} 0.4 & 0.6 \\ 0.8 & 0.2 \end{pmatrix}$$

$$p^{(1)} = p^{(0)} M = (0.8 \quad 0.2)$$

$$p^{(2)} = p^{(1)} M = (0.48 \quad 0.52)$$

$$p^{(3)} = p^{(2)} M = (0.608 \quad 0.392)$$

$$p^{(4)} = p^{(3)} M = (0.557 \quad 0.443)$$

The required probability is 0.443.

General formulation

There is a finite set of states known as the state space. If there are r of them, they can conveniently be numbered $1, \ldots, r$. The random process

$$\{X_t\} \quad \text{for } t = 0, 1, \ldots$$

is viewed at a discrete set of times, when it is in one of the states. The process can change states between these times. The Markov property is that the probability of it moving from state i to state j depends on those two states, but not on the history of the process before it arrived in state i. That is:

$$\Pr(X_{t+1} = j \mid X_t = i, \text{and states taken by } X_{t-1}, \ldots, X_0) = \Pr(X_{t+1} = j \mid X_t = i)$$

If we assume a stationary transition mechanism, these probabilities are constant one-step transition probabilities, which we will refer to as p_{ij}. The $r \times r$ matrix of transition probabilities, which we will denote by M, is known as the transition matrix. The two-step transition probabilities $p_{ij}^{(2)}$ are given by:

$$p_{ij}^{(2)} = \sum_{k=1}^{r} p_{ik} \, p_{kj}$$

because we must consider all possible intermediate states. They are the elements of M^2. The more general result

$$p_{ij}^{(n+m)} = \sum_{k=1}^{r} p_{ik}^{(n)} \, p_{kj}^{(m)}$$

is known as the Chapman–Kolmogorov equation. The probability vector at time t is the set of probabilities that X_t is in any of the r states:

$$p^{(t)} = (p_1^{(t)} \ldots p_r^{(t)})$$

$$p^{(t+1)} = p^{(t)} M = (p^{(t-1)} M) M = p^{(t-1)} M^2 = \cdots = p^{(0)} M^{t+1}$$

In a regular chain some power of the transition matrix has only positive elements. That is, there is some time at which it is possible to be in any of the states regardless of the starting state. All higher powers of the transition matrix must also be positive. In a regular chain the probabilities of being in the various states tend to constant values as t becomes large, irrespective of the initial probabilities. This stationary probability vector, v, must satisfy

$$v = vM \tag{8.2}$$

One way of finding v is to solve these equations, together with the constraint that the elements sum to one. An alternative is to compute $p^{(t)}$ until it no longer changes.

Example 8.16
For the coffee and tea transition matrix, the stationary probability vector is the solution of:

$$(v_1 \quad v_2) = (v_1 \quad v_2) \begin{pmatrix} 0.4 & 0.6 \\ 0.8 & 0.2 \end{pmatrix}$$

The two equations

$$v_1 = 0.4v_1 + 0.8v_2$$

$$v_2 = 0.6v_1 + 0.2v_2$$

are linearly dependent, and the constraint that

$$v_1 + v_2 = 1$$

is needed to obtain the unique solution:

$$v_1 = 0.571 \quad v_2 = 0.429$$

The physical interpretation is that 57% of customers choose coffee in the long run.

Probability chains

A natural extension of the Markov chain is to let transition probabilities depend on more than just the current state. A probability chain is said to be of order q if the probability that $X_{t+1} = j$ depends on the states taken at times $t, t-1, \ldots, t-q-1$. A probability chain of order 1 is a Markov chain. One way of fitting probability chains is nominal logistic regression with indicator variables for past states.

Consistency of pension fund performance

Brown *et al.* (1997) investigated the performance of 409 pension funds over seven years from 1986 until 1992. For each year, they ranked the funds according to their risk-adjusted performance using the Market Model. This is a nice application of regression analysis. Let R_{jt} be the return of fund j in period t and R_{mt} be the return on the market, as measured by the WM Pension Fund Index, for example. The coefficients α_j and β_j were estimated by regressing the quarterly returns for each fund on the market return for the seven-year period. That is, the model

$$R_{jt} = \alpha_j + \beta_j R_{mt} + E_{jt}$$

where E_{jt} are the error terms, was fitted for each fund. The abnormal return (AR_{jt}) on fund j in period t is defined by

$$AR_{jt} = R_{jt} - \beta_j R_{mt} = \alpha_j + E_{jt}$$

and is estimated by

$$\widehat{AR}_{jt} = R_{jt} - \hat{\beta}_j R_{mt}$$

The β_j represent risk, e.g. lower beta funds do better in a falling market. The abnormal returns for each year (τ) were combined into an annual percentage for each fund:

$$\widehat{AR}_{j\tau} = \left[\prod_{t=4\tau-3}^{4\tau} \{(\widehat{AR}_{jt}/100) + 1\} - 1 \right] \times 100\%$$

The funds could then be ranked on the basis of this index and divided into quartiles. Subsequently the transition matrix between quartiles from one year to the next, the states $1, \ldots, 4$ representing the upper quartile, $\ldots$, lower quartile respectively, could be estimated.

	1	2	3	4
1	0.30	0.28	0.22	0.19
2	0.23	0.29	0.28	0.21
3	0.26	0.26	0.25	0.23
4	0.22	0.17	0.24	0.37

According to the efficient market hypothesis, given free flows of information and capital, without any regulatory or cultural constraints, the price of assets should fully reflect all relevant information. Consequently an investor cannot consistently do better than the market average without taking extra risk, and investment fund managers may not be worth their fees. If the efficient market hypothesis holds, we cannot expect any consistency of performance of funds and all the entries in the transition matrix would be 0.25. There is evidence against this hypothesis and, in particular, the pension funds that display top quartile performance in one period have the highest estimated probability of displaying top quartile performance in the next. The apparent persistence effect is not dramatic, but the authors took care to demonstrate that the result was robust to alternative samples and methods of analysis. They estimated separate transition matrices for each year with qualitatively similar results, except for 1990/91 when markets fell. In order to allow for more consistency of performance than can be accounted for with a Markov chain, they looked at the number of funds which were in the top quartile for at least five out of seven years. Twelve out of 409 funds achieved this, which is significantly more than the 5.28 expected under the efficient market hypothesis.

Example 8.17

Lambin (1997) gives the following transition matrix (A) for brand switching in the heavy-duty truck market in Belgium. The source of the data is the MDA Consulting Group in Brussels. The transition probabilities are usually estimated through a survey or based on panel data. The brands are Daf (D), Mercedes (M), Renault (R), Scania (S), Volvo (V) and others (O).

$$
\begin{array}{c c c c c c c}
 & D & M & R & S & V & O \\
D & 0.562 & 0.153 & 0.013 & 0.023 & 0.114 & 0.135 \\
M & 0.082 & 0.595 & 0.024 & 0.023 & 0.111 & 0.165 \\
R & 0.090 & 0.092 & 0.530 & 0.050 & 0.012 & 0.226 \\
S & 0.081 & 0.133 & 0.000 & 0.656 & 0.061 & 0.069 \\
V & 0.165 & 0.129 & 0.012 & 0.017 & 0.600 & 0.077 \\
O & 0.117 & 0.171 & 0.048 & 0.029 & 0.103 & 0.532
\end{array}
$$

The market shares in period t were

$$p^{(t)} = (0.076, 0.163, 0.032, 0.033, 0.062, 0.634)$$

The predicted market shares for period $t+1$ are

$$p^{(t+1)} = p^{(t)}A$$

The matrix multiplication gives

$$p^{(t+1)} = [0.146, 0.232, 0.053, 0.048, 0.132, 0.389]$$

The stationary probability vector v is the solution of

$$v = vA$$

$$v = [0.205, 0.265, 0.046, 0.066, 0.200, 0.218]$$

If the transition matrix remained unchanged, after about 10 years Mercedes would have increased its market share to 26.5% mainly at the expense of O. Let $MS(t)$ be market share at time t. Then

$$MS(t+1) = \alpha MS(t) + \beta(1 - MS(t))$$

where α denotes the loyalty rate and β denotes the attraction rate. The loyalty rate is the proportion of buyers who, having bought brand X in period t, buy brand X in period $t+1$. The attraction rate is the proportion of buyers who, having bought a competing brand in period t, buy brand X in period $t+1$. For equilibrium

$$MS(t+1) = MS(t) = MS$$

and hence

$$MS = \beta/(1 + \beta - \alpha)$$

In the case of Daf we know MS is equal to 0.205 and α is 0.562. It follows that the loyalty rate β is 0.113. This is a weighted average of the transition probabilities from other brands to Daf, the weights being proportional to their long-run market shares.

8.7.2 Absorbing Markov chains

A Markov chain is *irreducible* if it is possible to move between any pair of states, though not necessarily in one step. A regular chain is irreducible, but the converse is not necessarily true, because there may be periodic states. A state *i* has a *period d* if when the chain starts in *i*, subsequent occupations of *i* can only occur at times which are multiples of *d*, i.e. *d*, 2*d*, 3*d*,.... A regular chain with no periodic states is called *aperiodic*. An aperiodic irreducible chain is regular. If an irreducible chain has periodic states, there is still a probability vector v satisfying equation (8.2), but $p^{(t)}$ does not tend to the limit v as *t* tends to infinity.

A state in a Markov chain is an *absorbing state* if it is impossible to leave it. A Markov chain is absorbing if it has at least one absorbing state. An absorbing chain is reducible, because the process must eventually end in an absorbing state. We will now find expressions for the probabilities of ending up in each of the absorbing states for each starting state, and the expected number of steps until absorption.

Canonical form of the transition matrix

Assume there are *r* states of which *s* are non-absorbing and (*r* − *s*) are absorbing. The transition matrix can be written in a partitioned form, known as the *canonical form*, as

$$M = \begin{pmatrix} I & O \\ R & Q \end{pmatrix}$$

where: I is the (*r* − *s*) × (*r* − *s*) identity matrix; O is the (*r* − *s*) × *s* zero matrix; R is an *s* × (*r* − *s*) matrix; and Q is an *s* × *s* matrix. Powers of this transition matrix will be of the form

$$M^t = \begin{pmatrix} I & O \\ * & Q^t \end{pmatrix}$$

where $*$ is some *s* × (*r* − *s*) matrix. The elements of Q^t are the *t*-step transition probabilities between the non-absorbing states. Since $Q^t \to 0$

$$I + Q + Q^2 + Q^3 + \cdots = (I - Q)^{-1}$$

This result is analogous to summing a geometric progression, and can easily be verified by pre-multiplying both sides by $(I - Q)$. The *fundamental matrix* is defined as:

$$N = (I - Q)^{-1}$$

Number of steps until absorption

Assume the process starts in the non-absorbing state *i*. Consider a non-absorbing state *j*, and define

$$u_{ij}^{(s)} = \begin{cases} 1 & \text{if the process is in state } j \text{ at end of step } s \\ 0 & \text{otherwise} \end{cases}$$

Then

$$E[u_{ij}^{(s)}] = 1 \times p_{ij}^{(s)} + 0 \times (1 - p_{ij}^{(s)}) = p_{ij}^{(s)}$$

Define $w_{ij}^{(t)}$ as the total number of times the process has been in state j, starting from i, at the end of t steps. Then,

$$w_{ij}^{(t)} = u_{ij}^{(0)} + u_{ij}^{(1)} + \cdots + u_{ij}^{(t)}$$

For example, suppose a chain goes: state 4; state 3; state 4; state 4; absorbing state. Then $u_{44}^{(0)} = 1$; $u_{44}^{(1)} = 0$; $u_{44}^{(2)} = 1$; $u_{44}^{(3)} = 1$; $u_{44}^{(4)} = 0$; and $w_{44}^{(4)} = 3$. Taking expectation gives

$$\mathrm{E}[w_{ij}^{(t)}] = \mathrm{E}[u_{ij}^{(0)}] + \mathrm{E}[u_{ij}^{(1)}] + \cdots + \mathrm{E}[u_{ij}^{(t)}] = p_{ij}^{(0)} + p_{ij}^{(1)} + \cdots + p_{ij}^{(t)}$$

which is the entry in the ith row and jth column of $I + Q + \cdots + Q^t$. It follows that the elements of N are given by the equation

$$n_{ij} = \lim_{t \to \infty} (\mathrm{E}[w_{ij}^{(t)}])$$

so the entries of N are the mean number of times in each non-absorbing state for each possible non-absorbing starting state. Adding the components of each row of N will give the mean number of steps before being absorbed for each possible non-absorbing starting state.

Mean first passage time for irreducible chain

In the pension fund example we can calculate the expected time until a lower quartile fund makes the upper quartile. This is done by changing 1 to an absorbing state, and calculating the mean number of steps until absorption for the newly constructed chain.

The modified transition matrix with the empty state made absorbing is:

$$
\begin{array}{c c c c c}
 & 1 & 2 & 3 & 4 \\
1 & \begin{pmatrix} 1 & 0 & 0 & 0 \\ 2 & 0.23 & 0.29 & 0.28 & 0.21 \\ 3 & 0.26 & 0.26 & 0.25 & 0.23 \\ 4 & 0.22 & 0.17 & 0.24 & 0.37 \end{pmatrix}
\end{array}
$$

Therefore

$$Q = \begin{pmatrix} 0.29 & 0.28 & 0.21 \\ 0.26 & 0.25 & 0.23 \\ 0.17 & 0.24 & 0.37 \end{pmatrix}$$

and

$$N = \begin{pmatrix} 2.09 & 1.14 & 1.11 \\ 1.02 & 2.06 & 1.09 \\ 0.95 & 1.09 & 2.03 \end{pmatrix}$$

The mean number of years until a lower quartile fund moves into the upper quartile is the sum of 0.95, 1.09 and 2.03 which equals 4.07.

Probabilities of ending in each absorbing state

Let b_{ij} be the probability that an absorbing chain will be absorbed in state j if it starts in the non-absorbing state i. We can go directly to j, or go via any of the non-absorbing states, i.e.

$$b_{ij} = p_{ij} + \sum p_{ik}b_{kj}$$

where the summation is over all the non-absorbing states. In matrix terms

$$B = R + QB$$
$$B - QB = R$$
$$B(I - Q) = R$$
$$B = (I - Q)^{-1}R = NR$$

Example 8.18
A communications satellite can be in any one of four states:

  0   failed
  1   badly positioned
  2   slight deviation from correct position
  3   correctly positioned

We take 0 and 3 as absorbing states. Corrective action can be taken from an earth station. The transition matrix between attempts at corrective action is

$$\begin{array}{cccc} & 0 & 3 & 1 & 2 \\ \begin{matrix} 0 \\ 3 \\ 1 \\ 2 \end{matrix} & \begin{pmatrix} 1 & 0 & 0 & 0 \\ 0 & 1 & 0 & 0 \\ 0.2 & 0 & 0.1 & 0.7 \\ 0.2 & 0.7 & 0 & 0.1 \end{pmatrix} \end{array}$$

in canonical form. Therefore

$$Q = \begin{pmatrix} 0.1 & 0.7 \\ 0 & 0.1 \end{pmatrix} \quad R = \begin{pmatrix} 0.2 & 0 \\ 0.2 & 0.7 \end{pmatrix}$$

and it follows that

$$N = \begin{pmatrix} 1.11 & 0.86 \\ 0 & 1.11 \end{pmatrix} \quad B = NR = \begin{pmatrix} 0.40 & 0.60 \\ 0.22 & 0.78 \end{pmatrix}$$

If the satellite starts out badly positioned, there is a 60% chance of correcting it.

Example 8.19
Two companies A and B are involved in a hostile advertising campaign. At each round A has a probability of 0.4 of putting B out of business and B has a probability of 0.2 of putting A out of business. What are the possible eventual outcomes and how likely are they? Define four states of which three are

absorbing: state 0 is both companies out of business, 1 is B out of business, 2 is A out of business, and 3 is both companies in business.

$$
\begin{array}{c@{\ }c@{\quad}c@{\quad}c@{\quad}c}
 & 0 & 1 & 2 & 3 \\
0 & 1 & 0 & 0 & 0 \\
1 & 0 & 1 & 0 & 0 \\
2 & 0 & 0 & 1 & 0 \\
3 & 0.08 & 0.32 & 0.12 & 0.48
\end{array}
$$

$$
\begin{aligned}
B &= (0.08 \quad 0.32 \quad 0.12)/0.52 \\
&= (0.15 \quad 0.62 \quad 0.23)
\end{aligned}
$$

There is a 15% chance of both going out of business and a 62% chance that A puts B out of business whilst remaining in business itself. You are asked to look at a three-company case in Exercise 8.27.

8.7.3 Markov processes

In a Markov chain, time is modelled as a sequence of discrete steps. For many applications, such as queueing processes and machine breakdown and repair cycles, it is more convenient to model time as continuous. A Markov process has a discrete state space and can change state in any time interval $(t, t + \delta t)$. The Markov property is that the probabilities of state transitions in the interval $(t, t + \delta t)$ depend on the state of the process at time t but not on its history. We will obtain a general form for the Markov process by applying a limiting argument to a Markov chain. For a Markov chain

$$
p^{(t+1)} = p^{(t)} M
$$

and hence

$$
p^{(t+1)} - p^{(t)} = p^{(t)}(M - I)
$$

Now suppose the time step has length δt and, for convenience of notation, lower the superscript:

$$
p(t + \delta t) - p(t) = p(t)(M - I)
$$

Divide by δt and take the limit as δt tends to zero to obtain

$$
\dot{p}(t) = p(t)L \tag{8.3}
$$

where

$$
\dot{p}(t) = \frac{dp(t)}{dt} \quad \text{and} \quad L = \lim_{\delta t \to 0} [(M - I)/\delta t]
$$

The matrix L is known as the rate matrix. The elements of M are $p_{ij}(\delta t)$ and, since $p_{ij}(0)$ is 1 if i equals j and 0 otherwise, the elements of I are $p_{ij}(0)$. It follows that a typical element of the rate matrix (λ_{ij}) is

$$\lambda_{ij} = \lim_{\delta t \to 0} [(p_{ij}(\delta t) - p_{ij}(0))/\delta t]$$

$$= \left[\frac{d}{dt} p_{ij}(t)\right]_{\text{evaluated at } t=0}$$

Conversely, for small δt,

$$p_{ij}(\delta t) \simeq \lambda_{ij} \delta t \quad \text{when } i \neq j$$

$$p_{ii}(\delta t) \simeq 1 + \lambda_{ii} \delta t \quad \text{when} \quad i = j$$

Since $\sum_j p_{ij}(\delta t) = 1$, it follows that

$$\sum_j \lambda_{ij} = 0$$

That is, the rows of the rate matrix must add to 0, which is a useful check in applications. A full solution for the system involves the solution of the linear simultaneous differential equations,

$$\dot{p}(t) = p(t)L$$

but practical applications are generally ergodic and the $p(t)$ tend to constant values as t increases. In such cases $\dot{p}(t)$ will tend to 0 and the steady-state v is given by the solution of the algebraic equation, known as the normal equations:

$$vL = 0 \tag{8.4}$$

together with the constraint $\sum v_i = 1$. The elements of v are the proportion of time the process spends in the corresponding states over a long period.

Example 8.20
A machine is required to operate continuously but sometimes fails. It is therefore always in one of two states: working (1) or under repair (0). We can represent the process by $X(t)$ where

$$X(t) = \begin{cases} 0 & \text{under repair} \\ 1 & \text{working} \end{cases}$$

The time until failure has an exponential distribution with a mean $1/\alpha$, and the repair time has an exponential distribution with a mean $1/\beta$ (Section 3.2.2). Alternatively we can specify the failure and repair rates as α and β. Let δt be a small time increment during which the machine can change state once or remain in its present state. For instance, if it is working now, the probability of failure is $\alpha \delta t$. We can ignore the possibility of two changes of state within time δt, failure and repair for example, because the corresponding probability would be of order $(\delta t)^2$. In the limit as δt tends to zero the following relationships become exact:

$$p_0(t + \delta t) = p_0(t)(1 - \beta \delta t) + p_1(t)\alpha \delta t$$

$$p_1(t + \delta t) = p_0(t)\beta \delta t + p_1(t)(1 - \alpha \delta t)$$

These can be rearranged to give

$$\frac{p_0(t + \delta t) - p_0(t)}{\delta t} = -\beta p_0(t) + \alpha p_1(t)$$

$$\frac{p_1(t + \delta t) - p_1(t)}{\delta t} = \beta p_0(t) - \alpha p_1(t)$$

and when $\delta t \to 0$

$$\dot{\boldsymbol{p}}(t) = \boldsymbol{p}(t) \begin{pmatrix} -\beta & \beta \\ \alpha & -\alpha \end{pmatrix}$$

where

$$\boldsymbol{p}(t) = (p_0(t), p_1(t))$$

Since $p_0(t) + p_1(t) = 1$, we can substitute for $p_0(t)$ and obtain a differential equation for $p_1(t)$:

$$\dot{p}_1(t) = \beta(1 - p_1(t)) - \alpha p_1(t)$$

If the machine is working at time 0, the initial condition is $p_1(0) = 1$, and the solution is

$$p_1(t) = \frac{\beta}{\alpha + \beta} + \frac{\alpha \exp[-(\alpha + \beta)t]}{\alpha + \beta}$$

As t becomes large $p_1(t)$ tends towards the constant, $\beta/(\alpha + \beta)$. The process is ergodic because this does not depend on the initial condition. In the long term the machine will be working a proportion $\beta/(\alpha + \beta)$ of time. We could obtain this result (v_1) more quickly by solving

$$\boldsymbol{v} \begin{pmatrix} -\beta & \beta \\ \alpha & -\alpha \end{pmatrix} = 0$$

with

$$v_0 + v_1 = 1$$

Example 8.21
The notation M/M/1 queue stands for exponential arrival times, exponential service times and 1 server. The proprietor of a small music shop, which sells CDs and sheet music, is the sole assistant. Customers arrive according to an exponential distribution at a rate α per minute. Service times are exponentially distributed with a mean of $1/\beta$ minutes. Define the Markov process $X(t)$ as the number of people in the shop at time t. Then

$$p_0(t + \delta t) = p_0(t)(1 - \alpha \delta t) + p_1(t)\beta \delta t$$

$$p_1(t + \delta t) = p_0(t)\alpha \delta t + p_1(t)(1 - \alpha \delta t - \beta \delta t) + p_2(t)\beta \delta t$$

$$p_2(t + \delta t) = p_1(t)\alpha \delta t + p_2(t)(1 - \alpha \delta t - \beta \delta t) + p_3(t)\beta \delta t$$

$$\vdots$$

Rearrangement gives the rate matrix

$$L = \begin{pmatrix} -\alpha & \alpha & 0 & 0 & 0 & \dots \\ \beta & -(\alpha+\beta) & \alpha & 0 & 0 & \dots \\ 0 & \beta & -(\alpha+\beta) & \alpha & 0 & \dots \\ 0 & 0 & \beta & -(\alpha+\beta) & \alpha & \dots \end{pmatrix}$$

The normal equations are:

$$-\alpha v_0 + \beta v_1 = 0$$
$$\alpha v_0 - (\alpha + \beta)v_1 + \beta v_2 = 0$$

and so on. The solution is

$$v_i = (\alpha/\beta)^i v_0$$

The constraint $\sum v_i = 1$, and the formula for the sum of a geometric progression (Appendix 1), give

$$v_0 = 1 - \alpha/\beta$$

provided $\alpha < \beta$. This is simply a requirement that the service rate is faster than the arrival rate. The average number of people in the shop is

$$\sum_{i=0}^{\infty} i v_i = \alpha/(\beta - \alpha)$$

It follows that the average waiting time, before service, is $\alpha/[\beta(\beta - \alpha)]$, and that the average time spent in the shop is $1/(\beta - \alpha)$. Some other standard queueing and machine repair problems are given in Exercises 8.29 and 8.30.

The assumptions of exponential distributions may not always be realistic, but the models tend to give a worst-case scenario for the given means of the distributions. Generally the higher the relative variability the longer the queues. To take an extreme example, if customers arrived at exactly one-minute intervals and service took exactly one minute there would be no queue. As soon as random variation is introduced into the model the queue will become infinite if the process continues indefinitely. Theoretical results for the case of one of the exponential distributions in queueing models being replaced by a general distribution are available. For instance, $M/G/r$ queueing models have exponential arrivals, a general distribution of service times with some specified mean and standard deviation, and r servers. Bunday (1996) gives a clear explanation of the derivation of these results, as well as many interesting applications of queueing theory. However, the analytic models are all rather idealised and simulations may give more realistic, if less elegant, solutions for practical systems.

8.8   Summary

1. An index number is a measure of changes in price over time of a group of items. The RPI is constructed to measure changes in the cost of living.

2. A time series is a sequence of measurements of a variable, usually considered to be continuous, at a discrete set of times.
3. The usual objectives of time-series analysis include the following.

(i) To make short-term forecasts. For example, we need to predict sales before planning production and setting work schedules.
(ii) To make longer-term forecasts for strategic planning. Simulation studies for different scenarios will give an indication of the range within which future values of the variable are likely to lie.
(iii) To summarise past performance.

4. Time series may include a trend and a seasonal pattern. These features often account for a substantial part of the variability in the series.
5. A stochastic process is stationary if the mean and variance–covariance structure do not change over time. If a time series is seasonally adjusted, and has any trend subtracted, it can reasonably be considered a realisation of a stationary process.
6. ARMA processes express a variable as a linear combination of its past values and a sequence of independent random variables (DWN). ARMA processes often provide a plausible model for the stationary process generating a time series. A potential ARMA model can be identified from the sample acf and pacf. If the chosen ARMA model is suitable, the residuals should appear to be uncorrelated.
7. AR(p) processes can easily be fitted by multiple regression. Multiple regression is also a convenient way of including other time-series variables which may improve predictions. Another aspect of multivariate time-series analysis is the forecasting of several inter-correlated variables.
8. Non-linear models for time series can account for sudden jumps and chaotic behaviour.
9. In a Markov chain the state variable moves between a discrete set of states at a discrete set of times. The Markov property is that the probability of moving from one state to any other state depends only on the present state, and not on the history of the process. If there are no absorbing states, the proportions of time spent in the various states tend towards constant values over a long period. These values are the entries in the stationary probability vector.
10. A Markov process is the continuous time version of the Markov chain. It provides a useful model for queueing processes, including machine maintenance.

Exercises

8.1 The monthly RPI index values for January until December 1996 are: 150.2, 150.9, 151.5, 152.6, 152.9, 153.0, 152.4, 153.1, 153.8, 153.8, 153.9, 154.4. Calculate a yearly average value by

(i) assuming you buy 100 baskets of goods each month
(ii) assuming you spend 1000 units each month.

Compare these values with the ONS figure of 152.7. Which method do you think is the more appropriate?

8.2 The Irving–Fisher price index is $\sqrt{LI_t \times PI_t}$ and the Marshall–Edgeworth index is

$$\frac{\sum_{i=1}^{m}(q_{it} + q_{i0})p_{it}}{\sum(q_{it} + q_{i0})p_{i0}}$$

Calculate the value of these for the data in Section 8.2.2.

8.3 A Laspeyre quantity index gives the ratio of the value of production now to the value of production in the base year at base year prices. A small computer manufacturer has sold the following items over the past four years.

Item	1995 price	Sales			
		1995	1996	1997	1998
PC base	500	900	950	980	1030
PC monitor	200	800	1000	960	1050
PC keyboard	100	400	500	600	800
Cables	2	1000	950	870	540

Calculate a quantity index weighted according to 1995 prices, taking 1995 as a base year of 100.

8.4 By considering $\text{var}(X_t + X_{t+k}) \geq 0$, and $\text{var}(X_t - X_{t+k}) \geq 0$, show that $-1 \leq \rho(k) \leq 1$.

8.5 (i) Minitab includes a routine to fit the Pearl–Reed logistic model

$$Y_t = 1/(\beta_0 + \beta_1(\beta_2^t))$$

Show how parameters are related to a, b and c in the form

$$Y_t = a/(1 + be^{-ct})$$

(ii) Minitab includes a routine to fit the exponential growth trend model

$$Y_t = \beta_0\beta_1^t + E_t$$

Show how this differs from fitting a regression line

$$\ln(y_t) = a + bt$$

8.6 The following data are sales (1000 litres) of kiwi fruit sorbet over the past five years on a small island. There are two seasons, cool and hot, and sales are higher in the latter. There has been an extensive advertising campaign over the past five years.

$$17 \quad 148 \quad 86 \quad 285 \quad 97 \quad 213 \quad 72 \quad 375 \quad 4 \quad 301$$

Analyse the data using CMA variant 1 and the Minitab Decomposition algorithm, with and without the last datum.

8.7 Sales of a robot lawn-mower over the past four years have been 2, 4, 8 and 8 thousand units respectively. Fit a cubic trend curve,

$$y = \beta_0 + \beta_1 t + \beta_2 t^2 + \beta_3 t^3$$

to the data:

t	1	2	3	4
y	2	4	8	8

Plot your curve, y against t, for t from 0 to 5. Is the forecast for next year $(t = 5)$ reasonable?

8.8 Consider the AR(2) process

$$X_t = \alpha_1 X_{t-1} + \alpha_2 X_{t-2} + E_t$$

This is stationary if, and only if, θ defined by

$$\theta = \alpha_1 \pm \sqrt{(\alpha_1^2 + 4\alpha_2)}/2$$

is such that $|\theta| < 1$. Show that if θ satisfies this condition, and θ is real, then

$$\alpha_1 + \alpha_2 < 1 \quad \text{and} \quad \alpha_1 - \alpha_2 > -1$$

Show that if θ satisfies this condition, and θ is complex, then

$$\alpha_2 > -1$$

8.9 Find the acf of the AR(2) process,

$$X_t = X_{t-1} - 0.5X_{t-2} + E_t$$

Find the variance of $\{X_t\}$ in terms of the variance of $\{E_t\}$ by taking variance of both sides of the defining equation and using the value of $\rho(1)$.

8.10 Show that the MA(1) models

(a) $X_t = Z_t + \beta Z_{t-1}$

and

(b) $X_t = Z_t + (1/\beta)Z_{t-1}$

have the same acf. Now express Z_t in terms of $X_t, X_{t-1}, \ldots$

Show that if $|\beta| < 1$ the series for model (a) converges. Deduce that the estimation procedure described in Section 8.3.5 will naturally lead to model (a). Model (a) is said to be invertible. The general definition of invertibility for an MA(q) process is that the roots of

$$\theta(B) = (\beta_0 + \beta_1 B + \cdots + \beta_q B^q) = 0$$

lie outside the unit circle. Verify that this is consistent with model (a) being invertible when $|\beta| < 1$.

8.11 Write the ARMA(2, 2) process,

$$X_t = \tfrac{3}{2}X_{t-1} - \tfrac{1}{2}X_{t-2} + Z_t - Z_{t-2}$$

in the form

$$\phi(B)X_t = \theta(B)E_t$$

Factorise the polynomials and show that this ARMA(2, 2) process is equivalent to the ARMA(1, 1) process

$$X_t = \tfrac{1}{2}X_{t-1} + Z_t + Z_{t-1}$$

This ARMA(2, 2) model exhibits redundancy. With some data sets it is possible to identify near redundant models using Minitab. All the coefficients of some high-order model can be statistically significant although the standard deviation of the errors is no smaller than it is for a simpler model.

8.12 Let $x_{i:n}$ be the ith largest in a random sample of n from a distribution with cdf $F(\ )$. Then, to a reasonable approximation,

$$F(x_{i:n}) = (i - 0.4)/(n + 0.2)$$

(a) If F is the Weibull cdf, i.e.

$$F(x) = 1 - e^{-(x/b)^a}$$

show that a line through a plot of $\ln(x_{i:n})$ against $\ln(-\ln(1 - (i - 0.4)/(n + 0.2)))$ will have gradient $1/a$ and an intercept of $\ln b$.

(b) The mean and variance of the Weibull distribution are

$$\mu = b\Gamma(1 + 1/a)$$

$$\sigma^2 = b^2[\Gamma(1 + 2/a) - \Gamma^2(1 + 1/a)]$$

Compare the mean and variance of the Weibull distribution fitted to the positive residuals in Figure 8.10(a) with their mean and standard deviation of 35.71 and 34.69 respectively.

(c) Ten pieces of electric cable were subjected to an accelerated life test for 100 days. Two lasted 100 days, but the rest failed after

$$30 \quad 39 \quad 46 \quad 51 \quad 58 \quad 66 \quad 80 \quad 97$$

days respectively. Use graphical methods to fit a Weibull distribution.

8.13 Suppose

$$Y_t = a + bt + E_t$$

Show that ∇Y_t is MA(1) with mean b, whereas $Y_t - (a + bt)$ is DWN.

8.14 Show that summing

$$\sum_1^t$$

is the inverse of differencing, ∇, if x_0 is taken to be 0 by demonstrating:

(i) $\nabla\left(\sum_{i=1}^t x_i\right) = x_t$

(ii) $\sum_{i=1}^t (\nabla x_i) = x_t$

8.15 An ARIMA$(2, 1, 1)(0, 1, 1)12$ model was fitted to the sweet white wine series. The Minitab parameter estimates were: AR1 -1.4387, AR2 -0.4722, MA1 -0.9772, SMA12 -0.7316, Constant -0.885.
Express the fitted model in the forms

(i) $\phi(B)(1 - B)(1 - B^{12})Y_t = \theta(B)\theta_{12}(B^{12})E_t$

(ii) $y_t = \ldots$

8.16 Consider the model

$$Y_t = m_t + E_t$$

where E_t is zero mean DWN with variance σ_E^2, $E_0 = 0$ and m_t are monthly seasonal effects, i.e. m_1 represents the January effect, $\ldots$, m_{12} represents the December effect, and $m_{i+12} = m_i$. Then

$$\nabla_{12} Y_t = E_t - E_{t-12}$$

and

$$Y_{12} = Y_0 + E_{12}$$

$$Y_{24} = Y_{12} + E_{24} - E_{12} = Y_0 + E_{24}$$

and so on.

Now suppose that the model has been identified from a fairly short time series as

$$\nabla_{12} Y_t = E_t - 0.9E_{t-12}$$

Show that

$$Y_{36} = Y_0 + E_{36} + 0.1E_{24} + 0.1E_{12}$$

What is the variance of Y after n years?

8.17 Let X be a variable with a mean μ and variance σ^2, and Y be $f(X)$. The first three terms of a Taylor expansion about μ give:

$$Y \simeq f(\mu) + f'(\mu)(X - \mu) + f''(\mu)(X - \mu)^2/2$$

Taking expectation gives

$$E[Y] \simeq f(\mu) + f''(\mu)\sigma^2/2$$

Use this result to show that if $Y = X^{\frac{1}{2}}$ then

$$E[Y] \approx \mu^{\frac{1}{2}} - \sigma^2/(8\mu^{\frac{3}{2}})$$

8.18 Fit a linear trend to the spend at 1989 prices in Table 8.4. Calculate the residuals and use an EWMA to forecast the stationary series for $t = 10$ and 11. Add these forecasts to the projected trend and compare your results with those in Example 8.9.

8.19 Take random numbers from $N(0, 1)$ and investigate realisations from the model

$$X_t = (0.6 + 0.3E_t)X_{t-1} + E_t$$

Repeat using random numbers from an $M(1)$ distribution (exponential mean 1) with 1 subtracted.

8.20   Refer to Section 8.6.2.

   (a)   If $y^3 = x + kwy$, and w is treated as a constant, show that $dx/dy = 0$ when $y^2 = kw/3$. Find the corresponding values of x.
   (b)   Show that the range of values taken by y in the equation

$$x = (3y^3 - 3kwy)/(3 + kw)$$

   is more realistic than in the equation of (a).

8.21   Write a program to calculate the logistic mapping

$$y_{t+1} = ky_t(1 - y_t)$$

from $y_1 = 0.1$ as far as y_{1000}, for any value of k between 0 and 4, and to plot y_t against t, and y_{t+1} against y_t. By setting $y_{t+1} = y_t$ find the value of the fixed point, in terms of k, for $0 < k < 3$. Check your answers by running the program. Investigate the behaviour of the mapping for k from 3 to 4 in steps of 0.1.

8.22   Goodwin (1990). Let v_t be employment and u_t be labour cost. The von Neumann model in a dimensionless form is

$$v_{t+1}/v_t = a(1 - v_t)v_t - u_t$$

$$u_{t+1}/u_t = 3.258v_t$$

where u_t, v_t are between 0 and 1. Investigate this system by computer simulation for $a = 2$, 2.8, 3.4, 3.5, 4.0. The output of your program should include a plot of v_t against u_t, and a plot of both u_t and v_t against t.

8.23   (After Guttorp, 1995.) The integral of a Wiener process is written formally as

$$X(t) = \int_0^t W(s)\, dW(s)$$

   Approximate $X(t)$ by the sum

$$S_n = \sum_{i=1}^n W(t_{i-1})(W(t_i) - W(t_{i-1}))$$

   (i)   Prove the identity

$$\sum_{i=1}^n W(t_{i-1})(W(t_i) - W(t_{i-1})) = \frac{W^2(t_n)}{2} - \frac{1}{2}\sum(W(t_i) - W(t_{i-1}))^2$$

   (ii)   Explain why

$$\sum_1^n (W(t_i) - W(t_{i-1}))^2 \simeq \sigma^2 t$$

   where σ^2 is the diffusion coefficient of $W(t)$.
   (iii)   Deduce that

$$X(t) = W^2(t)/2 - \sigma^2 t/2$$

(iv) Take $\sigma^2 = 1$, and interpret the differential notation form of (iii)

$$d(W^2(t)) = 2W(t)\,dW(t) + dt$$

(v) The Taylor expansion for a twice differentiable function f can be written

$$df(x) = f'(x)\,dx + f''(x)(dx)^2/2$$

Take $f(x) = x^2$ and $x = W(t)$ and show that

$$d(W^2(t)) = 2W(t)\,dW(t) + (dW(t))^2$$

Deduce that $(dW(t))^2 = dt$ and notice that the square of the stochastic differential element is not negligible.

(vi) Suppose that $Y(t)$ satisfies the stochastic differential equation

$$dY = \mu(Y)\,dt + \sigma(Y)\,dW$$

Let f be a twice differentiable function and $X(t) = f(Y(t))$. Show that X satisfies the stochastic differential equation

$$dX = (f'(Y)\mu(Y) + \tfrac{1}{2}f''(Y)\sigma^2(Y))\,dt + f'(Y)\sigma(Y)\,dW$$

This is referred to as Ito's formula.

8.24 A dealer can order mini-computers at the end of one week for delivery at the beginning of the next. If he has 0 or 1 computer in stock at the end of a week he places an order for 2, otherwise no order is placed. Let D_k be the demand for computers during the kth week. Assume that the demand in any week is independent of the demand in previous weeks, and that

$$\Pr\{D_k = 0\} = 0.4$$
$$\Pr\{D_k = 1\} = 0.2$$
$$\Pr\{D_k = 2\} = 0.2$$
$$\Pr\{D_k = 3\} = 0.2$$

Note that demand can exceed the stock. The dealer starts with three mini-computers in stock.

Let X_k be the number of computers in stock at the end of the kth week. The sequence $\{X_k\}$ is a Markov chain.

(i) Write down the one-step transition matrix.
(ii) Find the probability that he will have no computers at the end of the third week.
(iii) Find the stationary probability vector and deduce that the long-run proportion of times that he has no computers in stock at the end of the week is $1/3$.

8.25 A car manufacturer offers three models, a saloon, an open top sports car and an estate. In order to balance the work content of the assembly line the following rules are applied.

(i) Estates are separated by at least one other model.
(ii) A sports car is followed by a saloon.

Set up a suitable transition matrix using letters for unspecified probabilities. Suggest values for these probabilities that would give 50% saloons, 10% sports cars and 40% estates.

8.26 A small business can be in one of four states at the end of the financial year: expanding (3), satisfactory (2), in difficulties (1), bankrupt (0). Expanding and bankrupt will be considered absorbing states. The transition matrix is

$$
\begin{array}{c}
0 \\ 1 \\ 2 \\ 3
\end{array}
\begin{pmatrix}
1 & 0 & 0 & 0 \\
0.25 & 0.25 & 0.50 & 0 \\
0 & 0.25 & 0.25 & 0.50 \\
0 & 0 & 0 & 1
\end{pmatrix}
$$

Find the probabilities it will become bankrupt if (i) it is now in state 1, (ii) it is now in state 2, (iii) it is somewhere between states 1 and 2.

8.27 A cartel of three car manufacturers, A, B and C have fallen out. They are now fighting an advertising campaign. At each round A has probability $\frac{1}{2}$ of putting a rival out of business by comparative advertisements. The corresponding probabilities for B and C are $\frac{1}{3}$ and $\frac{1}{6}$ respectively. Each company attacks the strongest opponent still in business. Find the probabilities that A, B and C will stay in business.

8.28 The Chapman–Kolmogorov equations for a Markov process are

$$ p_{ij}(t+\tau) = \sum p_{ik}(t)p_{kj}(\tau) $$

Differentiate both sides partially with respect to τ, set τ to 0, and hence deduce equation (8.3).

8.29 Consider these examples of M/M/1 and M/M/2 queues with bounded storage. A small ship repair yard has two dry docks for two ships including the one being worked on. Local ships need repair at a rate of λ per day. There is one repair team, and the repair time is exponentially distributed with mean $1/\nu$ days. If a ship needs repair and the yard has a spare dry dock the ship will be sent to the yard, but if there is no free dock it will be sent elsewhere. Show that the proportion of time that both docks are in use is

$$ (\lambda/\nu)^2 / [1 + (\lambda/\nu) + (\lambda/\nu)^2] $$

and deduce the average number of lost repair jobs per day. Now suppose the ship repair yard has expanded. There are two repair teams and four dry docks. Explain why the rate matrix is:

$$
\begin{array}{c}
 \\ 0 \\ 1 \\ 2 \\ 3 \\ 4
\end{array}
\begin{array}{ccccc}
0 & 1 & 2 & 3 & 4 \\
\end{array}
\begin{pmatrix}
-\lambda & \lambda & 0 & 0 & 0 \\
\nu & -(\nu+\lambda) & \lambda & 0 & 0 \\
0 & 2\nu & -(2\nu+\lambda) & \lambda & 0 \\
0 & 0 & 2\nu & (-2\nu+\lambda) & \lambda \\
0 & 0 & 0 & 2\nu & -2\nu
\end{pmatrix}
$$

8.30 There are N identical machines in a factory. The machines are expected to operate continuously, but breakdowns occur. Each machine has an operating time that is exponentially distributed with mean $1/\theta$. Machines are also independent of each other. There is one repair crew, and the time taken to repair a machine is exponentially distributed with mean $1/\lambda$. A repair is completed before work starts on the next machine waiting for repair.

(i) Write down the rate matrix L.

(ii) Write down the normal equation

$$vL = 0$$

and verify that the solution is

$$v_i = \frac{1}{i!}\left(\frac{\lambda}{\theta}\right)^i v_0 \quad \text{for } i = 0, 1, \dots, N$$

where

$$v_0 = \left[\sum (\lambda/\theta)^i / i!\right]^{-1}$$

(iii) This is known as the Erlang telegraph formula. If N is the number of lines from a telephone exchange, interpret θ and λ in this context. What does v_0 represent? [Agner Erlang (1878–1929) was a Danish mathematician who worked for the Copenhagen Telephone Company.]

9

Optimisation strategies

Many optimisation problems are highly structured, and there are very efficient methods for their solution. The use of trial-and-error methods, including stochastic variants such as genetic algorithms, would be absurdly inefficient, at best, and quite impractical with a large number of variables.

9.1 Linear programming

A linear programming problem is to maximise, or minimise, a linear combination of variables subject to linear inequalities.

9.1.1 Graphical method

The specification for Enco engine oil is: specific gravity (SG) of at least 0.88, viscosity of at most 32 and a sulphur content below 0.35%. A manager intends blending oils X and Y with the company's stock of 100 thousand barrels. The relevant properties of the three oils are given below.

	SG	Viscosity	S%	Availability
Stock	0.89	20	0.30	100
X	0.92	44	0.45	Unlimited
Y	0.86	40	0.25	90

Let x and y be the amount, in thousand barrels, of oil X and Y blended with stock. The objective is to maximise the total volume

$$V = 100 + x + y \qquad (9.1)$$

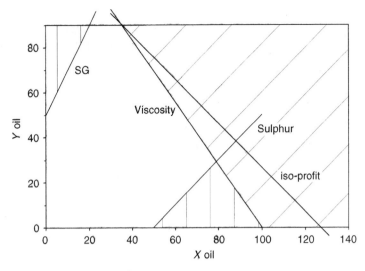

Figure 9.1   Graphical solution of oil-blending problem

subject to the constraints imposed by the specification. For example, the viscosity requirement is

$$\frac{20 \times 100 + 44x + 40y}{100 + x + y} \leq 32$$

and this is equivalent to

$$3x + 2y \leq 300 \tag{9.2}$$

The *SG* and sulphur content requirements are equivalent to:

$$-2x + y \leq 50 \tag{9.3}$$

$$x - y \leq 50 \tag{9.4}$$

The availability requirement is

$$y \leq 90 \tag{9.5}$$

and the physical requirement that volumes are non-negative is

$$0 \leq x, y \tag{9.6}$$

The inequalities define a feasible region (Figure 9.1). The manager requires the pair (x, y) that lies within this region and gives a maximum volume. Equation (9.1) can be written as

$$y = -x + (V - 100)$$

which is a straight line with gradient -1 and intercept $(V - 100)$, known as the iso-profit line. The maximum intercept of 130 is obtained when the line passes through the point $(40, 90)$ as shown in Figure 9.1. Therefore the optimal solution is to blend 40 thousand barrels of X and all 90 thousand barrels of Y with the 100 thousand barrels of oil in stock. This yields 230 thousand barrels of blended oil. Since the objective function and inequalities are linear in x and y, the solution must lie at a corner of the feasible region. There is a complication if the iso-profit line is parallel to a critical constraint. In such cases, any point on the segment of the critical constraint bounding the feasible region is an optimal solution. Two variable problems are easy to solve graphically but practical applications usually involve more variables. The restriction that the objective function and constraints are linear ensures that the solution lies at a vertex of the simplex enclosed by the hyperplanes representing the constraints. The simplex algorithm for linear programming is a routine for searching for the optimal vertex which gives an improvement at every iteration and indicates when the optimum solution is found, provided it exists. It may be that there is no solution satisfying a particular set of requirements. The simplex algorithm for linear programming is quite different from the simplex method for searching for the maximum of a general function, described in Appendix 3.

9.1.2   The simplex algorithm for linear programming

The algorithm is summarised in Table 9.1. The first step is to make the inequalities into equalities by adding non-negative slack variables, s_i. The oil-blending problem is to maximise:

$$V = 100 + x + y$$

subject to the constraints that all variables are non-negative and

$$3x + 2y + s1 = 300$$
$$-2x + y + s2 = 50$$
$$x - y + s3 = 50$$
$$y + s4 = 90$$

If an equality specified that a linear combination of x and y should exceed some constant, then a surplus variable would be subtracted to give an equation. The surplus variable would be non-negative.

A feasible solution is any set of values of the variables x, y, $s1$, ..., $s4$ that satisfies the constraints. If there are m constraints and n variables, a basic solution is obtained by setting $(n-m)$ variables to zero and solving the m equations for the remaining m variables provided these equations have a unique solution. The variables which are set equal to zero are called the non-basic variables. The other variables are called the basic variables and form a basis. The key result for the linear programming algorithm is that the basic feasible solutions correspond to the vertices of the feasible region. The optimal solution will be at a vertex of the feasible region. It follows that we only need to investigate basic feasible solutions.

The simplex algorithm starts from the origin of the x–y plane, i.e. x and y are set equal to 0 and the slack variables form the basis. This is summarised as Tableau 1 in Table 9.2(a). The main body of the table, including the right-hand column, gives the equations which represent the constraints. The initial

Table 9.1 Flow chart for simplex method, maximisation problem (after Dunn and Ramsing, 1981). (Reprinted by permission of Prentice-Hall, Inc, Upper Saddle River, NJ.) RHS is 'right-hand side' of equations

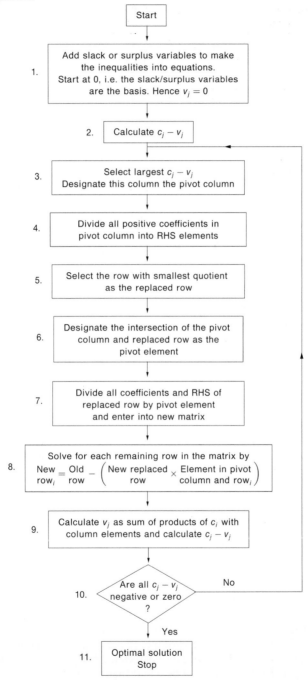

Start

1. Add slack or surplus variables to make the inequalities into equations. Start at 0, i.e. the slack/surplus variables are the basis. Hence $v_j = 0$

2. Calculate $c_j - v_j$

3. Select largest $c_j - v_j$
Designate this column the pivot column

4. Divide all positive coefficients in pivot column into RHS elements

5. Select the row with smallest quotient as the replaced row

6. Designate the intersection of the pivot column and replaced row as the pivot element

7. Divide all coefficients and RHS of replaced row by pivot element and enter into new matrix

8. Solve for each remaining row in the matrix by
$$\frac{\text{New}}{\text{row}_i} = \frac{\text{Old}}{\text{row}} - \left(\frac{\text{New replaced}}{\text{row}} \times \frac{\text{Element in pivot}}{\text{column and row}_i} \right)$$

9. Calculate v_j as sum of products of c_i with column elements and calculate $c_j - v_j$

10. Are all $c_j - v_j$ negative or zero? — No

Yes

11. Optimal solution Stop

Table 9.2 Simplex tableaux for the oil-blending problem

(a) Tableau 1

c_j	solution	1	1	0	0	0	0	
c_i	variables	x	y	$s1$	$s2$	$s3$	$s4$	
0	$s1$	3	2	1	0	0	0	300
0	$s2$	−2	1	0	1	0	0	50
0	$s3$	1	−1	0	0	1	0	50
0	$s4$	0	1	0	0	0	1	90
	v_j	0	0	0	0	0	0	0
	$c_j - v_j$	1	1	0	0	0	0	

(b) Tableau 2

c_j	solution	1	1	0	0	0	0	
c_i	variables	x	y	$s1$	$s2$	$s3$	$s4$	
0	$s1$	7	0	1	−2	0	0	200
1	y	−2	1	0	1	0	0	50
0	$s3$	−1	0	0	1	1	0	100
0	$s4$	2	0	0	−1	0	1	40
	v_j	−2	1	0	1	0	0	50
	$c_j - v_j$	3	0	0	−1	0	0	

(c) Tableau 3

c_j	solution	1	1	0	0	0	0	
c_i	variables	x	y	$s1$	$s2$	$s3$	$s4$	
0	$s1$	0	0	1	1.5	0	−3.5	60
1	y	0	1	0	0	0	1	90
0	$s3$	0	0	0	0.5	1	0.5	120
1	x	1	0	0	−0.5	0	0.5	20
	v_j	1	1	0	−0.5	0	1.5	110
	$c_j - v_j$	0	0	0	0.5	0	−1.5	

(d) Tableau 4

c_j	solution	1	1	0	0	0	0	
c_i	variables	x	y	$s1$	$s2$	$s3$	$s4$	
0	$s2$	0	0	0.67	1	0	2.33	40
1	y	0	1	0	0	0	1	90
0	$s3$	0	0	−0.33	0	1	−1.25	100
1	x	1	0	0	0	0	−0.67	40
	v_j	1	1	0	0	0	0.33	130
	$c_j - v_j$	0	0	0	0	0	−0.33	

solution is to set x and y at 0, and hence $s1$, $s2$, $s3$ and $s4$ at 300, 50, 50 and 90 respectively. The c_j are the coefficients of the variables in the objective function, those for the slack variables being 0 because they contribute nothing to V. The c_i are the coefficients of the basic variables in the objective function. The v_j are the sum of the products of the c_i with the coefficients of the jth variable in the ith equation. The v_j in Tableau 1 will all be 0 and their interpretation becomes apparent from Tableau 2 onwards. Suppose the variable heading column j is increased by 1 unit. This will lead to changes in the other variables which will have the effect of decreasing V by v_j, However, V is also increased by c_j so the net effect is an increase by $c_j - v_j$. Hence the next variable to enter the basis is the one corresponding to the largest $c_j - v_j$. If two values of $c_j - v_j$ are equal either variable could be chosen. The column with the largest value of $c_j - v_j$, corresponding to the new variable to enter the basis, is known as the pivot column. The positive coefficients in the pivot column are divided into the numbers on the right-hand sides of the equations to find the limiting constraint. This is the row for which the quotient is smallest, row i say. The coefficient in the ijth position is called the pivot element. The other steps in the algorithm are elementary row operations on the set of equations that result in the basic variable having a coefficient of 1 in row i, and 0 in the other rows. For the first tableau, the pivot column could be either x or y. I chose to take y and the $s2$ row has the smallest positive quotient. The pivot element is at the intersection of row $s2$ and column y. The solution can be read from the final tableau: $s1$ and $s4$ are 0, and hence $s2$, y, $s3$ and x are 40, 90, 100 and 40 respectively. The interpretation of $s2$ and $s3$ is that the sulphur and SG requirements are not critical. The maximum value of $x + y$ is 130, and the value of V is found by adding the constant, 100, to obtain 230. Software for solving linear programming problems is available on disks provided with many operations research textbooks, e.g. Erikson and Hall (1989) and Taha (1997). Bunday and Garside (1987) is another useful source, and Press et al. (1992) give an algorithm.

In some problems, such as the following, the slack and surplus variables do not provide a basic feasible solution.

$$\text{Maximise} \quad V = 2y + x$$
$$\text{subject to} \quad -x + y \geq 1$$
$$y \leq 3$$
$$x + y \leq 4$$

The inequalities can be made into equations by introducing a non-negative surplus variable s_1 and slack variables s_2 and s_3.

$$-x + y - s1 = 1$$
$$y + s2 = 3$$
$$x + y + s3 = 4$$

However, setting x and y equal to 0 does not give a basic feasible solution because $s1$ would be -1, infringing the requirement that all variables take non-negative values. Introduction of an artificial variable u removes this difficulty.

$$-x + y - s1 \qquad +u = 1$$
$$y \qquad +s2 \qquad = 3$$
$$x + y \qquad +s3 \qquad = 4$$

There is now a basic feasible solution with x, y and $s1$ equal to 0, $u = 1$, $s2 = 3$ and $s3 = 4$. It is necessary to modify the objective function by adding $-Mu$, where M is a very large number, so that the optimum solution excludes u. In this case,

$$V = 2y + x - Mu$$

You need to introduce an artificial variable to answer Exercise 9.3.

The simplex algorithm can also be used for a minimisation problem. The difference is that the variable with the most negative $(c_j - v_j)$ enters the basis. The optimal solution is reached when all the $(c_j - v_j)$ are zero or positive. Alternatively, the objective function can be multiplied by -1 and then maximised. Negative valued variables can be accommodated as the difference between two non-negative valued variables. Some other aspects of linear programming are covered in Exercises 9.1 up to 9.4.

9.1.3   Chance constrained programming

A stochastic variant of the linear programming problem is to maximise

$$z = \sum_{j=1}^{n} c_j x_j$$

subject to the constraints that

$$\Pr\left\{ \sum_{j=1}^{n} a_{ij} x_j \le b_i \right\} \ge 1 - \alpha_i \quad \text{for } i = 1, \ldots, m$$

$$0 \le x_j \quad \text{for } j = 1, \ldots, n$$

If b_i is normally distributed with mean $E[b_i]$ and variance $\text{var}(b_i)$, and the a_{ij} are known constants, the problem is identical to the deterministic linear programming problem with

$$\sum_{j=1}^{n} a_{ij} x_j \le E[b_i] - z_{\alpha_i} \sqrt{\text{var}(b_i)}$$

Example 9.1

The manager at Enco knows that an estimate of sulphur content, made by an independent party from a tin of blended oil, will be unbiased but have a standard deviation of 0.02. The manager is confident that the company's measurements of sulphur content of the oils X and Y are far more precise. Then a requirement that

$$\Pr (\text{estimated sulphur content} < 0.35\%) = 0.9$$

would be equivalent to:

$$\frac{100 \times 0.30 + x \times 0.45 + y \times 0.25}{100 + x + y} \le 0.35 - 1.28 \times 0.02$$

If the a_{ij} are also random variables the problem is no longer linear because the standard deviation of the linear combination

$$\sum_{j=1}^{n} a_{ij} x_j$$

is

$$\left\{ \sum_{j=1}^{n} \mathrm{var}(a_{ij})x_j^2 + \sum_{k \neq j} \sum_{j=1}^{n} \mathrm{cov}(a_{ij}a_{ik})x_j x_k \right\}^{1/2}$$

The following strategy could be used to solve the problem with standard linear programming software. First solve the problem with the variance and co-variances set at 0. Use the values of x_j at the optimal solution to calculate an approximate standard deviation of the linear combination, and solve the resultant linear programming problem. This process could be iterated, although it is possible that it might not converge satisfactorily if several vertices of the simplex are near the optimum.

9.2 PERT

9.2.1 Introduction

An Italian has bought a rather dilapidated restaurant which she intends converting to a pizzeria. The steps involved in the project are shown in a flow diagram (Figure 9.2). If each activity is to be identified uniquely by two node numbers it is necessary to introduce a dummy activity which takes no time. This is shown by a broken line. The program evaluation and review technique (PERT) is a stochastic version of critical path analysis (CPA). PERT requires an optimistic (L), most likely (M), and pessimistic time (U) for each activity. A rough interpretation of a pessimistic time is that there should be only a 1 in 1000 chance of its being exceeded, and an optimistic time can be interpreted in a complementary way. From these figures expected times (μ_i), roughly based on the mean of a triangular distribution with these limits and a mode M (Exercise 9.5), can be calculated for each activity (i). That is,

$$\mu_i = (L + 4M + U)/6$$

Standard deviations are also calculated, roughly based on a normal distribution, as

$$\sigma_i = (U - L)/6$$

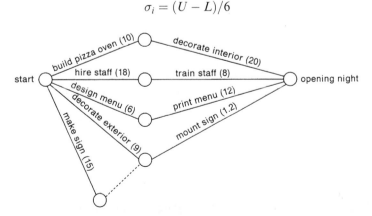

Figure 9.2 Flow diagram for pizzeria project

Table 9.3   Timings for pizzeria project

Activity	Optimistic time L	Most likely time M	Pessimistic time U
Build pizza oven	7	10	13
Decorate interior	15	20	25
Hire staff	9	16	35
Train staff	6	8	10
Design menu	5	6	7
Print menu	6	12	18
Decorate exterior	6	8	16
Make sign	13	15	17
Mount sign	1	1	2

The critical path is now found for the case in which activities take their expected times. The three time estimates for each activity for the pizzeria project are given in Table 9.3 and the expected times are shown in Figure 9.2. In this simple example the critical path is easily seen to be the building of the oven followed by the interior decorating, but rules are needed in order to program a general computer algorithm. Dunn and Ramsing (1981) suggest the following method.

9.2.2   Finding the critical path

Each node is uniquely identified by a number, and has two times associated with it. These are the earliest start time (EST) for the next activity and the latest finishing time (LFT) for the last activity. The EST is the earliest time at which an activity can start, assuming that all preceding activities have been started and completed as soon as possible. The LFT is the latest time at which an activity can start if it is not to delay completion of the project. In a flow diagram, the node number, EST and LFT are shown in the lower semi-circle, upper left quadrant and upper right quadrant of the circles representing the nodes. The steps to find the critical path follow.

1. Go through the flow diagram and calculate the EST for all nodes, including the final one even though there is no next activity.
2. Set the LFT of the final node equal to its EST.
3. Work back from the final node and calculate LFTs.
4. Identify any arrow for which the EST equals the LFT on the node at its start and the EST equals the LFT on the node at its end. This is a necessary condition for the arrow to be on the critical path.
5. If the condition in (4) is met and the difference between the EST at the ends of the arrow equals the expected time for the activity, then the arrow is on the critical path.

9.2.3   Probability of completion

Let T be the total time along the critical path, and assume times taken for each activity vary independently about their expected values. The mean and variance of T are given by

$$\mu = \sum_{\substack{\text{critical} \\ \text{path}}} \mu_i$$

$$\sigma^2 = \sum_{\substack{\text{critical} \\ \text{path}}} \sigma_i^2$$

T will be approximately normally distributed, by the central limit theorem, and, for example, T has a 99% chance of being less than $\mu + 2.33\sigma$. The limitation of this analysis is that T need not be the time until completion, especially if a near critical path has a large standard deviation associated with it. An improvement on relying on the distribution of T only is to find the pessimist's critical path when all activities take their pessimistic times, repeat the calculation for the total time along this path, and then take the longer of the two estimates.

Example 9.2
The owner plans to have an opening night and asks for an upper 99% estimate of the time to complete the pizzeria project. It is clear from Figure 9.2 that the critical path is building the oven and decorating the interior.

$$\text{oven:} \qquad\qquad\qquad \mu_{\text{oven}} = 10 \quad \sigma_{\text{oven}} = 1$$
$$\text{interior decoration:} \quad \mu_{\text{int}} = 20 \quad \sigma_{\text{int}} = 1.67$$

Hence $\mu = 30$ and $\sigma = 1.95$

$$T \sim N(30, (1.95)^2)$$
$$\Pr(T < 34.5) = 0.99$$

The pessimist's critical path is hiring staff and training them. Let the associated time be $T_{\text{pess}}(\ )$

$$\text{hire staff:} \quad \mu_{\text{hire}} = 18 \quad \sigma_{\text{hire}} = 4.33$$
$$\text{train staff:} \quad \mu_{\text{train}} = 8 \quad \sigma_{\text{train}} = 0.67$$
$$T_{\text{pess}} \sim N(26, (4.38)^2)$$

and the upper 1% of this distribution is 36.2. The opening night should be scheduled after 37 working days.

   In practice we will usually try to reduce the variability and expected times of critical activities. It may be possible, for example, to divert resources from some of the activities with slack time.

9.3   Portfolio analysis

Ranne (1999) describes an investment optimisation problem and lists an APL routine for its solution. There are n different investment categories which might include various shares, variable interest bank deposits, fixed interest government stock and other debentures. The investment returns per monetary unit have expected values μ_i, standard deviations σ_i and correlation coefficients ρ_{ij}. The standard deviation is a measure of risk of the investment. Returns on shares may be highly correlated. A proportion of the initial capital, p_i, is to be invested in each category. The mean μ and standard deviation of the whole portfolio are:

$$\mu = \sum_i \mu_i p_i$$

$$\sigma = \left(\sum_i \sum_j \sigma_i \sigma_j \rho_{ij} p_i p_j \right)^{1/2}$$

The basic problem specification is to find the portfolio with the minimum standard deviation corresponding to a specified expected return μ. An alternative is to find the maximum expected return given a specified standard deviation. However, many fund managers consider that the solutions to these optimisation problems tend to give poorly balanced portfolios, and therefore impose minimum and maximum proportions for each investment category. The fact that most of the expected values, variances and covariances on which the optimisation is based are just estimates based on past performance, is another good reason for imposing constraints on the proportions. Thus the problem for a specified expected return is:

$$\text{minimise} \quad \sum \sum \sigma_i \sigma_j \rho_{ij} p_i p_j$$

$$\text{subject to} \quad \sum p_i = 1$$

$$\sum \mu_i p_i = \mu$$

$$m_i \le p_i \le M_i \quad \text{for } i = 1, \ldots, n$$

It is assumed that the μ_i, σ_i, σ_j and ρ_{ij} are known, and that the μ, m_i and M_i are specified. The method of Lagrange multipliers, augmented by consideration of points at the ends of the specified ranges (Exercise 9.6), leads to the following set of equations:

$$2 \sum_i \sigma_i \sigma_j \rho_{ij} p_i + a + b \mu_j = \lambda_j \quad \text{for } j = 1, \ldots, n$$

$$\sum_i p_i - 1 \quad = 0$$

$$\sum_i \mu_i p_i - \mu = 0$$

where in each investment category one of the following conditions is true: either $\lambda_i = 0$ or $p_i = m_i$ or $p_i = M_i$. For each possible combination of these conditions, the system of equations is linear and can be solved very quickly. The number of possible conditions is less than n^3 because certain combinations, such as $p_i = m_i$ for all i, cannot occur. The program listed in Ranne's paper made use of an APL function for solving a set of linear equations and is very succinct. The computing language J has similar features and can be evaluated for two weeks, free of charge. It is available from:

http://www.jsoftware.com.

You are asked to solve a four-component portfolio in Exercise 9.8. It is quite feasible to do so without any special software.

9.4   Stochastic dynamic programming

9.4.1   Introduction

A small cooperative makes soft toys. At the end of each year the business is in one of two states: thriving (1) or floundering (2). The state transitions form a Markov chain with transition matrix P,

$$P = \begin{array}{c} 1 \\ 2 \end{array}\begin{array}{cc} 1 & 2 \\ \begin{pmatrix} 0.50 & 0.50 \\ 0.30 & 0.70 \end{pmatrix} \end{array}$$

There is a reward structure associated with the changes in state over the year. If the business continues to thrive sales provide a net income of 9 monetary units (mu), but if it moves from the thriving state to the floundering state the reduced sales during the transition period provide a net income of only 3 mu. If the business moves from floundering to thriving the sales again provide an income of 3 mu, but if it remains in a floundering state the revenue from sales does not cover outgoings and the net outcome is a loss of 4 mu. These rewards can be summarised in a matrix R:

$$R = \begin{bmatrix} 9 & 3 \\ 3 & -4 \end{bmatrix}$$

The rewards take no account of likely future earnings, and the transition probabilities are independent of the history of the process.

It is possible to increase the chances of moving to a thriving state by spending more time on design work, but this will have a cost and the rewards are reduced by 4 mu. The transition matrix would become

$$P(\text{des}) = \begin{pmatrix} 0.75 & 0.25 \\ 0.70 & 0.30 \end{pmatrix} \qquad R(\text{des}) = \begin{pmatrix} 5 & -1 \\ -1 & -8 \end{pmatrix}$$

Suppose the lease on the business premises expires at the end of the next year and the members of the cooperative have agreed to split up and do other things. They must decide whether to spend time on design work, and they base their decision on expected monetary values (EMV). There are four cases to consider:

(i)   present state 1 with no design work

$$\text{EMV} = 0.50 \times 9 + 0.50 \times 3 = 6$$

(ii)   present state 2 with no design work

$$\text{EMV} = 0.30 \times 3 + 0.70 \times (-4) = -1.9$$

(iii) present state 1 with design work

$$\text{EMV} = 0.75 \times 5 + 0.25 \times (-1) = 3.5$$

(iv) present state 2 with design work

$$\text{EMV} = 0.70 \times (-1) + 0.30 \times (-8) = -3.1$$

The decision is taken to do without design work, regardless of the state the company is in now. However, the decisions are less straightforward if there are two years left on the lease. It might now be worthwhile spending time on design in the first year so as to increase the probability of obtaining the benefits from starting the final year in a thriving state. We can analyse this by drawing two decision trees, one for each of the two possible present states, but relying on decision trees for more years, more states and more options is not feasible. The roll-back procedure, that is, the method of working back from the final time, needs to be formalised as stochastic dynamic programming (SDP).

9.4.2 Value iteration

Assume an ergodic Markov chain with n states. Let t be the discrete time step for $t = 0, 1, \ldots, T$. The EMV over the remaining time steps, at time t, will depend on the state (i) and decision (k) at time t and is denoted by $v_t(i, k)$. The elements of the transition and reward matrices also depend on the decision and are written $p_{ij}(k)$ and $r_{ij}(k)$ respectively. Finally let $v_{t+1}^*(j)$ represent the EMV of the best sequence of decisions from state j over the remaining time steps until T. If we are in state i, the EMV of a certain move to j, with decision k, followed by application of the best sequence of decisions would be:

$$r_{ij}(k) + v_{t+1}^*(j)$$

Unfortunately we cannot guarantee a move to state j so we need to take expected value over all possible states. Thus

$$v_t(i, k) = \sum_{\substack{\text{states} \\ j}} p_{ij}(k)(r_{ij}(k) + v_{t+1}^*(j)) \quad \text{for } j = 1, \ldots, n$$

The best decision is that which maximises this expected value:

$$v_t^*(i) = \max_{\substack{\text{decisions} \\ k}} \left\{ \sum_{\substack{\text{states} \\ j}} p_{ij}(k)(r_{ij}(k) + v_{t+1}^*(j)) \right\} \tag{9.7}$$

The value iteration algorithm proceeds by working backwards from some specified final values for $v_T^*(j)$. In many cases 0 is appropriate for all the final values, but if it is desirable to finish in a particular state, l say, $v_T^*(l)$ can be made large.

Example 9.3
The cooperative has four years to run on the lease of the premises. All members agree to cease trading then, rather than look for new premises. The time steps will run from now, $t = 0$, up to $t = 4$. There is no financial advantage to be

gained from being in the thriving state when the lease expires, so

$$v_4^*(1) = v_4^*(2) = 0$$

The remaining calculations require more steps. The decisions will be abbreviated to 'no' for no design work and 'des' for design work.

For $t = 3$

$$
\begin{aligned}
v_3(1, \text{no}) &= 0.50 \times (9 + 0) &+ 0.50 \times (3 + 0) &= 6 \\
v_3(2, \text{no}) &= 0.30 \times (3 + 0) &+ 0.70 \times (-4 + 0) &= -1.9 \\
v_3(1, \text{des}) &= 0.75 \times (5 + 0) &+ 0.25 \times (-1 + 0) &= 3.5 \\
v_3(2, \text{des}) &= 0.70 \times (-1 + 0) + 0.30 \times (-8 + 0) &&= -3.1
\end{aligned}
$$

Hence the optimal decisions and values are

$$
\begin{aligned}
v_3^*(1) &= 6 &&\text{no design work} \\
v_3^*(2) &= -1.9 &&\text{no design work}
\end{aligned}
$$

For $t = 2$

$$
\begin{aligned}
v_2(1, \text{no}) &= 0.50 \times (9 + 6) &+ 0.50 \times (3 + (-1.9)) &= 8.05 \\
v_2(2, \text{no}) &= 0.30 \times (3 + 6) &+ 0.70 \times (-4 + (-1.9)) &= -1.43 \\
v_2(1, \text{des}) &= 0.75 \times (5 + 6) &+ 0.25 \times (-1 + (-1.9)) &= 7.52 \\
v_2(2, \text{des}) &= 0.70 \times (-1 + 6) + 0.30 \times (-8 + (-1.9)) &&= 0.53
\end{aligned}
$$

Hence the optimal decisions and values are

$$
\begin{aligned}
v_2^*(1) &= 8.05 &&\text{no design work} \\
v_2^*(2) &= 0.53 &&\text{design work}
\end{aligned}
$$

For $t = 1$

$$
\begin{aligned}
v_1(1, \text{no}) &= 0.50 \times (9 + 8.05) &+ 0.50 \times (3 + 0.53) &= 10.290 \\
v_1(2, \text{no}) &= 0.30 \times (3 + 8.05) &+ 0.70 \times (-4 + 0.53) &= 0.886 \\
v_1(1, \text{des}) &= 0.75 \times (5 + 8.05) &+ 0.25 \times (-1 + 0.53) &= 9.905 \\
v_1(2, \text{des}) &= 0.70 \times (-1 + 8.05) + 0.30 \times (-8 + 0.53) &&= 2.694
\end{aligned}
$$

Hence the optimal decisions and values are

$$
\begin{aligned}
v_1^*(1) &= 10.290 &&\text{no design work} \\
v_1^*(2) &= 2.694 &&\text{design work}
\end{aligned}
$$

For $t = 0$

$$
\begin{aligned}
v_0(1, \text{no}) &= 0.50 \times (9 + 10.29) &+ 0.50 \times (3 + 2.694) &= 12.492 \\
v_0(2, \text{no}) &= 0.30 \times (3 + 10.29) &+ 0.70 \times (-4 + 2.694) &= 3.073 \\
v_0(1, \text{des}) &= 0.75 \times (5 + 10.29) &+ 0.25 \times (-1 + 2.694) &= 11.891 \\
v_0(2, \text{des}) &= 0.70 \times (-1 + 10.29) + 0.30 \times (-8 + 2.694) &&= 4.911
\end{aligned}
$$

Hence the optimal decisions and values are

$$v_0^*(1) = 12.492 \quad \text{no design work}$$

$$v_0^*(2) = 4.911 \quad \text{design work}$$

The value iteration algorithm is already close to reaching a steady state. If the cooperative owned the premises and hoped to stay in business indefinitely, T would be large and the steady-state policy would be to spend time on design next year if the current year ends in the floundering state, Also, the increase in $v_t^*(i)$ between t equal to τ and t equal to $\tau - 1$ tends to a constant value, known as the gain (g), which represents the average value of remaining in business for a year. It is the same for all i, so in this example it will be between:

$$v_0^*(1) - v_1^*(1) = 12.492 - 10.290 = 2.202$$

and

$$v_0^*(2) - v_1^*(2) = 4.911 - 2.694 = 2.217$$

We can find the exact value from the following argument. Let the fixed probability vector of the transition matrix corresponding to the optimal decisions be ω. This gives the proportion of time in each state. The gain is the sum of the products of the proportions of time spent in each state with the expected reward over a single transition from the state. The expected reward from state i is

$$q_i = \sum_{\substack{\text{states} \\ j}} p_{ij} r_{ij} \quad \text{where } j = 1, \ldots, n \tag{9.8}$$

and the gain is

$$g = \sum_{\substack{\text{states} \\ i}} \omega_i q_i \quad \text{where } i = 1, \ldots, n \tag{9.9}$$

The transition matrix for the cooperative's optimal policy is

$$P_{\text{opt}} = \begin{pmatrix} 0.50 & 0.50 \\ 0.70 & 0.30 \end{pmatrix}$$

and

$$\omega = (0.5833, 0.4167)$$

$$g = 0.5833 \times (0.5 \times 9 + 0.5 \times 3) + 0.4167 \times (0.7 \times (-1) + 0.3 \times (-8)) = 2.208$$

The existence of a constant gain in the steady state is the essence of the policy iteration algorithm for problems in which the time to go is indefinite.

9.4.3 Policy iteration procedure

For any chosen policy

$$v_t(i) = \sum_{\substack{\text{states} \\ j}} p_{ij}(r_{ij} + v_{t+1}(j)) \quad j = 1, \ldots, n \tag{9.10}$$

Also for any ergodic Markov chain there is a constant gain, g, such that:

$$v_t(i) = g(T - t) + v_i \quad \text{for } T \text{ large relative to } t \tag{9.11}$$

A consequence of equation (9.11) is that

$$v_t(i) = g + v_{t+1}(i)$$

as explained in the preceding section. Equation (9.11) also implies that the specific value of starting in state i, v_i, becomes negligible in percentage terms over a long time. Substitution of equation (9.11) into equation (9.10) gives the value determination operation (VDO):

$$g + v_i = \sum_j p_{ij}r_{ij} + \sum_j p_{ij}v_j \tag{9.12a}$$

The policy improvement routine (PIR) (Howard, 1960) is: for each state i find the alternative policy that maximises

$$\sum p_{ij}^+ r_{ij}^+ + \sum p_{ij}^+ v_j \tag{9.12b}$$

where p_{ij}^+ and r_{ij}^+ are the transition probabilities and rewards for this new policy and v_j are obtained from the solution of (9.12a). The policy iteration procedure revolves around these two equations. It is convenient to start with equation (9.12a) using any policy, for example that which maximises the immediate rewards (q_i). Equation (9.12a) is a set of n linear equations in $n + 1$ unknowns, $v_1, \ldots, v_n$ and g. Unique solutions are obtained by setting v_1, for example, equal to 0. Then the other v_i represent the relative values of starting in state i rather than state 1. The next step is to apply the PIR criteria (equation (9.12b)). The PIR will not give the optimal policy at this stage, unless the starting policy was an exceptionally good guess, because the v_j from equation (9.12a) do not correspond to the optimal policy, but Howard proves it can only give an improvement. The procedure stops when the policies on two successive iterations are identical. You could try the algorithm on the cooperative's problem.

9.4.4   SDP for reservoir systems

Value iteration

The following example demonstrates the essential features of a typical SDP algorithm for a single reservoir system. The basic atom (unit) of water is a small fraction of the reservoir's total capacity, e.g. a unit of water is $\frac{1}{50}$ of total capacity and the reservoir can hold 50 units. Time intervals are typically one month. Let s_t represent the amount of water in the reservoir at the beginning of time interval t so, for example, s_t^j represents j units in the reservoir. Let i_t represent the inflow to the reservoir during time period t. This inflow is unknown but its probability distribution can be estimated from flow records and is typically assumed to depend on the inflow (i_{t-1}) during the past time period. The discrete probabilities of the form $p_{k,l}$ represent the probability of l units inflow during time t if there were k units of inflow during time period $(t - 1)$. The state of the system at time t is denoted by the pair (s_t^i, i_{t-1}^k).

Now let u_t represent the releases during time period t, that is, u_t is the decision variable. In practice u_t would probably have several components, e.g. some of: industrial demand, domestic demand, hydroelectric power requirements, irrigation and sustaining a minimum flow for navigation or fish. The remaining notation is V_t for the value of the optimum strategy at the beginning of time period t and G_t for the immediate gain during time period t obtained by releasing u_t. Then the SDP algorithm is:

$$V_t(s_t^i, i_{t-1}^k) = \max_{u_t}\left\{ G_t(u_t) + \sum p_{k,l} V_{t+1}(s_{t+1}^j, i_t^l) \right\} \qquad (9.13)$$

where the summation is over all possible inflows during time period t, and

$$s_{t+1}^j = \begin{cases} s_t^i + i_t^l - u_t & \text{if this is less than } s \text{ max} \\ s_{\max} & \text{the maximum capacity of the reservoir otherwise} \end{cases}$$

The value iteration (VI) algorithm proceeds by working backwards from some specified final values for V_N. A convenient choice is:

$$V_N(s_N^i, i_{N-1}^k) = 0 \quad \text{for all } i, k \qquad (9.14)$$

If the algorithm is iterated enough times a steady state which is independent of the final V_N will be found. This is known as value iteration.

So far no explicit mention has been made of seasonal variation. The immediate gains (G_t) and transition probabilities $(p_{k,l})$ would be expected to vary seasonally. This does not have any substantial effect on the value iteration algorithm, although the computer code will be somewhat more complicated and convergence will only be to a steady-state yearly policy, the within-year decisions depending on the season.

An indication of the size of the computation is given by assuming, as an example, six possible inflows, i.e. $0, \ldots, 5$ units, and 6×6 release decisions, i.e. $0, \ldots, 5$ for supply and $0, \ldots, 5$ to maintain flow in the river. We then have $50 \times 6 = 300$ states and 36 possible decisions to consider for each state. In addition, for each state and decision we have to allow for the probability distribution of inflows which itself has six terms.

1. Select a state pair (s_t^i, i_{t-1}^k).
2. Consider a release decision u_t.
3. Calculate for all possible i_t^l:

   (i) The probability distribution of inflows conditional on the last inflow. That is:

   $$p_{k,l} = \Pr(i_t^l \mid i_{t-1}^k)$$

   (ii) The state at the end of the time period, and equivalently at the beginning of period $t + 1$, for each possible inflow from:

   $$s_{t+1}^j = s_t^i + i_t^l$$

   or the maximum capacity of the reservoir should the above exceed it.
   (iii) The immediate gain $G(u_t)$.

(iv) $\{G_t(u_t) + \sum p_{k,l} V_{t+1}(s_{t+1}^j, i_t^l)\}$ remembering that V_{t+1} is available from the previous step.
4. Repeat from (2) over all possible release decisions.
5. Select the release decision which maximises the expression in 3(iv). This is $V_t(s_t^i, i_{t-1}^k)$.
6. Repeat for all possible states.

Policy iteration

A policy is a rule which specifies release decisions for any given state. An initial policy might be chosen as the releases which maximise the immediate gains. The policy iteration algorithm assumes that the Markov chain is ergodic for any chosen policy. It is intuitively plausible, and can be formally proved, that for any given policy:

$$V_{N-t}(s_t^i, i_{t-1}^k) = (N - t)g + V(k) \tag{9.15}$$

for a large number of steps to go $(N - t)$. This is a statement that the long-term gain only depends on the initial state through an additive constant. It follows that the effect of the initial state becomes negligible, in relative terms, after a long time.

A consequence of equation (9.15) is that

$$V_t(s_t^i, i_{t-1}^k) = g + V_{t+1}(s_t^j, i_{t-1}^k) \tag{9.16}$$

and substitution of equation (9.16) into equation (9.13) gives, for every possible state indexed by k

$$g + V_{t+1}(s_{t+1}^j, i_t^k) = G_t + \sum p_{kl} V_{t+1}(s_{t+1}^j, i_t^l) \tag{9.17}$$

Equation (9.17) can be solved for g and the $V_{t+1}(s_{t+1}^j, i_t^l)$ relative to an arbitrary value for any one l. This is known as the value determination operation. The significance of the $V_{t+1}(,)$ is that they indicate an improved policy. This is the policy improvement operation. The procedure is iterated. It is a neat algorithm which avoids the necessity to check every possible policy for every possible state. However, the crucial assumption of ergodicity does not hold for seasonally varying parameters.

SDP for a multi-reservoir system

The multi-reservoir problem has interested research workers for at least 50 years (e.g. Massé, 1946). As computing power increases so will the number of reservoirs that can be handled by SDP without recourse to special variants. However, the number of reservoirs in large water supply systems is still likely to exceed the number that can be handled with standard SDP. Turgeon (1981) proposed a method for solving the problem for n reservoirs in series. The method breaks down the problem into $n - 1$ problems of two-state variables: the content of the reservoir under consideration (i), and the total contents of the downstream reservoirs. It follows that the release policy for reservoir i

depends on its water content and the total contents of the reservoirs which lie downstream. Turgeon (1980) had devised a similar sub-optimal solution for the problem of reservoirs in parallel.

Archibald *et al.* (1997) adapted the work of Turgeon (1980, 1981) to more general multi-reservoir systems. They required that the ratio of working volume to capacity be the same for all the reservoirs (equally full heuristic) and showed that their operating policies were close to optimal for a three-reservoir system. They also applied it to an example with 17 reservoirs.

9.5   Summary

1. Linear programming is used for maximising, or minimising, a linear combination of variables which are subject to linear constraints. The solution lies at a corner of the simplex. The stochastic version replaces deterministic constraints with probability statements.
2. PERT is critical path analysis which allows for variability in the specification of times for activities.
3. The portfolio problem is the minimisation of a quadratic function subject to linear constraints.
4. Stochastic dynamic programming is an algorithm that formalises the roll-back principle used to analyse decision trees. The value iteration algorithm is used when there is a finite number of steps to go. If a steady-state solution is required the policy iteration algorithm is an alternative method. However, the value iteration algorithm will converge to a steady state if it is run back enough times.

Exercises

9.1 Verify that the basic feasible solutions for the oil-blending problem of Section 9.1 do correspond to vertices of the feasible region.

9.2 (After Taha, 1997.) A soft drinks manufacturer sells Dandelion and Burdock (DB) and a drink called Bettabru. One thousand bottles of DB uses 6 kg of dandelions and 1 kg of burdock, and 1 thousand bottles of Bettabru uses 4 kg of dandelions and 2 kg of burdock. The maximum daily availability of dandelions is 24 kg, whereas burdock is limited to a maximum of 6 kg. The profit per thousand bottles is 5 monetary units (mu) for DB, and 4 mu for Bettabru. The sales manager requires that daily production of Bettabru is kept below 2 thousand bottles, and that production of Bettabru does not exceed production of DB by more than 1 thousand bottles. Find the optimal production and corresponding daily profit.

9.3 A linear programming problem, the primal problem (P), is defined as follows:

$$\text{maximise } z = \sum_{j=1}^{n} c_j x_j$$

subject to

$$\sum_{j=1}^{n} a_{ij}x_j \leq b_i \quad \text{for } i = 1, \ldots, m$$

$$0 \leq x_j \quad \text{for } j = 1, \ldots, n$$

The corresponding dual problem (D) is defined as:

$$\text{minimise } w = \sum_{i=1}^{m} b_i y_i$$

subject to

$$\sum_{i=1}^{m} a_{ij} y_i \geq c_j \quad \text{for } j = 1, \ldots, n$$

$$0 \leq y_i \quad \text{for } i = 1, \ldots, m$$

The maximum value of z will be the same as the minimum value of w. The non-zero values of $c_j - v_j$ in the final tableau of the dual problem are the solution values for the variables in the primal problem. If the kth constraint in the primal problem is not binding then the kth dual variable is zero.

(a) Show that if problem 9.1 is taken as P the dual problem is

$$\text{minimise } w = 24y_1 + 6y_2 + y_3 + 2y_4$$

subject to

$$6y_1 + y_2 - y_3 \geq 5$$

$$4y_1 + 2y_2 + y_3 + y_4 \geq 4$$

$$0 \leq y_1, y_2, y_3, y_4$$

(b) Show that the optimal solution to (a) is $w = 21$, and is obtained when $y_1 = 0.75$, $y_2 = 0.5$ and $y_3 = y_4 = 0$. Compare the optimal value of z in Exercise 9.2 with that for w.

(c) In the general context of resource allocation, P has n economic activities and m resources, and c_j is profit per unit of activity j. Resource i, whose maximum availability is b_i, is used at a rate of a_{ij} per unit of activity j. From the fact that the optimal solutions for P and D are equal, deduce that the dual variables represent worth per unit of resource i.

(d) A primal problem is:

$$\text{max } z = c^T x \quad \text{subject to } Ax \leq b$$

Its dual is

$$\text{min } w = b^T y \quad \text{subject to } A^T y \geq c$$

Prove that $w \geq z$.

9.4 Let x and y be units of water released from a dam for hydroelectricity generation and irrigation respectively. The objective is to maximise benefits

B given by

$$B = x + 5y$$

subject to the following constraints.

(i)   The irrigation release must exceed 4 units.
(ii) The total release must be less than 24 units.
(iii) The irrigation release must not exceed the hydroelectric release by more than 12 units.

(a) Write down the constraints as inequalities in x and y.
(b) Solve the problem graphically.
(c) Introduce slack variables and an artificial variable and hence solve the problem using the simplex algorithm.
(d) Adapt the problem so that the origin is a basic feasible solution and solve the adapted problem using the simplex algorithm.
(e) Write down the dual of the adapted problem. Solve this dual problem graphically and by the simplex algorithm.

9.5 A triangular distribution has a lower bound L, a mode M, and an upper bound U. Find the pdf, and hence the mean and standard deviation in terms of L, M and U.

9.6 Verify that application of the algorithm of Section 9.2.2 to the data in Table 9.3 leads to the critical path: building the oven, followed by decorating the interior.

9.7 Find the stationary points of

$$y = 4 + 2(x-1)^3 - 3(x-1)^2$$

Find the greatest value of the function over the domain $0 \le x \le 4$.

9.8 Four investments have expected returns and standard deviations of returns as shown below.

Investment (i)	μ_i	σ_i
1	0.10	0.02
2	0.06	0.00
3	0.20	0.08
4	0.40	0.20

Investments 3 and 4 are correlated with equal ρ to 0.70 but all other correlations are 0. The proportions of each investment must lie between 0.1 and 0.6. Find, by any means, the optimum portfolio with μ equal to 0.15.

Appendix 1   Discounting and compound interest

A1.1   Discounting

Suppose we are certain to receive B dollars in n years' time. If interest rates are constant at $\alpha \times 100\%$ per year, an amount of A dollars invested now would be worth B dollars after n years if

$$A(1 + \alpha)^n = B$$

If we make A the subject of the formula we have the present value of the promise of B in the future. This process is known as discounting values, and is given by:

$$A = B(1 + \alpha)^{-n}$$

In this context the assumed interest rate is also known as the discount rate. The time (n) can be reckoned in months, for example, in which case $\alpha \times 100\%$ is the monthly interest rate. The practical limitation of such calculations is the volatility of interest rates.

Example A1.1
A company has offered to pay a consultant 200 000 dollars for a replacement networked system of personal computers, once it has been set up satisfactorily. The consultant estimates that this will take 8 months, and she will have to borrow some money to pay for hardware at the beginning of the project. Before deciding whether or not to accept the contract, she discounts the payment at a rate of 1.5% per month over 9 months. The extra month is to allow for a delay in receiving the final payment. On this basis the present value is

$$200\,000 \times (1.015)^{-9} = 174\,918$$

A1.2   Mortgage repayment and geometric progressions

A company has borrowed M dollars to expand the business. The loan will be repaid over n years by annual payments of A. If we assume an interest rate of $\alpha \times 100\%$ per annum we have the following balances at ends of years.

end of year 1 $M(1+\alpha) - A$

end of year 2 $(M(1+\alpha) - A)(1+\alpha) - A$

end of year 3 $((M(1+\alpha) - A)(1+\alpha) - A)(1+\alpha) - A$

$\vdots$

end of year n $M(1+\alpha)^n - A((1+\alpha)^{n-1} + \cdots + 1)$

If the M dollars are to be repaid at the end of year n, then setting the final balance equal to 0 gives:

$$A = \frac{M(1+\alpha)^n}{1 + \cdots + (1+\alpha)^{n-1}}$$

The denominator is an example of a geometric progression.

Geometric progression

In general, let the first term be a and the common ratio be r. Then the sum of n terms, S_n, is

$$S_n = a + ar + ar^2 + \cdots + ar^{n-1}$$

Now multiply both sides by r

$$rS_n = ar + ar^2 + ar^3 + \cdots + ar^n$$

and subtract to obtain

$$S_n = \frac{a(1 - r^n)}{1 - r}$$

provided r is not equal to 1. A very important special case is the sum to infinity which is only convergent if r is less than 1. Then

$$S_\infty = \frac{a}{1 - r} \quad \text{for } |r| < 1$$

If the formula for S_n is used for the denominator of the annual repayment we obtain

$$A = \frac{\alpha M(1+\alpha)^n}{(1+\alpha)^n - 1}$$

Example A1.2

A loan of 1 000 000 dollars over 12 years at an annual interest rate of 10% would require annual repayments of

$$A = \frac{0.1 \times 1\,000\,000 \times (1.1)^{12}}{(1.1)^{12} - 1} = 137\,554$$

A1.3 Force of interest

We are offered a loan at an apparent annual interest rate of $\alpha \times 100\%$, but the small print states that this is applied monthly and compounded up. So, if we borrow 100 monetary units at the end of one year we will owe $100 \times (1 + \alpha/12)^{12}$ rather than $100 \times (1 + \alpha)$. If $\alpha \times 100\%$ is 20%, the actual annual interest rate would be 21.94%. Now suppose the interest is applied continuously. At the end of one year we would owe

$$\operatorname*{limit}_{n \to \infty} 100 \times (1 + \alpha/n)^n$$

This is an important limit which occurs in other contexts, including the derivation of the Poisson distribution given in this book. If we apply a Taylor series expansion we have:

$$\left(1 + \frac{\alpha}{n}\right)^n = 1 + n\frac{\alpha}{n} + \frac{n(n-1)}{2!}\left(\frac{\alpha}{n}\right)^2 + \frac{n(n-1)(n-2)}{3!}\left(\frac{\alpha}{n}\right)^3 + \cdots$$

$$= 1 + \alpha + \left(1 - \frac{1}{n}\right)\frac{\alpha^2}{2!} + \left(1 - \frac{1}{n}\right)\left(1 - \frac{2}{n}\right)\frac{\alpha^3}{3!} + \cdots$$

Now let n tend to infinity, in which case the condition for convergence of the Taylor series ($|\alpha/n| < 1$) will be satisfied for any α, to obtain

$$\operatorname*{limit}_{n \to \infty} \left(1 + \frac{\alpha}{n}\right)^n = 1 + \alpha + \frac{\alpha^2}{2!} + \frac{\alpha^3}{3!} + \cdots$$

$$= e^\alpha$$

If α is 0.2, e^α equals 1.2214. A continuous interest rate of 0.2, known as a force of interest, is therefore equivalent to 22.14% per annum when time is reckoned in years. A force of interest of 0.1823 is equivalent to a true 20% per annum. If an amount A is borrowed for a period t at a force of interest α, the amount owed will be $A e^{\alpha t}$.

Appendix 2   Summary of introductory statistics course

A2.1   Probability

Probability is measured on a scale from 0 to 1. However we define probability, the same rules for calculating probabilities of complex events hold.

Addition rule

$$\Pr(A \text{ or } B) = \Pr(A) + \Pr(B) - \Pr(A \text{ and } B) \tag{A2.1}$$

In mathematics 'or' conventionally includes the possibility of both. If $\Pr(A \text{ and } B) = 0$, we say A and B are *mutually exclusive*.

Complement

$$\Pr(\text{not } A) = 1 - \Pr(A)$$

This useful result is proved from the addition rule by substituting $\bar{A}$ (not A) for B.

Conditional probability

$$\Pr(A \mid B) = \Pr(A \text{ and } B) / \Pr(B) \tag{A2.2}$$

The vertical line is read as 'conditional on' or 'given that'.

Multiplicative rule

$$\Pr(A \text{ and } B) = \Pr(A \mid B) \Pr(B) \quad \text{and} \quad = \Pr(B \mid A) \Pr(A) \tag{A2.3}$$

A and B are *independent* if, and only if,

$$\Pr(A \text{ and } B) = \Pr(A) \Pr(B)$$

A2.2 Sample statistics and population parameters

Suppose we have a simple random sample (SRS) $\{x_i\}$ of size n from a finite population of size N.

	Sample estimate of population parameter	Population parameter
Mean	$\bar{x} = \sum x_i/n$	$\mu = \sum x_i/N$ or $\int xf(x)\,dx$
Variance	$s^2 = \sum(x_i - \bar{x})^2/(n-1)$	$\sigma^2 = \sum(x_i - \mu)^2/N$
Standard deviation	$s = \sqrt{s^2}$	$\sigma = \sqrt{\sigma^2}$
Skewness	$\hat{\gamma} = [\sum(x_i - \bar{x})^3/(n-1)]/s^3$	$\gamma = [\sum(x_i - \mu)^3/N]/\sigma^3$
Kurtosis	$\hat{\kappa} = [\sum(x_i - \bar{x})^4/(n-1)]/s^4$	$\kappa = [\sum(x_i - \mu)^4/N]/\sigma^4$
Coefficient of variation	$\widehat{CV} = s/\bar{x}$	$CV = \sigma/\mu$

Expected value

If a population is infinite a line chart for a discrete variable will tend towards a probability mass function $P(x)$. A histogram for a continuous variable will tend towards a probability density function (pdf) $f(x)$, and the cumulative frequency polygon tends toward the cumulative distribution function (cdf) $F(x)$. The pdf and cdf are related by

$$\frac{dF(x)}{dx} = f(x) \quad \text{and conversely} \quad F(x) = \int_{-\infty}^{x} f(\theta)\,d\theta$$

Expected value is a population average. The mean (μ) of a variable X is

$$E[X] = \sum xP(x) \quad \text{or} \quad \int xf(x)\,dx$$

for the discrete or continuous case respectively. The variance for a continuous variable is

$$\sigma^2 = E[(X - \mu)^2] = \int(x - \mu)^2 f(x)\,dx$$

and the expected value of any function $\phi(X)$ is

$$E[\phi(X)] = \int \phi(x)f(x)\,dx$$

Definitions for a discrete variable use $P(x)$ instead of $f(x)\,dx$. Two useful results are:

(i) $E[(X - \mu)] = E[X] - \mu = \mu - \mu = 0$
(ii) $\sigma^2 = E[(X - \mu)^2] = E[X^2 - 2\mu X + \mu^2] = E[X^2] - 2\mu E[X] + \mu^2$

If this is rearranged we have

$$E[X^2] = \mu^2 + \sigma^2$$

A2.3   Binomial distribution

A Bernoulli trial is an experiment with precisely two possible outcomes, generally referred to as success and failure. Let X be the number of successes in n such trials with a constant probability of success p. The constant probability of success implies that the outcomes of the trials are independent. Then

$$\Pr(X = x) = P(x) = {}_nC_x p^x(1 - p)^{n - x} \quad \text{for } x = 0, \ldots, n$$

where ${}_nC_x$ is the number of ways of choosing x items from $n(n!/(x!(n - x)!))$. The variable X has a binomial distribution:

$$X \sim \text{Bin}(n, p)$$

The mean and variance of X are np and $np(1 - p)$ respectively. The mean can be obtained as follows:

$$\mu = \text{E}[X] = \sum_{x=0}^{n} xP(x) = \sum x \, {}_nC_x p^x(1 - p)^{n - x}$$

$$= \sum_{x=1}^{n} x \, {}_nC_x p^x(1 - p)^{n - x} = np \sum_{y=0}^{n-1} {}_{n-1}C_y p^y(1 - p)^{n - 1 - y} = np$$

The variance can be found from a similar argument.

A2.4   Uniform distribution

The variable X has a uniform distribution on $[a, b]$ if

$$f(x) = 1/(b - a) \quad \text{for } a \le x \le b \tag{A2.4}$$

and we use the notation

$$X \sim U[a, b]$$

The mean and variance of X are easily found from the defining integral, and equal $(a + b)/2$ and $(b - a)^2/12$ respectively. Random numbers with a uniform distribution on $[0, 1]$ are the basis for stochastic simulations.

A2.5   Normal distribution

The pdf is

$$f(x) = \frac{1}{\sigma\sqrt{2\pi}} \, e^{-((x-\mu)/\sigma)^2/2} \quad \text{for } -\infty < x < \infty \tag{A2.5}$$

The parameters μ and σ are the mean and standard deviation of the distribution. The cdf

$$F(x) = \int_{-\infty}^{x} \frac{1}{\sigma\sqrt{2\pi}} e^{-((\theta-\mu)/\sigma)^2/2} \, d\theta \tag{A2.6}$$

has to be evaluated numerically. This leads to the definition of the *standard normal distribution* as a normal distribution with a mean of 0 and a standard deviation of 1. It is conventional to use z for a standard normal variable and the Greek letter ϕ for its distribution, so its pdf is written

$$\phi(z) = \frac{1}{\sqrt{2\pi}} e^{-z^2/2}$$

and its cdf, Table A9.1, is

$$\Phi(z) = \frac{1}{\sqrt{2\pi}} \int_{-\infty}^{z} e^{-\theta^2/2} \, d\theta$$

The *inverse* cdf (Φ^{-1}) is defined by:

$$z = \Phi^{-1}(p) \quad \text{if and only if} \quad \Phi(z) = p$$

The upper $\alpha \times 100\%$ percentage point is the value above which a proportion α of the distribution lies. That is

$$\Phi(z_\alpha) = 1 - \alpha$$

We write Z for a standard normal variable, so

$$\frac{X - \mu}{\sigma} \sim Z$$

where $Z \sim N(0, 1)$.

A2.6 Moment-generating function

A continuous probability distribution can be characterised by either its pdf or its cdf, and a discrete distribution can similarly be defined by its probability function or the cumulated probabilities. An alternative representation is the *moment-generating function* (mgf). The mgf of a variable X is defined by

$$M_X(\theta) = E[e^{\theta x}] \tag{A2.7}$$

which holds for both discrete and continuous distributions. If it exists, it determines the distribution uniquely. The mgf of a binomial distribution is

$$M_X(\theta) = [(1 - p) + p e^{\theta}]^n$$

The mgf of a normal distribution is

$$M_X(\theta) = e^{\mu\theta + (1/2)\sigma^2\theta^2}$$

Moments of a distribution

It is easy to show, from the definition and the Maclaurin series expansion for $e^{\theta x}$, that

$$E[X^k] = \left(\frac{d^k}{d\theta^k} M_X(\theta) \right)_{\theta = 0}$$

The moments about the mean

$$E[(X - \mu)^k]$$

follow by expanding $(X - \mu)^k$ and using the expressions for $E[X^j]$ for j up to k.

Moment-generating function of a sum of independent variables

Let $X_1, \ldots, X_n$ be independent variables with mgf $M_{X_1}(\theta), \ldots, M_{X_n}(\theta)$ respectively. Define the sum T by

$$T = X_1 + \cdots + X_n$$

Then the mgf of T is the product of the mgfs of $X_1, \ldots, X_n$. The proof follows from the definition and the factorisation of $f(x_1, \ldots, x_n)$ as $f(x_1) \ldots f(x_n)$ since the variables are independent.

  This result can be used to prove the central limit theorem (see Metcalfe, 1994, for example).

A2.7   Multivariate probability distributions

If we have two variables we can draw a bivariate histogram with a total volume of one. As the sample size increases we imagine this tending towards a smooth surface given by the formula

$$z = f(x, y)$$

such that $f(x, y) \geq 0$ and

$$\int \int f(x, y) \, dx \, dy = 1$$

This is known as a bivariate pdf. The bivariate cdf is $F(x, y)$ where

$$F(x, y) = \Pr(X < x \text{ and } Y < y)$$

$$= \int_{-\infty}^{x} \int_{-\infty}^{y} f(\zeta, \eta) \, d\zeta \, d\eta$$

It follows that

$$\frac{\partial^2 F}{\partial x \, \partial y} = f(x, y)$$

The variables X and Y are *independent* if and only if

$$f(x, y) = f_X(x) f_Y(y)$$

Marginal distributions

The *marginal distribution* of X is just the distribution of X, i.e. the information about Y is ignored.

$$f_X(x) = \int f(x, y) \, dy$$

The distribution of Y is obtained in a similar fashion, by integrating over x. The subscript is used if it is needed to distinguish it from other pdfs.

Expected value

The expected value of some function ϕ of X and Y is given by

$$E[\phi(X, Y)] = \int\int \phi(x, y)f(x, y)\,dx\,dy$$

In particular *covariance*, which is a measure of linear association, is defined by

$$\text{cov}(X, Y) = E[(X - \mu_X)(Y - \mu_Y)] = \int\int (x - \mu_X)(y - \mu_Y)f(x, y)\,dx\,dy$$

The *correlation* is a dimensionless measure defined by $\rho = \text{cov}(X, Y)/(\sigma_X \sigma_Y)$. If X and Y are independent $\text{cov}(X, Y)$ equals zero. The converse is not necessarily true; for example, there may be a quadratic relationship or the values that can be taken by X may depend on the value of Y.

Conditional distributions

The conditional distribution of Y given x is defined by

$$f(y \mid x) = f(x, y)/f(x)$$

Bivariate normal distribution and regression

Suppose that (X, Y) has a bivariate normal distribution. Then if W and Z are defined by

$$W = (X - \mu_X)/\sigma_X \quad \text{and} \quad Z = (Y - \mu_Y)/\sigma_Y$$

(W, Z) has a standardised bivariate normal distribution and the pdf is

$$f(w, z) = \frac{1}{2\pi\sqrt{(1 - \rho^2)}} \exp\left\{\frac{-1}{2(1 - \rho^2)}(w^2 - 2\rho wz + z^2)\right\}$$

The marginal distributions are standard normal.
    Some straightforward algebra leads to the conditional distribution

$$f(z \mid w) = \frac{1}{\sqrt{2\pi}\sqrt{(1 - \rho^2)}} \exp\left\{\frac{-1}{2(1 - \rho^2)}(z - \rho w)^2\right\}$$

This is a normal distribution with mean, $E[Z \mid w]$, equal to ρw and a variance of $(1 - \rho^2)$. The regression line of z on w is

$$z = \rho w$$

This result can be rescaled so it is explicitly in terms of Y and x. The regression line of Y on x is

$$y = \mu_Y + \rho\frac{\sigma_Y}{\sigma_X}(x - \mu_X)$$

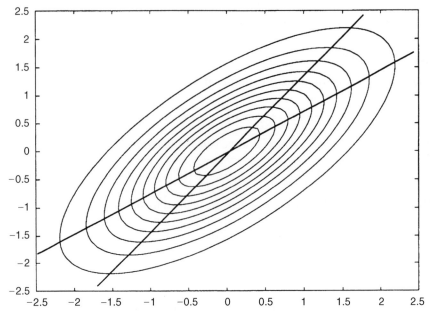

Figure A2.1 Bivariate normal distribution and regression lines

An identical argument leads to the regression line of X on y which is

$$x = \mu_X + \rho \frac{\sigma_X}{\sigma_Y}(y - \mu_Y)$$

The two regression lines are not the same. Since $|\rho| < 1$, the regression line of y on x is less steep than the major axis of the elliptical contours of the bivariate distribution, whereas the regression line of x on y is steeper (Figure A2.1).

Multivariate normal distribution

Let $x = (x_1, \ldots, x_k)^T$

$$f(x) = (2\pi)^{-k/2}|V|^{-1/2}\exp(-\tfrac{1}{2}(x - \mu)^T V^{-1}(x - \mu)) \qquad \text{(A2.8)}$$

where μ is the mean and V is the variance–covariance matrix, often just called the covariance matrix. If the elements of V are denoted by σ_{ij}, then $\sigma_{ii} = \text{var}(x_i)$ and $\sigma_{ij} = \text{cov}(x_i x_j)$ if $i \neq j$.

A2.8 Sample correlation

The population correlation, ρ, is estimated by the sample correlation, r. If we have n data pairs (x_i, y_i), the sample covariance is

$$\widehat{\text{cov}} = \sum(x_i - \bar{x})(y_i - \bar{y})/(n - 1) \qquad \text{(A2.9a)}$$

and

$$r = \widehat{\text{cov}}/(s_x s_y) \tag{A2.9b}$$

A $(1 - \alpha) \times 100\%$ confidence interval for ρ is

$$\tanh[\text{arctanh}(r) \pm z_{\alpha/2}/\sqrt{(n-3)}]$$

A2.9   Functions of variables

Linear function

Let X and Y be variables with means μ_X, μ_Y and variances σ_X^2, σ_Y^2, respectively. Now construct a variable which is a linear combination of X and Y:

$$W = aX + bY \tag{A2.10a}$$

where a and b are constants. Then

$$\mu_W = a\mu_X + b\mu_Y \tag{A2.10b}$$

$$\sigma_W^2 = \text{E}[(W - \mu_W)^2] = \text{E}[((aX + bY) - (a\mu_X + b\mu_Y))^2]$$

$$= \text{E}[(a(X - \mu_X) + b(Y - \mu_Y))^2]$$

$$\sigma_W^2 = a^2\sigma_X^2 + b^2\sigma_Y^2 + 2ab\,\text{cov}(X, Y) \tag{A2.10c}$$

Notice that the final term disappears if X and Y are independent. The result extends to any number of variables.

Non-linear function

Let ϕ be an arbitrary function of X and Y and expand ϕ about their mean values. Then

$$\phi(X, Y) = \phi(\mu_X, \mu_Y) + \frac{\partial \phi}{\partial x}(X - \mu_X) + \frac{\partial \phi}{\partial y}(Y - \mu_Y)$$

$$+ \frac{1}{2!}\left(\frac{\partial^2 \phi}{\partial x^2}(X - \mu_X)^2 + 2\frac{\partial^2 \phi}{\partial x\,\partial y}(X - \mu_X)(Y - \mu_Y)\right.$$

$$\left. + \frac{\partial^2 \phi}{\partial y^2}(Y - \mu_Y)^2\right) + \cdots$$

where all the partial derivatives are evaluated at (μ_X, μ_Y). We start by considering the question of bias. Take expectation of both sides to obtain the approximate result that

$$\text{E}[\phi(X, Y)] = \phi(\mu_X, \mu_Y) + \frac{1}{2}\frac{\partial^2 \phi}{\partial x^2}\sigma_X^2 + \frac{1}{2}\frac{\partial^2 \phi}{\partial y^2}\sigma_Y^2 + \frac{\partial^2 \phi}{\partial x\,\partial y}\text{cov}(X, Y)$$

If the variables are independent the covariance term is zero and can be dropped. An approximation to the variance of $\phi(X, Y)$ is usually based on the linear terms only. If X and Y are independent

$$\mathrm{var}(\phi(X, Y)) \simeq \left(\frac{\partial \phi}{\partial x}\right)^2 \sigma_X^2 + \left(\frac{\partial \phi}{\partial y}\right)^2 \sigma_Y^2$$

Unbiased estimators and bias

An estimator X for a parameter θ is unbiased if

$$\mathrm{E}[X] = \theta$$

Bias is defined as the difference between $\mathrm{E}[X]$, which is its mean μ, and θ. Mean squared error (MSE) is

$$\mathrm{E}[(X - \theta)^2] = \mathrm{E}[((X - \mu) + (\mu - \theta))^2]$$
$$= \mathrm{E}[(X - \mu)^2] + (\mu - \theta)^2$$
$$= \mathrm{var}(X) + (\mathrm{bias})^2$$

A biased estimator is acceptable if the bias is small compared with its variance. For example, S is a slightly biased estimator of σ. An *estimator* is said to be consistent if the bias and variance tend to zero as the sample size tends to infinity.

A2.10   Distribution of the sample mean

Let $\{X_i\}$ be a random sample of size n from any distribution with a mean μ and variance σ^2. Define the sample total

$$T = X_1 + \cdots + X_n$$

Using the results for linear functions

$$\mu_T = \mu + \cdots + \mu = n\mu$$

and

$$\sigma_T^2 = \sigma^2 + \cdots + \sigma^2 = n\sigma^2$$
$$\sigma_T = \sigma\sqrt{n}$$

since randomisation makes an assumption of independence valid. Now

$$\bar{X} = T/n$$

and it follows that

$$\mu_{\bar{X}} = n\mu/n = \mu \quad \sigma_{\bar{X}}^2 = n\sigma^2/n^2 = \sigma^2/n$$

and

$$\sigma_{\bar{X}} = \sigma/\sqrt{n}$$

The practical consequence of the central limit theorem is that T and $\bar{X}$ are approximately normally distributed if the sample is reasonably large. This is an example of a *sampling distribution*, so called because it is the hypothetical distribution of $\bar{X}$ that would be obtained if we took an infinite number of samples of size n. The approximation is usually adequate if n exceeds 30, and if the original population is itself near normal any value of n will suffice. If we are sampling from a finite population of size N, the variance of $\bar{X}$ is reduced by a factor of $(1 - n/N)$, known as the *finite population correction*.

$$\bar{X} \sim N\left(\mu, \left(\frac{\sigma}{\sqrt{n}}\right)^2\right)$$ (A2.11)

leads to a confidence interval for μ

$$\Pr\left(-z_{\alpha/2} < \frac{\bar{X} - \mu}{\sigma/\sqrt{n}} < z_{\alpha/2}\right) = 1 - \alpha$$

$$\Pr(\bar{X} - z_{\alpha/2}\sigma/\sqrt{n} < \mu < \bar{X} + z_{\alpha/2}\sigma/\sqrt{n}) = 1 - \alpha$$

A $(1 - \alpha) \times 100\%$ confidence interval for μ is

$$\bar{x} \pm z_{\alpha/2}\sigma/\sqrt{n}$$

The limitation of this result is that it assumes the population standard deviation is known. W.S. Gosset introduced Student's t-distribution to allow for the use of the sample estimator of the population standard deviation.

Student's *t*-distribution

Student's t-distribution makes allowance for the use of the sample estimator of the population standard deviation. That is:

$$\frac{\bar{X} - \mu}{S/\sqrt{n}} \sim t_{n-1}$$

where t_{n-1} is Student's t-distribution with $n - 1$ degrees of freedom. The degrees of freedom is the divisor used when calculating s, in this context the sample size minus 1. A $(1 - \alpha) \times 100\%$ confidence interval for μ when σ is estimated by s is,

$$\bar{x} \pm t_{n-1, \alpha/2}s/\sqrt{n}$$

where $t_{n-1, \alpha/2}$ is the upper $\alpha/2 \times 100\%$ point of the distribution. The t-distribution tends towards the standard normal distribution as the degrees of freedom increase.

A2.11 The chi-square distribution

The special cases of the gamma distribution with $c = \nu/2$ and $\lambda = 1/2$ are known as chi-square distributions with ν degrees of freedom (χ_ν^2). The mgf of a gamma distribution is given by

$$M(\theta) = \int_0^\infty e^{\theta x} \lambda^c x^{c-1} (\Gamma(c))^{-1} e^{-\lambda x}\, dx$$

and substituting $y = x(\lambda - \theta)$ gives

$$M(\theta) = (1 - \theta/\lambda)^{-c} \quad \text{for } \theta < \lambda$$

A chi-square distribution with ν degrees of freedom has mgf

$$M(\theta) = (1 - 2\theta)^{-\nu/2}$$

The mean and variance are ν and 2ν. It follows from the form of $M(\theta)$ that the sum of independent chi-square random variables with ν_1 and ν_2 degrees of freedom have a chi-square distribution with $(\nu_1 + \nu_2)$ degrees of freedom.

A2.12   Sampling distribution of sample variance

If $Z \sim N(0, 1)$ and $Y = Z^2$ then $Y \sim \chi_1^2$. Now suppose $\{X_i\}$ is a random sample from $N(\mu, \sigma^2)$. Then

$$\sum \left(\frac{X_i - \mu}{\sigma} \right)^2 \sim \chi_n^2$$

If μ is replaced by $\bar{X}$, it can be shown that

$$\sum \left(\frac{X_i - \bar{X}}{\sigma} \right)^2 \sim \chi_{n-1}^2$$

and this is equivalent to

$$\frac{(n - 1)S^2}{\sigma^2} \sim \chi_{n-1}^2$$

The result that $E[S^2] = \sigma^2$ can be deduced from the fact that the mean of a χ^2 distribution equals its degrees of freedom or, more directly, by taking expected value of the left-hand and right-hand expressions in the following identities:

$$\sum (X_i - \mu)^2 = \sum (X_i - \bar{X} + \bar{X} - \mu)^2 = \sum (X_i - \bar{X})^2 + n(\bar{X} - \mu)^2$$

We can use the chi-square distribution to construct a $(1 - \alpha) \times 100\%$ confidence interval for σ^2, and hence σ

$$\Pr \left(\chi_{n-1, 1-\alpha/2}^2 < \frac{(n - 1)S^2}{\sigma^2} < \chi_{n-1, \alpha/2}^2 \right) = 1 - \alpha$$

Rearrangement gives

$$\Pr \left(\frac{(n - 1)S^2}{\chi_{n-1, \alpha/2}^2} < \sigma^2 < \frac{(n - 1)S^2}{\chi_{n-1, 1-\alpha/2}^2} \right) = 1 - \alpha$$

and hence a $(1 - \alpha) \times 100\%$ confidence interval for σ^2 of

$$\left[\frac{(n - 1)s^2}{\chi_{n-1, \alpha/2}^2}, \frac{(n - 1)s^2}{\chi_{n-1, 1-\alpha/2}^2} \right]$$

A corresponding confidence interval for σ is obtained by taking the square root of the interval for σ^2. Percentage points of the chi-square distribution are given in Table A9.3.

A2.13 The *F*-distribution

If U and V are independent random variables having chi-square distributions with ν_1 and ν_2 degrees of freedom, then the distribution of

$$W = \frac{U/\nu_1}{V/\nu_2}$$

has an *F*-distribution with ν_1 and ν_2 degrees of freedom. If W has an *F*-distribution with ν_1 and ν_2 degrees of freedom, its reciprocal has an *F*-distribution with ν_2 and ν_1 degrees of freedom:

$$\text{If} \quad W \sim F_{\nu_1,\nu_2} \quad \text{then} \quad W^{-1} \sim F_{\nu_2,\nu_1}$$

Therefore only the upper tail is tabled (Table A9.4, for example).

In particular, if we have independent random samples of sizes n_A and n_B from normal populations with variances σ_A^2, σ_B^2 respectively then

$$\frac{S_A^2}{S_B^2} \times \frac{\sigma_B^2}{\sigma_A^2} \sim F_{n_A-1,n_B-1}$$

It follows that a $(1-\alpha) \times 100\%$ confidence interval for σ_A^2/σ_B^2 is given by

$$[(s_A^2/s_B^2)F_{n_B-1,\,n_A-1,\,1-\alpha/2}, \; (s_A^2/s_B^2)F_{n_B-1,\,n_A-1,\,\alpha/2}]$$

with

$$F_{n_B-1,\,n_A-1,\,1-\alpha/2} = 1/F_{n_A-1,\,n_B-1,\,\alpha/2}$$

A2.14 Differences in means

Independent samples

Populations A and B have means μ_A and μ_B, and standard deviations σ_A and σ_B, respectively. Suppose we have two independent random samples of sizes n_A and n_B from these populations. We rely on the central limit theorem, which is a good approximation if the sample is large or the population near-normal, to state that

$$\bar{X}_A \sim N(\mu_A, \sigma_A^2/n_A) \quad \text{and} \quad \bar{X}_B \sim N(\mu_B, \sigma_B^2/n_B)$$

Therefore, since the samples are independent,

$$\bar{X}_A - \bar{X}_B \sim N(\mu_A - \mu_B, \sigma_A^2/n_A + \sigma_B^2/n_B)$$

By the usual argument, a $(1-\alpha) \times 100\%$ confidence interval for $\mu_A - \mu_B$ is given by

$$\bar{x}_A - \bar{x}_B \pm z_{\alpha/2}(\sigma_A^2/n_A + \sigma_B^2/n_B)^{1/2}$$

Usually σ_A^2 and σ_B^2 are unknown, so they are replaced by their sample estimates to obtain an approximate $(1-\alpha) \times 100\%$ confidence interval for $\mu_A - \mu_B$ as

$$\bar{x}_A - \bar{x}_B \pm t_{\nu,\alpha/2}(s_A^2/n_A + s_B^2/n_B)^{1/2}$$

where

$$\nu = \left(\frac{s_A^2}{n_A} + \frac{s_B^2}{n_B}\right)^2 \bigg/ \left(\frac{(s_A^2/n_A)^2}{n_A - 1} + \frac{(s_B^2/n_B)^2}{n_B - 1}\right)$$

This is unlikely to be an integer but interpolation in Table A9.2 can be used.

Paired comparisons

A sample of n paired units is available. Treatment A is applied to a randomly selected unit from each pair, and treatment B is applied to the other. The results for pair i are x_{Ai} and x_{Bi}. The differences are:

$$d_i = (x_{Ai} - x_{Bi})$$

and the mean and standard deviation of $\{d_i\}$ are $\bar{d}$ and s_d. A $(1 - \alpha) \times 100\%$ confidence interval for the mean of the population of all such differences is

$$\bar{d} \pm t_{n-1,\,\alpha/2} s_d / \sqrt{n}$$

In some cases both treatments can be applied to the same experimental unit, in which case only n units are required.

A2.15   Proportions

Confidence interval for a proportion

Let p represent the proportion of defectives in the population. Take a random sample of size n from this population. Let X be the number of defectives in the sample.

$$X \sim \text{Bin}(n, p)$$

Provided both np and $n(1 - p)$ exceed about five, X is approximately normally distributed. That is, to a good approximation:

$$X \sim N(np, np(1 - p)) \quad \text{and hence} \quad \frac{X}{n} \sim N(p, p(1 - p)/n))$$

An approximate $(1 - \alpha) \times 100\%$ confidence interval for p is

$$\frac{x}{n} \pm z_{\alpha/2} \sqrt{\frac{x}{n}\left(1 - \frac{x}{n}\right) \bigg/ n}$$

Confidence interval for difference in proportions

Consider two large (or infinite) populations A, B and let the proportion of defectives be p_A, p_B respectively. Draw random samples of sizes n_A, n_B from the populations and let X and Y represent the number of defectives in each sample.

Provided n_A, n_B exceed at least 30, and $n_A p_A$, $n_A(1 - p_A)$, $n_B p_B$ and $n_B(1 - p_B)$ all exceed 5, an approximate $(1 - \alpha) \times 100\%$ confidence interval for $p_A - p_B$ is given by

$$(x/n_A - y/n_B) \pm z_{\alpha/2}\sqrt{(x/n_A)(1 - x/n_A)/n_A + (y/n_B)(1 - y/n_B)/n_B}$$

A2.16 Hypothesis testing

A null hypothesis (H_0) is a supposition put forward as a basis for decision making. If there is no evidence against H_0, in favour of H_1, we act as if H_0 is true. The test statistic is chosen so that it is sensitive to the assumption H_0. The *P*-value associated with the calculated value of the test statistic is the probability of obtaining a value as extreme as, or more extreme than, that observed, if H_0 is true. The result is statistically significant at, or beyond, the $\alpha \times 100\%$ level if the *P*-value is less than α. The smaller the *P*-value the stronger the evidence against H_0. For example, consider testing a null hypothesis that a mean equals μ_0, from a random sample of size n.

$$H_0: \mu = \mu_0$$

$$H_1 \text{ (two-sided): } \mu \neq \mu_0$$

The test statistic is:

$$\frac{\bar{X} - \mu_0}{S/\sqrt{n}}$$

If H_0 is true the test statistic has a *t*-distribution with $n - 1$ degrees of freedom. Let the calculated value of the test statistic be t_{calc}. The *P*-value with the two-sided alternative hypothesis is:

$$\Pr(t_{n-1} < -|t_{\text{calc}}| \text{ or } |t_{\text{calc}}| < t_{n-1})$$

If we test at the $\alpha \times 100\%$ level we reject H_0 if the *P*-value is less than α. An immediate consequence of the construction of a $(1 - \alpha) \times 100\%$ confidence interval for μ:

$$\bar{x} \pm t_{n-1, \alpha/2}s/\sqrt{n}$$

is that we reject H_0 if this interval does not include 0.

A one-sided alternative hypothesis could be $H_1: \mu < \mu_0$, in which case the *P*-value is

$$\Pr(t_{n-1} < t_{\text{calc}})$$

A2.17 Goodness-of-fit tests

The chi-square goodness-of-fit test

This test compares observed (O_k) and expected (E_k) frequencies over c class intervals. If a simple random sample is taken the observed frequency is roughly approximated as a Poisson variable with mean E_k. The variance of a

Poisson variable equals its mean so, provided the expected value exceeds about 5, we have, approximately,

$$\frac{O_k - E_k}{\sqrt{E_k}} \sim N(0, 1)$$

and

$$\frac{(O_k - E_k)^2}{E_k} \sim \chi_1^2$$

Their sum

$$W = \sum_{k=1}^{c} \frac{(O_k - E_k)^2}{E_k} \sim \chi_{c-1-p}^2$$

We lose one degree of freedom for each parameter of the distribution which we estimate from the data, and one more for the constraint that the sum of the observed frequencies equals the sum of the expected frequencies.

Appendix 3 Pseudo-random number generation

A3.1 Uniform random numbers and random digits

There are electronic devices, such as noise diodes, which use the thermal agitation of electrons (Johnson noise) to approximate white noise. Such devices could be adapted to generate uniform random numbers. However, simulation work usually relies on mathematical algorithms which, although deterministic, produce sequences of numbers that appear to be distributed as independent uniform random variables. These algorithms are properly known as pseudo-random number generators, but this is usually abbreviated by dropping the 'pseudo'. Cooke *et al.* (1990) describe more of the background. The following is a simple example of a mixed congruential generator, using division modulo 6075

$$I_{j+1} = 106I_j + 1283 \quad (\text{mod } 6075)$$

where the instruction (mod 6075) means: let I_{j+1} equal the remainder after dividing $(106I_j + 1283)$ by 6075. It will generate a sequence of integers $\{I_j\}$ between 0 and 6074. These integers are divided by 6075 to give a sequence of decimals between 0 and 1

$$r_j = I_j/6075$$

The $\{r_j\}$ appear to be independent uniformly distributed random variables, except for the fact that the sequence must repeat itself after each run of 6075 numbers – if not before. The generator is called 'mixed' because it involves addition as well as multiplication, and 'congruential' because of the modular division which results in numbers differing by multiples of 6075 being equivalent. To see how the calculation works let us start with I_0 equal to 127. Then

$$I_1 = 106 \times 127 + 1283 = 14\,745(\text{mod } 6075) = 2595$$

$$I_2 = 276\,353(\text{mod } 6075) = 2978$$

$$I_3 = \cdots = 1051$$

$$I_4 = \cdots = 3339 \text{ etc.}$$

The corresponding $\{r_j\}$ sequence starts: 0.427 161, 0.490 206, 0.173 004, 0.549 630, ... A sequence of random digits $\{n_j\}$ can be obtained by multiplying the $\{r_j\}$ by 10 and taking the integer part. The first few are 4, 4, 1, 5.

Fortran and most other languages have an MOD function which will do the division modulo 6075 for you, but you can obtain the same effect by:

$$(i/k - \text{integer part of } i/k) \times k$$

Remember to use integer arithmetic. The uniform distribution of the $\{r_j\}$ can be checked by a histogram, but there are many tests that can be applied to help justify an assumption that the $\{r_j\}$ are independent. For example:

(i)   plot r_{j+1} against r_j and check that the points are spread all over the square;
(ii) calculate the correlations between $\{r_j\}$ and $\{r_{j+k}\}$ for k from 1 upwards, and check that they are within sampling variation of 0;
(iii) check the proportions of two-digit combinations in $\{n_j\}$.

In practice you will want a reliable generator with a much longer cycle length, and you will not wish to test it yourself. Reputable mathematical or statistical software should provide good generators, but it is often convenient to write your own within a program.

Ross (1997) suggests multiplicative congruential generators of the form

$$I_{j+1} = aI_j(\text{mod } m)$$

where for a 32-bit word machine $m = 2^{31} - 1$ and $a = 7^5$, and for a 36-bit word machine $m = 2^{35} - 31$ and $a = 5^5$. Wichmann and Hill (1982) suggest combining numbers from several multiplicative generators and their algorithm is used widely.

A3.2   Wichmann's and Hill's algorithm

This is a brief description of Wichmann's and Hill's (1982) algorithm. It relies on the fact that the fractional part of the sum of n independent uniform random numbers remains uniform, despite the fact that their sum tends to normality. To see why, suppose $R_1 \sim U[0,1]$ and add any number (a) between 0 and 1, to R_1. Then

$$R_1 + a \sim U[a, 1 + a]$$

Now, the interval $[a, 1 + a]$ can be split up as $[a, 1) + [1, 1 + a)$. If we take the fractional part of $(R_1 + a)$ the corresponding interval will be

$$[a, 1) + [0, a)$$

which is $[0, 1]$. Therefore, the fractional part of $R_1 + a$ has a $U[0,1]$ distribution. In particular, a could be the value taken by an independent variable

$$R_2 \sim U[0, 1]$$

The general result follows by induction. (Note the result would still apply if a has any value, and R_2 could have any distribution provided R_2 is independent of R_1.) The reason for combining numbers from several generators is that none is perfect and the combination is an improvement with a longer cycle length.

The algorithm is: I_0, J_0 and K_0 can be arbitrarily set at any integer values within the range $1-30\,000$:

$$I_{m+1} = 171I_m \quad (\mathrm{mod}\ 30\,269)$$

$$J_{m+1} = 172J_m \quad (\mathrm{mod}\ 30\,307)$$

$$K_{m+1} = 170K_m \quad (\mathrm{mod}\ 30\,323)$$

which must be written in integer arithmetic

$$S_{m+1} = I_{m+1}/30\,269 + J_{m+1}/30\,307 + K_{m+1}/30\,323$$

$$R_{m+1} = S_{m+1} - \text{integer part of } S_{m+1}$$

An adjustment to the algorithm to avoid the possibility of obtaining exactly zero, within the precision of the machine, is given by McLeod (1985). Although each of the three simple multiplicative congruential generators has a prime number for its modulus, the cycle it produces does not include the value zero. Each cycle length is one less than the corresponding modulus, and the cycle length of the sum of the simple generators is the lowest common multiple of the three cycle lengths. This is approximately 0.7×10^{13}.

A3.3 Generating random numbers from discrete distributions

Suppose a discrete random variable X can take integer values from 0 upwards, and has a probability mass function $P(x)$. The cumulative distribution function $F(x)$ is defined by

$$F(x) = \sum_{i=0}^{x} P(i)$$

Now generate a uniform random number R from $U[0,1]$. Then a random number X from the discrete distribution is given by:

$$X = \begin{cases} 0 & \text{if } R \le F(0) \\ 1 & \text{if } F(0) < R \le F(1) \\ 2 & \text{if } F(1) < R \le F(2) \\ \vdots \end{cases}$$

Example A3.1
Generate random numbers from a Poisson distribution with mean 3.5. First calculate:

$$P(0) = e^{-3.5} \qquad\qquad = 0.0302$$

$$P(1) = 3.5P(0) \qquad = 0.1057$$

$$P(2) = 3.5P(1)/2 = 0.1850$$

$$P(3) = 3.5P(2)/3 = 0.2158$$

$$\vdots$$

$$P(13) = \qquad\qquad\qquad = 0.0001$$

Then calculate

$$F(0) = 0.0302$$
$$F(1) = 0.1359$$
$$F(2) = 0.3208$$
$$F(3) = 0.5366$$
$$F(4) = 0.7254$$
$$\vdots$$
$$F(13) = 1.0000$$

Now generate uniform random numbers, and locate the corresponding Poisson numbers. For example, the uniform random numbers on the left give the Poisson random numbers on the right:

$$0.192\,11 \text{ gives } 2$$
$$0.945\,20 \text{ gives } 7$$
$$0.709\,86 \text{ gives } 4$$
$$0.652\,49 \text{ gives } 4$$
$$\text{etc.}$$

Example A3.2
Generate random numbers from a negative binomial distribution with parameters r and p equal to 5.183 and 0.561 respectively. The same principle is applied, the only difference being in the probability calculations. It is not necessary to evaluate gamma functions because of cancellations.

$$P(0) = \frac{\Gamma(r)}{\Gamma(r)0!}p^r = p^r \qquad\qquad = 0.0500$$

$$P(1) = \frac{\Gamma(r+1)}{\Gamma(r)1!}p^r(1-p) = rp^r(1-p) \qquad\qquad = 0.1137$$

$$P(2) = \frac{\Gamma(r+2)}{\Gamma(r)2!}p^r(1-p)^2 = (r+1)rp^r(1-p)^2/2 = 0.1544$$

and so on.

A3.4   Generating random numbers from continuous distributions

A3.4.1   Inverse cumulative distribution function method

Suppose a random variable X has a cdf $F(x)$ and we want to generate random numbers from this distribution. For any a and b

$$\Pr(a < X < b) = F(b) - F(a)$$

Both $F(a)$ and $F(b)$ are probabilities, and must therefore be between 0 and 1. So, if R is distributed $U[0, 1]$ then

$$\Pr(F(a) < R < F(b)) = F(b) - F(a)$$

as well. Now apply the inverse cdf (F^{-1}) to the inequality to obtain the equivalent probability statement

$$\Pr(a < F^{-1}(R) < b) = F(b) - F(a)$$

Since a and b were arbitrary numbers $F^{-1}(R)$ and X have the same distribution with cdf F. All we need do is generate a sequence of uniform random numbers and then apply F^{-1}.

Example A3.3
Port authorities use Monte Carlo simulation methods for planning the numbers of berths, cranes and other facilities. Suppose bulk carriers arrive, according to a Poisson process, at an average rate of 6.4 per 24 hours and we wish to generate random times between arrivals.

Let T be the time in hours between arrivals. The distribution of T is exponential, $M(\lambda)$, with a rate parameter (λ) of 0.267 per hour, and the cdf is

$$F(t) = 1 - e^{-0.267t}$$

If

$$F(t) = r$$

then

$$t = -\ln(1 - r)/0.267$$

The sequence of uniform random numbers on the left would correspond to the times on the right:

> 0.733 36 gives 4.95 hours
>
> 0.444 51 gives 2.20 hours
>
> 0.038 17 gives 0.15 hours
>
> 0.796 77 gives 5.97 hours
>
> etc.

A3.4.2   Acceptance–rejection algorithm

This description applies to the generation of random numbers from univariate continuous distributions, but the acceptance–rejection (AR) algorithm can be used in many other situations. In particular it is an essential component of the Metropolis–Hastings algorithm (Chib and Greenberg, 1995). Suppose we wish to generate random numbers from a distribution with pdf $f(x)$ for which the inverse cdf cannot be expressed as an algebraic formula in terms of elementary functions. The AR algorithm relies on finding some other pdf $g(x)$ which does have a convenient algebraic form of inverse cdf that can be expressed as a formula and which, when multiplied by some constant k, dominates $f(x)$. That is:

$$f(x) \leq kg(x) \quad \text{for all } x$$

and, for the algorithm to be efficient, k should be as small as possible. The algorithm proceeds as follows:

- *Step 1*   Generate a random number W from the distribution with pdf $g(x)$.
- *Step 2*   Generate a uniform random number R from $U(0, 1)$.
- *Step 3*   Calculate

$$c = \frac{f(W)}{kg(W)}$$

- *Step 4*   If $R \leq c$ set

$$X = W$$

and accept X as a random number from the distribution with pdf $f(x)$. If $c < R$ reject W and return to Step 1. The process is repeated, by returning to Step 1, until the required sample size is obtained. The number of iterations until a number is accepted has a geometric distribution with a mean k. The algorithm works because the probability of being in a small interval is proportional to the height of the pdf.

Example A3.4
Generate random numbers from a standard folded normal distribution, i.e.

$$f(x) = (2/\pi)^{1/2} \exp(-x^2/2) \quad \text{for } 0 \leq x$$

A suitable $g(x)$ is the exponential with $\lambda = 1$, i.e.

$$g(x) = \exp(-x)$$

The smallest value for k is $\sqrt{2e/\pi}$, and the pdf touch when $x = 1$. On average 1.32 iterations will be required for each random number.

Example A3.5
Generate random numbers from a beta $Be(2, 4)$ distribution, i.e.

$$f(x) = 20x(1 - x)^3 \quad \text{for } 0 < x < 1$$

A suitable $g(x)$ is $U[0, 1]$, and the minimum k is $135/64$. On average 2.11 iterations will be required for each random number.

Example A3.6
Generate random numbers from a gamma distribution. The probability density function for a gamma-distributed random variable X is

$$f(x) = \frac{\lambda(\lambda x)^{c-1}}{\Gamma(c)} e^{-\lambda x} \quad \text{for } 0 \leq x$$

For small integer values of c, random variates can conveniently be generated by adding independent variates from an exponential distribution with mean $1/\lambda$.

However, the AR algorithm can be used for non-integer values of c. A possible choice for $g(x)$ is an exponential distribution with the same mean: if c is

$$g(x) = \lambda e^{-\lambda x} \quad \text{for } 0 \le x$$

The mode of the gamma distribution is the value of x for which $f'(x) = 0$, and it is straightforward to verify that this occurs when $x = (c-1)/\lambda$. Therefore the minimum value for k is $(c-1)^{c-1}/\Gamma(c)$. It follows that the choice of an exponential distribution for $g(x)$ would become very inefficient for c greater than about 3.

A3.5   Generating random numbers from a normal distribution

A3.5.1   Univariate normal distribution

Sum of uniform variates

A consequence of the central limit theorem is that the distribution of the sum of n uniform random variables rapidly approaches normality as a normal distribution n tends to infinity. In practice a value of 12 for n will usually suffice. If R_i are independent and

$$R_i \sim U[0, 1]$$

then to a very good approximation

$$\sum_{i=1}^{12} R_i \sim N(6, (1)^2)$$

Random numbers which are produced in this way can be scaled to any other normal distribution: subtract 6; multiply by the required standard deviation; and then add the required mean.

Box–Muller method

Let R_1 and R_2 be independent variables from $U[0, 1]$. A point with coordinates $(\cos(2\pi R_2), \sin(2\pi R_2))$ is uniformly distributed on the unit circle. It is also straightforward to show that

$$-2\ln(R_1) \sim \chi_2^2$$

Now if Z_1 and Z_2 are independent $N(0, 1)$ then

$$W = Z_1^2 + Z_2^2$$

has a χ_2^2 distribution. The point $(Z_1/W^{1/2}, Z_2/W^{1/2})$ is uniformly distributed on a unit circle. Hence an independent pair of standard normal variables is given by

$$Z_1 = (-2\ln(R_1))^{1/2} \sin(2\pi R_2)$$
$$Z_2 = (-2\ln(R_1))^{1/2} \cos(2\pi R_2)$$

Polar method

The Polar method is based on the same principles as the Box–Muller method but it is more computationally efficient because it avoids the need to calculate sine and cosine functions (Ross, 1997). The algorithm is

- *Step 1* Generate R_1 and R_2 from $U[0, 1]$.
- *Step 2* Set $U_1 = 2R_1 - 1$, $U_2 = 2R_2 - 1$ and $W = U_1^2 + U_2^2$.
- *Step 3* If $W > 1$ return to Step 1, otherwise return two independent standard normal variables

$$Z_1 = [-2\ln(W)/W]^{1/2} U_1$$

$$Z_2 = [-2\ln(W)/W]^{1/2} U_2$$

A3.5.2   Multivariate normal distribution

Let X be a $k \times 1$ vector which has a multivariate normal distribution with mean μ and variance–covariance matrix V. Then there is a matrix Q such that

$$V = Q^2$$

The matrix Q is found from the eigenvalue analysis of V. That is

$$V = M\Lambda M^{-1}$$

where Λ is a diagonal matrix with the eigenvalues along its diagonal and M is a matrix with columns given by corresponding eigenvalues. Then $\Lambda^{1/2}$ is the diagonal matrix obtained by taking the square roots of the eigenvalues. Let $z = (z_1, \ldots, z_k)^T$ be a vector of independent standard normal random numbers. A random deviate from the original multivariate normal distribution is given by:

$$X = Qz$$

This means of generating random numbers from a multivariate normal distribution is often referred to as the Choleski approach. Since V is a symmetric matrix, Q is too, and Q^2 is equivalent to $Q^T Q$.

A3.6   Metropolis–Hastings algorithm

An irreducible Markov chain (MC) with $N \times N$ transition matrix P has a stationary probability vector v which satisfies

$$v_j = \sum_{i=1}^{N} v_i p_{ij}$$

The process is said to be time reversible if and only if

$$v_i p_{ij} = v_j p_{ji}$$

This property corresponds to a physical description that is symmetric with respect to past and future, and is also known as the law of detailed balance

(Guttorp, 1995). It is straightforward to construct a matrix P which does not correspond to a time-reversible process because a doubly stochastic matrix (rows and columns add to 1) is only time reversible if it is symmetric (Bhattacharya and Waymire, 1990).

The Metropolis–Hastings algorithm is a method for constructing an MC with a given stationary probability vector v. The motivation is that it may be easier to run the chain than to sample directly from the distribution represented by v. Ross (1997) gives some examples from combinatorial and queueing problems, and the continuous version of the algorithm is often used (Chib and Greenberg, 1995). However, the explanation is simpler for the discrete case. Take any irreducible $N \times N$ transition matrix Q, with elements q_{ij}, and define

$$p_{ij} = q_{ij}\alpha_{ij} \quad \text{if } j \neq i$$
$$p_{ii} = q_{ii} + \sum_{k \neq i} q_{ik}(1 - \alpha_{ik}) \tag{A3.1}$$

where the α_{ij} have yet to be defined. The MC will be reversible and have stationary probability vector v if

$$v_i p_{ij} = v_j p_{ji} \quad \text{for } j \neq i$$

This condition is satisfied if

$$\alpha_{ij} = \min\left(\frac{v_j q_{ji}}{v_i q_{ij}}, 1\right)$$

which follows from substituting for p_{ij} and p_{ji} from equation (A3.1). Notice that if α_{ij} is less than 1 then α_{ji} equals 1 and vice versa. An important practical feature of the result is that only ratios of the form v_j/v_i are needed and v does not have to have been scaled so that its elements add to 1.

Algorithm (discrete form after Ross, 1997)

- *Step 0* Choose an irreducible matrix Q.
- *Step 1* Let $n = 0$ and $X_0 = i$.
- *Step 2* Generate Y such that $\Pr(Y = j) = q_{ij}$ and R from $U[0, 1]$.
- *Step 3* Suppose $Y = k$. If $R < v_k q_{ki}/(v_i q_{ik})$ then $X_{n+1} = Y$, otherwise $X_{n+1} = X_n$.
- *Step 4* $n = n + 1$, $i = X_{n+1}$ and return to Step 2.

Algorithm (continuous form after Chib and Greenberg, 1995)

- *Step 0* Choose a convenient candidate-generating probability density $q(x, y)$.
- *Step 1* Let $n = 0$ and $X_0 = x$.
- *Step 2* Generate Y from $q(y|x)$ and R from $U[0, 1]$.
- *Step 3* Suppose $Y = y$. If $R < v(y)q(y, x)/(v(x)q(x, y))$ then $X_{n+1} = Y$, otherwise $X_{n+1} = X_n$.
- *Step 4* $n = n + 1$, $x = X_{n+1}$ and return to Step 2.

A choice of symmetric candidate-generating density, i.e. $q(x, y) = q(y, x)$, is an important special case for which the probability of a move reduces to $v(y)/v(x)$.

Example A3.7
The M–H algorithm can be used to generate random numbers from a multi-variate normal distribution: variable $X = (X_1, \ldots, X_k)^T$ with mean μ and variance–covariance matrix V. A symmetric candidate-generating density is

$$y = \mu - (x - \mu) + u$$

where $u = (U_1, \ldots, U_k)^T$ and the U_i are independent $U[-1, 1]$. Then

$$\alpha(x, y) = \min \left\{ \frac{\exp[-\frac{1}{2}(y - \mu)^T v^{-1} (y - \mu)]}{\exp[-\frac{1}{2}(x - \mu)^T v^{-1} (x - \mu)]}, 1 \right\}$$

Chib and Greenberg (1995) compare the results with the Choleski approach and give other examples of the use of the Metropolis–Hastings algorithm.

Appendix 4 Bayes' linear estimator

This account is based on O'Hagan (1994). y and x are scalar and vector random variables.

We wish to minimise the expected value of the squared error

$$D = E[((a + b^T x) - y)^2]$$
$$= E[a^2 + 2ab^T x + b^T xx^T b - 2ay - 2b^T xy + y^2]$$
$$= a^2 + 2ab^T E[x] + b^T E[xx^T]b - 2aE[y] - 2b^T E[xy] + E[y^2]$$

Now

$$\text{var}(x) = E[xx^T] - E[x]E[x^T]$$

and

$$\text{cov}(x, y) = E[xy] - E[x]E[y]$$

and so

$$D = (a + b^T E[x] - E[y])^2 + b^T \text{var}(x)b - 2b^T \text{cov}(x, y) + \text{var}(y)$$

This can be written as

$$D = (a + b^T E[x] - E[y])^2 + (b - b^*)^T \text{var}(x)(b - b^*) + D^*$$

where

$$b^* = \text{var}(x)^{-1} \text{cov}(x, y)$$
$$D^* = \text{var}(y) - \text{cov}(x, y) \text{var}(x)^{-1} \text{cov}(x, y)$$

The two leading terms on the right-hand side of the expression for D are non-negative. Therefore, D will be minimised if they are made zero by setting

$$b = b^*$$
$$a = -b^{*T} E[x] + E[y]$$

We will now show that

$$E[y] + \boldsymbol{b}^{*T}(\boldsymbol{x} - E[\boldsymbol{x}])$$

is the posterior mean for y, given $\boldsymbol{x}$. It is a particular case of the following result. We let w be an estimator of θ, and find the value of w which minimises

$$D = E[(w - \theta)^2] = E[w^2 - 2\theta w + \theta^2]$$

$$= w^2 - 2wE[\theta] + E[\theta^2]$$

We set dD/dw equal to zero

$$2w - 2E[\theta] = 0$$

to find that $w = E[\theta]$.

Appendix 5   Optimisation algorithms

A5.1   Introduction

The objective is to minimise some function f with respect to m parameters. Nothing is lost by referring only to minimisation because minimising $-f$ is equivalent to maximising f. The following optimisation methods can be used to search for a local minimum. They have two limitations. Firstly, when there are constraints imposed on parameter values the least value taken by a function need not be at a minimum, because the least value may occur on the boundary of the region over which the parameters vary. Secondly, a local minimum is not necessarily the global minimum. A plot of $f(t) = e^{2t} \sin t$ for t from 0 to 5.49π illustrates both of these remarks. If m is only one or two there may not be any practical problem, but difficulties often arise with larger values of m. *Simulated annealing* and *genetic algorithms* (Chapter 7) aim to reduce the problem by allowing some probability of changes in parameter values that lead to a local increase. Another strategy is to start deterministic algorithms from different guesses of parameter values.

A5.2   Simplex method

A *simplex* is a geometric figure formed by $m + 1$ points in an m-dimensional space, e.g. a triangle when $m = 2$, and a tetrahedron when $m = 3$. We begin with $(m + 1)$ parameter points $\boldsymbol{\theta}_1, \ldots, \boldsymbol{\theta}_{m+1}$ (each of which has m elements corresponding to the m parameters) and evaluate $f(\boldsymbol{\theta}_i)$ for each point. Find the greatest of these function values, and suppose it is at $\boldsymbol{\theta}_p$. We move to a point $\boldsymbol{\theta}_r$ which is the reflection of $\boldsymbol{\theta}_p$ along a line joining $\boldsymbol{\theta}_p$ to the centroid of the other points in the simplex $(\boldsymbol{\theta}_0)$. That is,

$$\boldsymbol{\theta}_r = \boldsymbol{\theta}_0 + (\boldsymbol{\theta}_0 - \boldsymbol{\theta}_p)$$

where

$$\boldsymbol{\theta}_0 = \sum_{\substack{i=1 \\ i \neq p}}^{m+1} \boldsymbol{\theta}_i / m$$

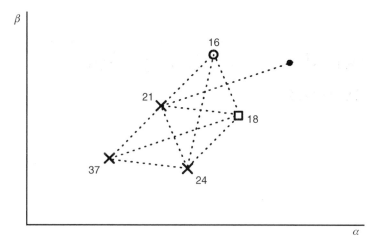

Figure A5.1   Simplex algorithm for minimisation. The X at which the function equals 37 is reflected to give a function value of 18 at the square. The cross (function value 24) is reflected to the circle (function value 16), the cross (function value 21) then reflects to the dot, and so on.

The routine is repeated for the simplex obtained by replacing θ_p with θ_r (Figure A5.1). When the procedure leads to a point θ_r for which $f(\theta_r)$ exceeds all the $f(\theta_i)$, excepting $i = p$, the size of the simplex is reduced. The usual convergence criterion is to stop when

$$\sum_{i=1}^{m+1} (f(\theta_i) - f(\theta_0))^2 < \varepsilon$$

for some suitably small ε. There are many variations on the rigid simplex algorithm, and one of the best known is the modified simplex method (MSM) of Nelder and Mead (1965) which allows an expansion or contraction at each step. For example, if $f(\theta_r)$ is less than all the $f(\theta_i)$ the simplex is expanded to

$$\theta_s = \theta_0 + 2(\theta_0 - \theta_p)$$

If $f(\theta_s)$ is less than all the $f(\theta_i)$ the program continues with the asymmetric simplex obtained by replacing θ_p with θ_s. If $f(\theta_s)$ is not less than all the $f(\theta_i)$ the program continues with the simplex obtained by replacing θ_p with θ_r.

A5.3   Gradient methods

Methods such as steepest descent, conjugate gradients and Newton–Raphson make use of the derivatives of the function. Details can be found in Everitt (1987) and Press *et al.* (1992), for example.

Appendix 6 Bootstrap methods for confidence intervals

The basic principle will be demonstrated with the following example. The ratios (r_i) of out-turn cost to predicted cost for eight water distribution schemes are:

$$1.16 \quad 0.90 \quad 1.09 \quad 0.95 \quad 0.75 \quad 1.50 \quad 1.48 \quad 1.67$$

The mean ratio ($\bar{r}$) is 1.29 and the standard deviation s_r is 0.34. The individual ratios are unlikely to be well approximated by a normal distribution, but the construction for a confidence interval for μ_r using a t-distribution is not very sensitive to this assumption. In particular, a 95% confidence interval for μ_r is given by

$$1.19 \pm 2.365 \times 0.33/\sqrt{8}$$

which gives $[0.91, 1.46]$. We will compare this interval with one obtained using a bootstrap procedure. The simplest method is:

- *Step 1*
  Use random numbers, to re-sample, with replacement, 8 data from the set of 8 ratios. For example, I took x_i from $U[0, 1]$ and let

  $$m_i = \text{round}(x_i \times 8 + 0.501)$$

  for $i = 1, \ldots, 8$. This gave me,

  $$3 \quad 2 \quad 3 \quad 2 \quad 5 \quad 8 \quad 4 \quad 6$$

  and the first *bootstrap sample* was

  $$1.09 \quad 0.90 \quad 1.09 \quad 0.90 \quad 0.75 \quad 1.67 \quad 0.95 \quad 1.50$$

  Calculate the mean ratio for the bootstrap sample, 1.11 in this case.
- *Steps 2 ... 1000*
  Repeat Step 1, a large number of times. A thousand will suffice.
- *Final step*
  Sort the 1000 bootstrap sample means into ascending order. For a 95% confidence interval take the 25th and 975th value. I obtained

  $$[0.97, 1.40]$$

This is narrower than the interval based on a t-distribution, as tends to be the case with small samples. It is closer to the interval we would have obtained if we had used $z_{0.025}$ instead of $t_{7, 0.025}$. An improvement is to use what are called *pivotal* methods (see, for example, Efron and Gong, 1983).

An important modification, which is often useful in more complicated situations, is the *parametric bootstrap*. The device is to estimate the parameters of an assumed distribution, from the original sample, and then to generate bootstrap samples from this distribution rather than by re-sampling the original sample. This idea can be extended to complex stochastic models.

Bootstrap methods are a means of assessing the precision of estimates, and require programming ability, rather than statistical expertise, for their implementation. Intuitively, it seems that they will tend to give confidence intervals that are somewhat narrower than they should be for a given level, but good enough for most management purposes. The methods can be justified by mathematical arguments. More details about practical applications, including Minitab code, can be found in Taffe and Garnham (1996).

Appendix 7 Expected value of the residual sum of squares in multiple regression

A7.1 Expected value of the sum of squared residuals

The multiple regression model can be written

$$Y = XB + E$$

where the errors are assumed to be independently distributed with mean 0 and variance σ^2.

The residual sum of squares (RSS) is given by the equation

$$RSS = (Y - X\hat{B})^T(Y - X\hat{B})$$

If we remember, from the derivation of $\hat{B}$, that

$$(X^T X)\hat{B} = X^T Y$$

we can rewrite RSS as

$$RSS = Y^T Y - \hat{B}^T X^T X \hat{B}$$

If we use result (iv) of Exercise 5.9:

$$E[RSS] = E[Y^T Y] - [B^T X^T X B + \text{tr}((X^T X)(X^T X)^{-1}\sigma^2)]$$

$$= E[Y^T Y] - B^T X^T X B - (k+1)\sigma^2$$

Finally, remembering the scalar version of the same result,

$$E[Y_i^2] = \text{var}(Y_i) + E[Y_i]^2$$

the first term on the right-hand side

$$E[Y^T Y] = n\sigma^2 + \sum_{i=1}^{n}(\beta_0 + \beta_1 x_{1i} + \cdots + \beta_k x_{ki})^2 = n\sigma^2 + B^T X^T X B$$

so

$$E[RSS] = (n - k - 1)\sigma^2$$

The estimator S^2 of σ^2 is $RSS/(n - k - 1)$, so it follows that

$$E[S^2] = \sigma^2$$

The expression $B^T(X^TX)B \geq 0$, because X^TX is the inverse of a covariance matrix multiplied by a positive constant. In later sections we will use the fact that S^2 is not correlated with $\hat{B}$. A formal proof will not be given, but there is no reason to suppose that $\hat{\beta}_k$ being above its mean β_k has any effect on the relationship of S^2 to its mean σ^2.

A7.2 Making predictions

The most common reason for fitting a multiple regression model is to make predictions for the dependent variable (response) when the values of the predictor variables are known. You should remember the distinction between confidence limits for the mean value of the dependent variable, and limits of prediction for a single value taken by it. For large samples $\pm 2s$ will give approximate, but rather too narrow, 95% limits of prediction. This approximation ignores the uncertainty in the estimation of parameters and therefore improves as the sample size increases.

Confidence interval for the mean value of Y given x

Let $x_p = (1 \quad x_{1p} \ldots x_{kp})^T$. The predicted mean value of Y given that $x = x_p$ is

$$E[Y \mid x = x_p] = x_p^T \hat{B}$$

The variance of this quantity is obtained from result (iv) of Exercise 5.9.

$$\text{var}(x_p^T \hat{B}) = x_p^T((X^TX)^{-1}\sigma^2)x_p$$

where $(X^TX)^{-1}\sigma^2$ is the covariance matrix of $\hat{B}$. Hence a $(1 - \alpha) \times 100\%$ confidence interval for $E[Y \mid x = x_p]$ is

$$x_p^T \hat{B} \pm t_{n-k-1, \, x/2}s\sqrt{x_p^T(X^TX)^{-1}x_p}$$

Limits of prediction for a single value of Y

$(1 - \alpha) \times 100\%$ limits of prediction for a single value of Y when $x = x_p$ are

$$x_p^T \hat{B} \pm t_{n-k-1, \, x/2}s\sqrt{1 + x_p^T(X^TX)^{-1}x_p}$$

These limits will be quite different if changing variances and different distributions are considered.

Appendix 8 Questionnaire on culture change

Answer each statement as accurately as you can.

Allocate points as follows:

The statement always applies .... 3 points
The statement usually applies .... 2 points
The statement rarely applies .... 1 point
The statement never applies .... 0 point

		Point
(1)	I am clear about the company's objectives.	....
(2)	Departmental structures are understood and people can relate to them.	....
(3)	Budgets are controlled very effectively.	....
(4)	Departments work well together and support each other.	....
(5)	The allocation of resources matches the needs and objectives of departments.	....
(6)	People have the right experience, skills and training for the job.	....
(7)	Work is always completed within the budget.	....
(8)	This organisation is quick to see market opportunities in the outside environment.	....
(9)	Management is skilful at putting over its case.	....
(10)	Departments work hard to cooperate with each other.	....
(11)	There are effective systems to convey information from top management down through the organisation.	....
(12)	What top management says is widely trusted.	....
(13)	Important information flows easily up the organisation.	....
(14)	Managers have effective systems to identify potential problems.	....
(15)	The organisation's objectives are communicated to its employees and they understand them.	....
(16)	Efficient work methods are used by individuals, by groups and between groups.	....

(17) The process of decision making is effective.   ....
(18) Groups get together and work on common problems.   ....
(19) The appropriate type of technology is used.   ....
(20) Managers take training and development seriously.   ....
(21) High performance levels are consistently achieved.   ....
(22) The organisation has succeeded in putting across a good public image to its customers.   ....
(23) Top management makes every effort to get its message across.   ....
(24) Adequate mechanisms exist to coordinate the work of different departments.   ....
(25) Management has direct lines of communication right down the organisation.   ....
(26) Management credibility is high.   ....
(27) Managers make great efforts to keep in touch with the workforce.   ....
(28) The formal communication system is so effective that people do not have to rely on the grapevine.   ....
(29) People know what is expected of them, and they do it.   ....
(30) Communications between management and staff are very effective.   ....
(31) Job achievements are systematically appraised against performance standards.   ....
(32) Meetings are generally very productive.   ....
(33) Departments never overspend on their budget.   ....
(34) All available talent is mobilised and used.   ....
(35) Equipment is in good condition and well maintained.   ....
(36) The organisation is quick to detect potential threats outside.   ....
(37) When changes are made, great pains are taken to explain the reasons.   ....
(38) There are effective procedures for integrating the work of specialist groups.   ....
(39) Everyone is regularly updated with news about the fortunes of the organisation.   ....
(40) Usually management information is accepted at face value.   ....
(41) Those lower down in the organisation feel that senior management fully understands their difficulties.   ....
(42) Everyone knows what decisions they can make, and to whom they must refer for authority.   ....
(43) Plans are effective and the priorities are clear.   ....
(44) Delegation is at the right level.   ....
(45) Managers understand the priorities of other departments.   ....
(46) Teams consciously look at ways of improving how they work.   ....
(47) There is no unused capacity, or under-utilised resource.   ....
(48) Risk takers are supported and innovation rewarded.   ....
(49) Quality standards are kept at a high level.   ....
(50) The organisation invests a lot of effort to find out what its customers actually want.   ....
(51) Managers can be relied upon to put forward clear proposals which are supported by strong arguments.   ....

(52) Effective action is taken to resolve inter-departmental problems. ....

(53) There are regular presentations to everyone on what is happening in the organisation. ....

(54) Top management is trusted to safeguard the interests of all employees. ....

(55) There are systems which ensure that any employee's ideas are carefully considered. ....

(56) Effective steps have been taken to ensure that people are not burdened by unnecessary information. ....

(57) Adequate time is spent planning for the future. ....

(58) Departments have a well-balanced workload. No department is overloaded while others have it relatively easy. ....

(59) Information on variances is fed back to managers for corrective action. ....

(60) Teams generally have shared aims and objectives. ....

(61) Pressure of work never leads to lower safety standards. ....

(62) People are encouraged to update their skills. ....

(63) Management is good at identifying threats and opportunities. ....

(64) People in this organisation constantly look outside for ideas to keep up to date. ....

(65) Managers are good at selling ideas. ....

(66) There are effective methods of coordinating the work of different departments. ....

(67) In the last month, I have been given a formal briefing from my manager about what is going on in the organisation. ....

(68) Managers do not blame others, but take full responsibility when things go wrong. ....

(69) Senior managers go out of their way to talk informally with people throughout the organisation. ....

(70) Unnecessary paperwork has been ruthlessly eliminated. ....

You may put your completed questionnaire in the envelope
supplied and please seal it before returning it to me.

ONCE AGAIN, THANK YOU FOR YOUR ASSISTANCE.

PROBLEM AREAS – SCORING

Write the score alongside the question number on the lists below.

						TOTAL
OBJECTIVES	1 –	15 –	29 –	43 –	57 –	....
ORGANISATION	2 –	16 –	30 –	44 –	58 –	....
CONTROL	3 –	17 –	31 –	45 –	59 –	....
TEAMWORK	4 –	18 –	32 –	46 –	60 –	....
RESOURCES	5 –	19 –	33 –	47 –	61 –	....
PEOPLE	6 –	20 –	34 –	48 –	62 –	....
EFFICIENCY	7 –	21 –	35 –	49 –	63 –	....
MARKETING	8 –	22 –	36 –	50 –	64 –	....
PERSUASIVE MANAGEMENT	9 –	23 –	37 –	51 –	65 –	....
INTEGRATING MECHANISM	10 –	24 –	38 –	52 –	66 –	....
DOWNWARD FLOW	11 –	25 –	39 –	53 –	67 –	....
HIGH TRUST	12 –	26 –	40 –	54 –	68 –	....
UPWARD FLOW	13 –	27 –	41 –	55 –	69 –	....
APT ADMINISTRATION	14 –	28 –	42 –	56 –	70 –	....

Appendix 9 Statistical tables

Table A9.1   Values of z, the standard normal variable, from 0.0 by steps of 0.01 to 3.9, showing the cumulative probability up to z. (Probability correct to 4 decimal places)

z	0.00	0.01	0.02	0.03	0.04	0.05	0.06	0.07	0.08	0.09
0.0	0.5000	0.5040	0.5080	0.5120	0.5160	0.5199	0.5239	0.5279	0.5319	0.5359
0.1	0.5398	0.5438	0.5478	0.5517	0.5557	0.5596	0.5636	0.5675	0.5714	0.5753
0.2	0.5793	0.5832	0.5871	0.5910	0.5948	0.5987	0.6026	0.6064	0.6103	0.6141
0.3	0.6179	0.6217	0.6255	0.6293	0.6331	0.6368	0.6406	0.6443	0.6480	0.6517
0.4	0.6554	0.6591	0.6628	0.6664	0.6700	0.6736	0.6772	0.6808	0.6844	0.6879
0.5	0.6915	0.6950	0.6985	0.7019	0.7054	0.7088	0.7123	0.7157	0.7190	0.7224
0.6	0.7257	0.7291	0.7324	0.7357	0.7389	0.7422	0.7454	0.7486	0.7517	0.7549
0.7	0.7580	0.7611	0.7642	0.7673	0.7704	0.7734	0.7764	0.7794	0.7823	0.7852
0.8	0.7881	0.7910	0.7939	0.7967	0.7995	0.8023	0.8051	0.8078	0.8106	0.8133
0.9	0.8159	0.8186	0.8212	0.8238	0.8264	0.8289	0.8315	0.8340	0.8365	0.8389
1.0	0.8413	0.8438	0.8461	0.8485	0.8508	0.8531	0.8554	0.8577	0.8599	0.8621
0.1	0.8643	0.8665	0.8686	0.8708	0.8729	0.8749	0.8770	0.8790	0.8810	0.8830
0.2	0.8849	0.8869	0.8888	0.8907	0.8925	0.8944	0.8962	0.8980	0.8997	0.9015
0.3	0.9032	0.9049	0.9066	0.9082	0.9099	0.9115	0.9131	0.9147	0.9162	0.9177
0.4	0.9192	0.9207	0.9222	0.9236	0.9251	0.9265	0.9279	0.9292	0.9306	0.9319
0.5	0.9332	0.9345	0.9357	0.9370	0.9382	0.9394	0.9406	0.9418	0.9429	0.9441
0.6	0.9452	0.9463	0.9474	0.9484	0.9495	0.9505	0.9515	0.9525	0.9535	0.9545
0.7	0.9554	0.9564	0.9573	0.9582	0.9591	0.9599	0.9608	0.9616	0.9625	0.9633
0.8	0.9641	0.9649	0.9656	0.9664	0.9671	0.9678	0.9686	0.9693	0.9699	0.9706
0.9	0.9713	0.9719	0.9726	0.9732	0.9738	0.9744	0.9750	0.9756	0.9761	0.9767
2.0	0.9772	0.9778	0.9783	0.9788	0.9793	0.9798	0.9803	0.9808	0.9812	0.9817
0.1	0.9821	0.9826	0.9830	0.9834	0.9838	0.9842	0.9846	0.9850	0.9854	0.9857
0.2	0.9861	0.9864	0.9868	0.9871	0.9875	0.9878	0.9881	0.9884	0.9887	0.9890
0.3	0.9893	0.9896	0.9898	0.9901	0.9904	0.9906	0.9909	0.9911	0.9913	0.9916
0.4	0.9918	0.9920	0.9922	0.9925	0.9927	0.9929	0.9931	0.9932	0.9934	0.9936
0.5	0.9938	0.9940	0.9941	0.9943	0.9945	0.9946	0.9948	0.9949	0.9951	0.9952
0.6	0.9953	0.9955	0.9956	0.9957	0.9959	0.9960	0.9961	0.9962	0.9963	0.9964
0.7	0.9965	0.9966	0.9967	0.9968	0.9969	0.9970	0.9971	0.9972	0.9973	0.9974
0.8	0.9974	0.9975	0.9976	0.9977	0.9977	0.9978	0.9979	0.9979	0.9980	0.9981
0.9	0.9981	0.9982	0.9982	0.9983	0.9984	0.9984	0.9985	0.9985	0.9986	0.9986
3.0	0.9987	0.9987	0.9987	0.9988	0.9988	0.9989	0.9989	0.9989	0.9990	0.9990
0.1	0.9990	0.9991	0.9991	0.9991	0.9992	0.9992	0.9992	0.9992	0.9993	0.9993
0.2	0.9993	0.9993	0.9994	0.9994	0.9994	0.9994	0.9994	0.9995	0.9995	0.9995
0.3	0.9995	0.9995	0.9995	0.9996	0.9996	0.9996	0.9996	0.9996	0.9996	0.9997
0.4	0.9997	0.9997	0.9997	0.9997	0.9997	0.9997	0.9997	0.9997	0.9997	0.9998
0.5	0.9998	0.9998	0.9998	0.9998	0.9998	0.9998	0.9998	0.9998	0.9998	0.9998
0.6	0.9998	0.9998	0.9999	0.9999	0.9999	0.9999	0.9999	0.9999	0.9999	0.9999
0.7	0.9999	0.9999	0.9999	0.9999	0.9999	0.9999	0.9999	0.9999	0.9999	0.9999
0.8	0.9999	0.9999	0.9999	0.9999	0.9999	0.9999	0.9999	0.9999	0.9999	0.9999
0.9	1.0000									

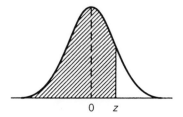

The curve is $\mathcal{N}(0,1)$, the standard normal variable. The table entry is the shaded area $\Phi(z) = \Pr(Z < z)$. For example when $z = 1.96$ the shaded area is 0.9750. Critical values of the standard normal distribution will be found in the bottom row of Table A9.2.

Table A9.2 Percentage points of Student's *t*-distribution

d.f.	P = 0.05	0.025	0.01	0.005	0.001	0.0005
1	6.314	12.706	31.821	63.657	318.31	636.62
2	2.920	4.303	6.965	9.925	22.327	31.598
3	2.353	3.182	4.541	5.841	10.214	12.924
4	2.132	2.776	3.747	4.604	7.173	8.610
5	2.015	2.571	3.365	4.032	5.893	6.869
6	1.943	2.447	3.143	3.707	5.208	5.959
7	1.895	2.365	2.998	3.499	4.785	5.408
8	1.860	2.306	2.896	3.355	4.501	5.041
9	1.833	2.262	2.821	3.250	4.297	4.781
10	1.812	2.228	2.764	3.169	4.144	4.587
11	1.796	2.201	2.718	3.106	4.025	4.437
12	1.782	2.179	2.681	3.055	3.930	4.318
13	1.771	2.160	2.650	3.012	3.852	4.221
14	1.761	2.145	2.624	2.977	3.787	4.140
15	1.753	2.131	2.602	2.947	3.733	4.073
16	1.746	2.210	2.583	2.921	3.686	4.015
17	1.740	2.110	2.567	2.898	3.646	3.965
18	1.734	2.101	2.552	2.878	3.610	3.922
19	1.729	2.093	2.539	2.861	3.579	3.883
20	1.725	2.086	2.528	2.845	3.552	3.850
21	1.721	2.080	2.518	2.831	3.527	3.819
22	1.717	2.074	2.508	2.819	3.505	3.792
23	1.714	2.069	2.500	2.807	3.485	3.767
24	1.711	2.064	2.492	2.797	3.467	3.745
25	1.708	2.060	2.485	2.787	3.450	3.725
26	1.706	2.056	2.479	2.779	3.435	3.707
27	1.703	2.052	2.473	2.771	3.421	3.690
28	1.701	2.048	2.467	2.763	3.408	3.674
29	1.699	2.045	2.462	2.756	3.396	3.659
30	1.697	2.042	2.457	2.750	3.385	3.646
40	1.684	2.021	2.423	2.704	3.307	3.551
60	1.671	2.000	2.390	2.660	3.232	3.460
120	1.658	1.980	2.358	2.617	3.160	3.373
∞	1.645	1.960	2.326	2.576	3.090	3.291

The last row of the table (∞) gives values of *z*, the standard normal variable.

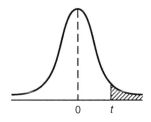

P is the shaded area

Table A9.3   Percentage points of the χ^2 distribution exceeded with probability P

				P			
d.f	0.995	0.975	0.050	0.025	0.010	0.005	0.001
1	3.9×10^{-5}	9.8×10^{-4}	3.84	5.02	6.63	7.88	10.83
2	0.010	0.051	5.99	7.38	9.21	10.60	13.81
3	0.071	0.22	7.81	9.35	11.34	12.84	16.27
4	0.21	0.48	9.49	11.14	13.28	14.86	18.47
5	0.41	0.83	11.07	12.83	15.09	16.75	20.52
6	0.68	1.24	12.59	14.45	16.81	18.55	22.46
7	0.99	1.69	14.07	16.01	18.48	20.28	24.32
8	1.34	2.18	15.51	17.53	20.09	21.96	26.13
9	1.73	2.70	16.92	19.02	21.67	23.59	27.88
10	2.16	3.25	18.31	20.48	23.21	25.19	29.59
11	2.60	3.82	19.68	21.92	24.73	26.76	31.26
12	3.07	4.40	21.03	23.34	26.22	28.30	32.91
13	3.57	5.01	22.36	24.74	27.69	29.82	34.53
14	4.07	5.63	23.68	26.12	29.14	31.32	36.12
15	4.60	6.26	25.00	27.49	30.58	32.80	37.70
16	5.14	6.91	26.30	28.85	32.00	34.27	39.25
17	5.70	7.56	27.59	30.19	33.41	35.72	40.79
18	6.26	8.23	28.87	31.53	34.81	37.16	42.31
19	6.84	8.91	30.14	32.85	36.19	38.58	43.82
20	7.43	9.59	31.41	34.17	37.57	40.00	45.32
21	8.03	10.28	32.67	35.48	38.93	41.40	46.80
22	8.64	10.98	33.92	36.78	40.29	42.80	48.27
23	9.26	11.69	35.17	38.08	41.64	44.18	49.73
24	9.89	12.40	36.42	39.36	42.98	45.56	51.18
25	10.52	13.12	37.65	40.65	44.31	46.93	52.62
26	11.16	13.84	38.89	41.92	45.64	48.29	54.05
27	11.81	14.57	40.11	43.19	46.96	49.64	55.48
28	12.46	15.31	41.34	44.46	48.28	50.99	56.89
29	13.12	16.05	42.56	45.72	49.59	52.34	58.30
30	13.79	16.79	43.77	46.98	50.89	53.67	59.70
40	20.71	24.43	55.76	59.34	63.69	66.77	73.40
50	27.99	32.36	67.50	71.42	76.16	79.49	86.66
60	35.53	40.48	79.08	83.30	88.38	91.95	99.61
70	43.28	48.76	90.53	95.02	100.43	104.22	112.32
80	51.17	57.15	101.88	106.63	112.33	116.32	124.84
90	59.20	65.65	113.15	118.14	124.12	128.30	137.21
100	67.33	74.22	124.34	129.56	135.81	140.17	149.44

For degrees of freedom $\nu > 100$, test $\sqrt{2\chi_\nu^2}$ as $N(\sqrt{2\nu - 1}, 1)$

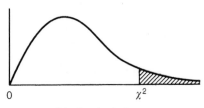

P is the shaded area

Table A9.4 Table of *F*-distribution. Upper 5% points

ν_2	ν_1 1	2	3	4	5	6	7	8	9	10	12	15	20	24	30	40	60	120	∞
1	161.4	199.5	215.7	224.6	230.2	234.0	236.8	238.9	240.5	241.9	243.9	245.9	248.0	249.1	250.1	251.1	252.2	253.3	254.3
2	18.51	19.00	19.16	19.25	19.30	19.33	19.35	19.37	19.38	19.40	19.41	19.43	19.45	19.45	19.46	19.47	19.48	19.49	19.50
3	10.13	9.55	9.28	9.12	9.01	8.94	8.89	8.85	8.81	8.79	8.74	8.70	8.66	8.64	8.62	8.59	8.57	8.55	8.53
4	7.71	6.94	6.59	6.39	6.26	6.16	6.09	6.04	6.00	5.96	5.91	5.86	5.80	5.77	5.75	5.72	5.69	5.66	5.63
5	6.61	5.79	5.41	5.19	5.05	4.95	4.88	4.82	4.77	4.74	4.68	4.62	4.56	4.53	4.50	4.46	4.43	4.40	4.36
6	5.99	5.14	4.76	4.53	4.39	4.28	4.21	4.15	4.10	4.06	4.00	3.94	3.87	3.84	3.81	3.77	3.74	3.70	3.67
7	5.59	4.74	4.35	4.12	3.97	3.87	3.79	3.73	3.68	3.64	3.57	3.51	3.44	3.41	3.38	3.34	3.30	3.27	3.23
8	5.32	4.46	4.07	3.84	3.69	3.58	3.50	3.44	3.39	3.35	3.28	3.22	3.15	3.12	3.08	3.04	3.01	2.97	2.93
9	5.12	4.26	3.86	3.63	3.48	3.37	3.29	3.23	3.18	3.14	3.07	3.01	2.94	2.90	2.86	2.83	2.79	2.75	2.71
10	4.96	4.10	3.71	3.48	3.33	3.22	3.14	3.07	3.02	2.98	2.91	2.85	2.77	2.74	2.70	2.66	2.62	2.58	2.54
11	4.84	3.98	3.59	3.36	3.20	3.09	3.01	2.95	2.90	2.85	2.79	2.72	2.65	2.61	2.57	2.53	2.49	2.45	2.40
12	4.75	3.89	3.49	3.26	3.11	3.00	2.91	2.85	2.80	2.75	2.69	2.62	2.54	2.51	2.47	2.43	2.38	2.34	2.30
13	4.67	3.81	3.41	3.18	3.03	2.92	2.83	2.77	2.71	2.67	2.60	2.53	2.46	2.42	2.38	2.34	2.30	2.25	2.21
14	4.60	3.74	3.34	3.11	2.96	2.85	2.76	2.70	2.65	2.60	2.53	2.46	2.39	2.35	2.31	2.27	2.22	2.18	2.13
15	4.54	3.68	3.29	3.06	2.90	2.79	2.71	2.64	2.59	2.54	2.48	2.40	2.33	2.29	2.25	2.20	2.16	2.11	2.07
16	4.49	3.63	3.24	3.01	2.85	2.74	2.66	2.59	2.54	2.49	2.42	2.35	2.28	2.24	2.19	2.15	2.11	2.06	2.01
17	4.45	3.59	3.20	2.96	2.81	2.70	2.61	2.55	2.49	2.45	2.38	2.31	2.23	2.19	2.15	2.10	2.06	2.01	1.96
18	4.41	3.55	3.16	2.93	2.77	2.66	2.58	2.51	2.46	2.41	2.34	2.27	2.19	2.15	2.11	2.06	2.02	1.97	1.92
19	4.38	3.52	3.13	2.90	2.74	2.63	2.54	2.48	2.42	2.38	2.31	2.23	2.16	2.11	2.07	2.03	1.98	1.93	1.88
20	4.35	3.49	3.10	2.87	2.71	2.60	2.51	2.45	2.39	2.35	2.28	2.20	2.12	2.08	2.04	1.99	1.95	1.90	1.84
21	4.32	3.47	3.07	2.84	2.68	2.57	2.49	2.42	2.37	2.32	2.25	2.18	2.10	2.05	2.01	1.96	1.92	1.87	1.81
22	4.30	3.44	3.05	2.82	2.66	2.55	2.46	2.40	2.34	2.30	2.23	2.15	2.07	2.03	1.98	1.94	1.89	1.84	1.78
23	4.28	3.42	3.03	2.80	2.64	2.53	2.44	2.37	2.32	2.27	2.20	2.13	2.05	2.01	1.96	1.91	1.86	1.81	1.76
24	4.26	3.40	3.01	2.78	2.62	2.51	2.42	2.36	2.30	2.25	2.18	2.11	2.03	1.98	1.94	1.89	1.84	1.79	1.73
25	4.24	3.39	2.99	2.76	2.60	2.49	2.40	2.34	2.28	2.24	2.16	2.09	2.01	1.96	1.92	1.87	1.82	1.77	1.71
26	4.23	3.37	2.98	2.74	2.59	2.47	2.39	2.32	2.27	2.22	2.15	2.07	1.99	1.95	1.90	1.85	1.80	1.75	1.69
27	4.21	3.35	2.96	2.73	2.57	2.46	2.37	2.31	2.25	2.20	2.13	2.06	1.97	1.93	1.88	1.84	1.79	1.73	1.67
28	4.20	3.34	2.95	2.71	2.56	2.45	2.36	2.29	2.24	2.19	2.12	2.04	1.96	1.91	1.87	1.82	1.77	1.71	1.65
29	4.18	3.33	2.93	2.70	2.55	2.43	2.35	2.28	2.22	2.18	2.10	2.03	1.94	1.90	1.85	1.81	1.75	1.70	1.64
30	4.17	3.32	2.92	2.69	2.53	2.42	2.33	2.27	2.21	2.16	2.09	2.01	1.93	1.89	1.84	1.79	1.74	1.68	1.62
40	4.08	3.23	2.84	2.61	2.45	2.34	2.25	2.18	2.12	2.08	2.00	1.92	1.84	1.79	1.74	1.69	1.64	1.58	1.51
60	4.00	3.15	2.76	2.53	2.37	2.25	2.17	2.10	2.04	1.99	1.92	1.84	1.75	1.70	1.65	1.59	1.53	1.47	1.39
120	3.92	3.07	2.68	2.45	2.29	2.17	2.09	2.02	1.96	1.91	1.83	1.75	1.66	1.61	1.55	1.50	1.43	1.35	1.25
∞	3.84	3.00	2.60	2.37	2.21	2.10	2.01	1.94	1.88	1.83	1.75	1.67	1.57	1.52	1.46	1.39	1.32	1.22	1.00

ν_1, ν_2 are numerator, denominator d.f. respectively.
Tabulated values are those exceeded with probability 0.05.

Table A9.4 (cont.)   Table of *F*-distribution. Upper 1% points

ν_2 \ ν_1	1	2	3	4	5	6	7	8	9	10	12	15	20	24	30	40	60	120	∞
1	4052	4999.5	5403	5625	5764	5859	5928	5981	6022	6056	6106	6157	6209	6235	6261	6287	6313	6339	6366
2	98.50	99.00	99.17	99.25	99.30	99.33	99.36	99.37	99.39	99.40	99.42	99.43	99.45	99.46	99.47	99.47	99.48	99.49	99.50
3	34.12	30.82	29.46	28.71	28.24	27.91	27.67	27.49	27.35	27.23	27.05	26.87	26.69	26.60	26.50	26.41	26.32	26.22	26.13
4	21.20	18.00	16.69	15.98	15.52	15.21	14.98	14.80	14.66	14.55	14.37	14.20	14.02	13.93	13.84	13.75	13.65	13.56	13.46
5	16.26	13.27	12.06	11.39	10.97	10.67	10.46	10.29	10.16	10.05	9.89	9.72	9.55	9.47	9.38	9.29	9.20	9.11	9.02
6	13.75	10.92	9.78	9.15	8.75	8.47	8.26	8.10	7.98	7.87	7.72	7.56	7.40	7.31	7.23	7.14	7.06	6.97	6.88
7	12.25	9.55	8.45	7.85	7.46	7.19	6.99	6.84	6.72	6.62	6.47	6.31	6.16	6.07	5.99	5.91	5.82	5.74	5.65
8	11.26	8.65	7.59	7.01	6.63	6.37	6.18	6.03	5.91	5.81	5.67	5.52	5.36	5.28	5.20	5.12	5.03	4.95	4.86
9	10.56	8.02	6.99	6.42	6.06	5.80	5.61	5.47	5.35	5.26	5.11	4.96	4.81	4.73	4.65	4.57	4.48	4.40	4.31
10	10.04	7.56	6.55	5.99	5.64	5.39	5.20	5.06	4.94	4.85	4.71	4.56	4.41	4.33	4.25	4.17	4.08	4.00	3.91
11	9.65	7.21	6.22	5.67	5.32	5.07	4.89	4.74	4.63	4.54	4.40	4.25	4.10	4.02	3.94	3.86	3.78	3.69	3.60
12	9.33	6.93	5.95	5.41	5.06	4.82	4.64	4.50	4.39	4.30	4.16	4.01	3.86	3.78	3.70	3.62	3.54	3.45	3.36
13	9.07	6.70	5.74	5.21	4.86	4.62	4.44	4.30	4.19	4.10	3.96	3.82	3.66	3.59	3.51	3.43	3.34	3.25	3.17
14	8.86	6.51	5.56	5.04	4.69	4.46	4.28	4.14	4.03	3.94	3.80	3.66	3.51	3.43	3.35	3.27	3.18	3.09	3.00
15	8.68	6.36	5.42	4.89	4.56	4.32	4.14	4.00	3.89	3.80	3.67	3.52	3.37	3.29	3.21	3.13	3.05	2.96	2.87
16	8.53	6.23	5.29	4.77	4.44	4.20	4.03	3.89	3.78	3.69	3.55	3.41	3.26	3.18	3.10	3.02	2.93	2.84	2.75
17	8.40	6.11	5.18	4.67	4.34	4.10	3.93	3.79	3.68	3.59	3.46	3.31	3.16	3.08	3.00	2.92	2.83	2.75	2.65
18	8.29	6.01	5.09	4.58	4.25	4.01	3.84	3.71	3.60	3.51	3.37	3.23	3.08	3.00	2.92	2.84	2.75	2.66	2.57
19	8.18	5.93	5.01	4.50	4.17	3.94	3.77	3.63	3.52	3.43	3.30	3.15	3.00	2.92	2.84	2.76	2.67	2.58	2.49
20	8.10	5.85	4.94	4.43	4.10	3.87	3.70	3.56	3.46	3.37	3.23	3.09	2.94	2.86	2.78	2.69	2.61	2.52	2.42
21	8.02	5.78	4.87	4.37	4.04	3.81	3.64	3.51	3.40	3.31	3.17	3.03	2.88	2.80	2.72	2.64	2.55	2.46	2.36
22	7.95	5.72	4.82	4.31	3.99	3.76	3.59	3.45	3.35	3.26	3.12	2.98	2.83	2.75	2.67	2.58	2.50	2.40	2.31
23	7.88	5.66	4.76	4.26	3.94	3.71	3.54	3.41	3.30	3.21	3.07	2.93	2.78	2.70	2.62	2.54	2.45	2.35	2.26
24	7.82	5.61	4.72	4.22	3.90	3.67	3.50	3.36	3.26	3.17	3.03	2.89	2.74	2.66	2.58	2.49	2.40	2.31	2.21
25	7.77	5.57	4.68	4.18	3.85	3.63	3.46	3.32	3.22	3.13	2.99	2.85	2.70	2.62	2.54	2.45	2.36	2.27	2.17
26	7.72	5.53	4.64	4.14	3.82	3.59	3.42	3.29	3.18	3.09	2.96	2.81	2.66	2.58	2.50	2.42	2.33	2.23	2.13
27	7.68	5.49	4.60	4.11	3.78	3.56	3.39	3.26	3.15	3.06	2.93	2.78	2.63	2.55	2.47	2.38	2.29	2.20	2.10
28	7.64	5.45	4.57	4.07	3.75	3.53	3.36	3.23	3.12	3.03	2.90	2.75	2.60	2.52	2.44	2.35	2.26	2.17	2.06
29	7.60	5.42	4.54	4.04	3.73	3.50	3.33	3.20	3.09	3.00	2.87	2.73	2.57	2.49	2.41	2.33	2.23	2.14	2.03
30	7.56	5.39	4.51	4.02	3.70	3.47	3.30	3.17	3.07	2.98	2.84	2.70	2.55	2.47	2.39	2.30	2.21	2.11	2.01
40	7.31	5.18	4.31	3.83	3.51	3.29	3.12	2.99	2.89	2.80	2.66	2.52	2.37	2.29	2.20	2.11	2.02	1.92	1.80
60	7.08	4.98	4.13	3.65	3.34	3.12	2.95	2.82	2.72	2.63	2.50	2.35	2.20	2.12	2.03	1.94	1.84	1.73	1.60
120	6.85	4.79	3.95	3.48	3.17	2.96	2.79	2.66	2.56	2.47	2.34	2.19	2.03	1.95	1.86	1.76	1.66	1.53	1.38
∞	6.63	4.61	3.78	3.32	3.02	2.80	2.64	2.51	2.41	2.32	2.18	2.04	1.88	1.79	1.70	1.59	1.47	1.32	1.00

ν_1, ν_1 are numerator, denominator d.f. respectively
Tabulated values are those exceeded with probability 0.01.

Table A9.5 Random digits

12005	84000	51051	92674	76575	35789	04180	75029	32490	39949
98859	09884	45275	09467	93026	32912	13941	23206	62419	67776
26604	95099	93751	00590	93060	64776	83565	69919	51623	27483
82984	65780	94428	30160	86023	52284	62463	70712	40687	92630
70888	14063	96700	83008	17579	71321	62664	51514	92195	46722
77803	61872	86245	68220	66267	01379	11304	01658	82404	46728
35228	49673	53552	51215	45611	83927	00772	99295	72154	24126
69965	74926	63366	47688	14279	42943	98863	86630	53925	22310
89716	61713	30650	49028	20285	37791	69149	41701	42403	64009
68348	85228	97590	90997	83339	95822	72969	14037	32379	96225
33821	41538	86376	71823	16285	92630	89531	59337	05421	17043
63162	18167	32088	41917	60942	63252	83886	54130	31841	04502
03431	44528	41760	68035	33731	43262	12789	40348	15532	95309
99198	35092	63655	23987	31112	88069	58720	41729	18757	96096
75535	45156	49477	10673	48262	78240	94031	06192	75221	13363
98554	52502	11780	04060	56634	58077	02005	80217	65893	78381
89725	00679	28401	79434	00909	22989	31446	76251	17061	66680
49221	37750	26367	44817	09214	82674	65641	14332	58211	49564
31783	96028	69352	78426	94411	38335	22540	37881	10784	84658
61025	72770	13689	21456	48391	00157	61957	11262	12640	17228
10581	30143	89214	52134	76280	77823	61674	96898	90487	43998
51753	56087	71524	64913	81706	33984	90919	86969	75553	87375
96050	08123	28557	04240	33606	10776	64239	81900	74830	92654
93998	95705	73353	26933	66089	25177	62387	34932	62021	34044
70974	45757	31830	09589	31037	91886	51780	21912	16444	52881
25833	71286	76375	43640	92551	46510	68950	60168	26349	04599
55060	28982	92650	71622	36740	05869	17828	29937	01020	90851
29436	79967	34383	85646	04715	80695	39283	50543	26875	94047
80180	08706	17875	72123	69723	52846	71310	72507	25702	33449
40842	32742	44671	72953	54811	39495	05023	61569	60805	26580
31481	16208	60372	94367	88977	35393	08681	53325	92547	31622
06045	35097	38319	17264	40640	63022	01496	28439	04197	63858
41446	12336	54072	47198	56085	25215	89943	41153	18496	76869
22301	07404	60943	75921	02932	50090	51949	86415	51919	98125
38199	09042	26771	15881	80204	61281	61610	24501	01935	33256
06273	93282	55034	79777	75241	11762	11274	41685	24117	98311
92201	02587	31599	27987	25678	69736	94487	41653	79550	92949
70782	80894	95413	36338	04237	19954	71137	23584	87069	10407
05245	40934	96832	33415	62058	87179	31542	18174	54711	21882
85607	45719	65640	33241	04852	87636	43840	42242	22092	28975
61175	56493	93453	90267	99471	04519	78694	17115	00371	64703
36079	22448	22686	31272	01245	66265	12670	29560	49346	20049
94688	39732	02785	73373	44876	39888	69352	40488	43849	95406
54047	85793	53994	28605	46114	91174	49646	85123	66246	72392
24997	69553	46802	24331	88523	89026	69776	55460	21984	76677

Table A9.6   Control chart factors for the sample range

	Lower percentage factors		Upper percentage factors	
Sample size	0.1%	2.5%	2.5%	0.1%
2	0.00	0.04	3.17	4.65
3	0.06	0.30	3.68	5.06
4	0.20	0.59	3.98	5.31
5	0.37	0.85	4.20	5.48
6	0.53	1.07	4.36	5.62
7	0.69	1.25	4.49	5.73
8	0.83	1.41	4.60	5.82
9	0.97	1.55	4.70	5.90
10	1.08	1.67	4.78	5.97

Multiply the estimate of the standard deviation by the tabled factors, which are based on a normal distribution.

Table A9.7   Gamma function (from Kreyszig, 1993)

α	$\Gamma(\alpha)$	α	$\Gamma(\alpha)$	α	$\Gamma(\alpha)$	α	$\Gamma(\alpha)$	α	$\Gamma(\alpha)$
1.00	1.000 000	1.20	0.918 169	1.40	0.887 264	1.60	0.893 515	1.80	0.931 384
1.02	0.988 844	1.22	0.913 106	1.42	0.886 356	1.62	0.895 924	1.82	0.936 845
1.04	0.978 438	1.24	0.908 521	1.44	0.885 805	1.64	0.898 642	1.84	0.942 612
1.06	0.968 744	1.26	0.904 397	1.46	0.885 604	1.66	0.901 668	1.86	0.948 687
1.08	0.959 725	1.28	0.900 718	1.48	0.885 747	1.68	0.905 001	1.88	0.955 071
1.10	0.951 351	1.30	0.897 471	1.50	0.886 227	1.70	0.908 639	1.90	0.961 766
1.12	0.943 590	1.32	0.894 640	1.52	0.887 039	1.72	0.912 581	1.92	0.968 774
1.14	0.936 416	1.34	0.892 216	1.54	0.888 178	1.74	0.916 826	1.94	0.976 099
1.16	0.929 803	1.36	0.890 185	1.56	0.889 639	1.76	0.921 375	1.96	0.983 743
1.18	0.923 728	1.38	0.888 537	1.58	0.891 420	1.78	0.926 227	1.98	0.991 708
1.20	0.918 169	1.40	0.887 264	1.60	0.893 515	1.80	0.931 384	2.00	1.000 000

Glossary

Accuracy
Accurate measurements are free from bias. A statistic is an accurate estimator of some population parameter if its expected value equals that parameter. Accuracy is distinct from precision which measures how close replicate measurements are to each other.

Action lines
Lines drawn on a control chart, such that only a few in a thousand points will lie outside them if the process is on target. Also called control limits.

Addition rule of probability
The probability of one or the other or both of two events occurring is: the sum of their individual probabilities of occurrence less the probability they both occur.

Aliases (design of experiments)
If there are a large number of factors to be investigated in an experimental programme a full factorial design may not be feasible. If a fraction of the full factorial is run, there will be some ambiguity over the cause of significant results. For example, an increase in strength might be due to the main effect of A or the interaction BDC or both. A and BDC are termed aliases. In practice, high-order interactions are often assumed to be negligible.

Analysis of variance (ANOVA)
The sum of squares, about the mean, for some variable is resolved into contributions which can be attributed to different sources. These sources can be different levels of variability, or predictor variables in regression, or treatments in a designed experiment.

ARIMA model
A variable is predicted from a linear combination of its past values, and the past forecasting errors.

Asset management plan (AMP)
A business plan for water companies. It includes estimates of the costs of maintaining, and where necessary improving, the water supply and sewerage systems over the following 20 years. UK companies have to produce an AMP every five years, and submit it to an independent government body.

Asymptotic result
A result which is proved by some limiting process, in which the sample size tends to infinity. Many such results still give reasonable approximations with small sample sizes. The central limit theorem is a well-known example.

Auto-covariance function (acvf), auto-correlation function (acf)
The acvf is the covariance between a variable at two times as a function of the time lag. The acf is the corresponding correlation.

Bayes' theorem
Bayes' theorem is used to update our knowledge, expressed in probabilistic terms, as we obtain new data. The posterior distribution is proportional to the product of the likelihood and the prior distribution.

Bayesian analysis
A statistical analysis which is formally based on the interpretation of probability as belief, and is modified in the light of additional data.

Bayes linear estimator
A minimum variance predictor of y from predictor variables which requires only variances and covariances to be specified.

Bernoulli trial
An experiment with two possible, mutually exclusive, outcomes of which one must occur.

Beta distribution
A continuous probability distribution defined over a domain from 0 to 1.

Beta function
The normalising constant for the beta distribution, which is a function of its parameters.

Bias
A systematic difference between the estimator and the parameter being estimated, i.e. the difference will persist when averaged over imaginary replicates of the sampling procedure. Formally, the difference between the mean of the sampling distribution and the parameter being estimated. If the bias is small by comparison with the standard deviation of the sampling distribution, the estimator may still be useful. For example, s^2 is unbiased for σ^2 and s is slightly biased for σ.

Bin
A shorter alternative to 'class interval' when grouping data.

Binomial distribution
The distribution of the number of successes in a set number of trials, with a constant probability of success.

Bivariate distribution
The distribution of data pairs.

Bootstrap confidence intervals
Confidence intervals based on the sampling distribution obtained from repeated re-sampling, with replacement, of the sample.

Box–Cox transform
A power transform scaled so that it includes natural logarithm as a special case.

Brownian motion
Motion of suspended particles, or molecules, in a fluid subject to random impacts from neighbouring particles or molecules.

Chi-square distribution
A special case of a gamma distribution which arises as the distribution of the sum of m independent standard normal variables. The parameter m is referred to as the degrees of freedom. The sample variance, in a random sample of size n from a normal distribution, has a distribution which is proportional to a chi-square distribution with $n - 1$ degrees of freedom.

Class intervals
Before a histogram is drawn, the data are grouped into classes which correspond to convenient divisions of the variable range. Each division is defined by its lower and upper limits, and the difference between them is the length of the class interval.

Coefficient
The factor for a predictor variable in a regression.

Coefficient of variation
The ratio of the standard deviation to the mean.

Common cause variability
Natural background variability in a process which is distinguished from special cause variability which has some temporary assignable cause that can be removed. A process is said to be in statistical control if only common cause variability is present.

Concomitant variables
Variables that can be monitored but not controlled.

Conditional distribution
The distribution of one variable given specific values for other associated variables.

Conditional probability
The probability of an event conditional on other events having occurred or an assumption they will occur. (All probabilities are conditional on the general context of the problem.)

Confidence interval
A 95%, or whatever, (frequentist) confidence interval for some parameter is an interval constructed in such a way that, on average, if you imagine millions of random samples of the same size, 95% of them will include the parameter. From a practical point of view we think there is a 95% chance the interval contains the parameter, and Bayesian confidence intervals are properly interpreted in this way.

Control limits
An alternative term for action lines.

Control variable
A variable that affects a process output and can be set to a specified value or changed in response to some signal.

Correlation
A dimensionless measure of linear association between two variables that lies between -1 and 1. Zero represents no association and negative values correspond to one variable increasing as the other decreases.

Cost accounting
The value of assets depreciates over time and provision is made for their eventual replacement.

Covariance
A measure of linear association between two variables, that equals the average value of the mean corrected products.

Cross-covariance function
The covariance between two different variables at different times, the time difference being known as the lag.

Cumulative distribution function (cdf)
A function which gives the probability a continuous variable is less than any value. It is the population analogue of the cumulative frequency polygon. Its derivative is the pdf.

Cumulative frequency polygon
A plot of the proportion, often expressed as a percentage, of data less than or equal to any value.

Degrees of freedom
The number of data used to calculate some statistic, less the number of physically independent parameters estimated from the data and used in the calculation. Each such estimate imposes a constraint. For example, if we estimate a population variance from a sample of n data we first estimate the population mean. Deviations from the sample mean are constrained to equal zero. Therefore we have $(n - 1)$ degrees of freedom for the sample variance. In multiple regression the number of constraints equals the number of coefficients to be estimated, which is one more than the number of explanatory variables (k). So, the estimate of the variance of the errors, based on the residuals, has $(n - k - 1)$ degrees of freedom.

Differencing
Successive values in a time series are subtracted to form a new series.

Discrete white noise (DWN)
A sequence of independent random variables.

Dow Jones index
The relative price of American stocks and shares, based on transactions for an agreed list of major companies.

Ensemble
An imaginary population of all possible time series that might be generated by the underlying random process.

Ergodic
A random process is ergodic in the mean, for example, if the time average from a single time series tends to the ensemble average (expected value) as the length of the time series increases.

Euclidian distance
The physical distance between two points in three-dimensional space is the square root of the sum of the squared differences in their coordinates (Pythagoras' theorem). The definition extends to any number of coordinates.

Expected monetary value
The sum of the products of possible financial outcomes, of some course of action, with their probabilities of occurrence.

Expected value
An average value in the population.

Explanatory variable
In a multiple regression the dependent variable, usually denoted by y, is expressed as a linear combination of the explanatory variables. In designed experiments it is helpful to subdivide the explanatory variables into control variables, whose values are chosen by the experimenter, and concomitant variables which can be monitored but not

preset. Alternative names for the explanatory variables include predictor variables. The dependent variable is often called the response, especially in an experimental context.

Exponential distribution
A continuous distribution of the times between occurrences, when occurrences are random and independent.

Extreme value distributions
Distributions that provide plausible models for the maximum value from repeated samples of the same notional size. For example, annual maximum wind speeds.

Factorial experiment
An experiment designed to examine the effects of two or more factors. Each factor is applied at two, or more, levels and all combinations of these factor levels are tried in a full factorial. This allows interactions to be investigated, and is also an efficient means of estimating the main effects. If the number of runs for the full factorial is prohibitive, fractional factorial designs, which retain the benefits, can be used.

Factorial function
The product $n \times (n - 1) \times \cdots \times 1$, for some integer n, which is written as $n!$.

Finite population correction (FPC)
A reduction in variance of statistics due to the fact that a substantial proportion, generally 10% or more, of the population has been sampled.

Fitted value
The predicted value of y corresponding to the x values for a datum used in fitting the model. The residual is the difference between the observed value of y and the fitted value of y.

Frequency
The number of times an event occurs. Also, when talking about harmonic functions (sines and cosines) the number of cycles per unit of time.

Frequentist analysis
A statistical analysis which is formally based on the relative frequency interpretation of probability, rather than treating probability as belief.

Gamma distribution
The distribution of the time until the rth occurrence in a Poisson process has a gamma distribution with the parameter 'c' equal to r. The pdf (multiplied by δt) can be written down directly as: the product of $r - 1$ arrivals in time t with the probability of an arrival in the interval $(t, t + \delta t)$. It generalises to non-integer c, and is often used in applications which are unrelated to Poisson processes because it gives a good empirical fit to data.

Gamma function
A generalisation of the factorial function to values other than positive integers.

Gaussian
A Gaussian distribution is an alternative name for the normal distribution.

Geometric distribution
The distribution of the number of failures before the first success in a sequence of Bernoulli trials.

Geometric sequence
A sequence in which the next term is obtained by multiplying the preceding term by a constant value known as the common ratio.

Highest density region
A domain which, for a chosen probability, includes that part of the probability density function which exceeds a certain height. The height depends on the probability, since the area under that part of the pdf which exceeds the height equals the probability.

Histogram
A chart consisting of rectangles drawn above class intervals with areas equal to the proportion of data in that interval. It follows that the heights of the rectangles equal the relative frequency density, and the total area equals one. If all the class intervals are of the same length the heights are proportional to the frequencies.

Hyper-parameter
A parameter of the distribution of parameters of some other distribution in a hierarchical Bayesian analysis.

Hyper-plane
A regression of Y on two predictor variables $x1$ and $x2$ can be represented by a plane in three dimensions. The corresponding geometric figure in more than three dimensions should be referred to as a hyper-plane, but plane is often used in a rather loose sense.

Hyper-surface
A regression of Y on two predictor variables $x1$ and $x2$, and functions of $x1$ and $x2$ which are commonly restricted to squares and cross-products, can be represented by a surface in three dimensions. The corresponding geometric figure in more than three dimensions should strictly be referred to as a hyper-surface, but surface is often used.

Imaginary infinite population
The population we are sampling from is often imaginary and arbitrarily large. A sample from a production line is thought of as a sample from the population of all items that will be produced if the process continues on its present settings. An estimate is thought of as a single value from the imaginary distribution of all possible estimates, so that we can give its precision.

Independent
Two events are independent if the probability one occurs does not depend on whether the other occurs.

Integer
A positive or negative whole number including 0.

Interaction
Two explanatory variables interact if the effect of one depends on the value of the other. Their product is then included as an explanatory variable in the regression. If this interaction depends on the value of some third variable a three-variable interaction exists, and so on.

Intercept
The constant term in a regression analysis. In the case of a regression of Y on a single predictor variable x, it is the intercept of the line on the y-axis.

Lag
In time series, the time between variables.

Laplace distribution
A distribution formed by identical back-to-back scaled exponential distributions.

Laspeyre index
A relative cost index based on a typical basket of goods that would have been bought in the base year.

Likelihood
The probability of obtaining the observed data, considered as a function of the unknown parameters.

Logistic regression
A regression with a logit as the dependent variable.

Logit
The natural logarithm of the ratio of a probability to its complement. The logit has a range from minus infinity to plus infinity.

Maclaurin series
A Taylor series about the point $(0, f(0))$.

Mahalanobis distance
A measure of the distance between two points calculated as a weighted sum of squared differences in coordinates, the weights being given by the inverse of the variance–covariance matrix.

Management by objectives (MBO)
Everyone in the company openly declares their objectives and times within which they aim to achieve them. Performance is then measured against these objectives.

Marginal distribution
The marginal distribution of a variable is the distribution of that variable. The 'marginal' refers to the fact that multivariate data are available, or being modelled, but information on the other variables has been ignored.

Markov property
'The future is independent of the past, given the present' (P. Guttorp).

Maximum likelihood estimation
A method of estimating parameters as those which maximise the likelihood function.

Meal
A mixture of powders used as raw material for a chemical process.

Mean
The sum of several quantities divided by their number. Also used as an alternative to 'average' in 'average value of . . .'.

Mean corrected
Data are mean corrected if their mean is subtracted from them. Mean-corrected data therefore have an average value of 0.

Median
The middle value if data are put into ascending order.

Method-of-moments estimation
A method of estimating parameters by equating sample moments to the corresponding population moments which are expressed in terms of the unknown parameters.

Minimax principle
Choose the strategy which minimises the maximum possible loss.

Mode
The most commonly occurring value. Also the value of the variable at which the pdf has its maximum.

Modulus
The distance of a point from the origin, also known as absolute value. In the case of real numbers it is the number without its sign.

Moment-generating function (mgf)
An integral transform of the pdf, which is an alternative characterisation of the distribution. It is a convenient method for finding moments and for obtaining the distribution of a sum of independent variables.

Monte Carlo methods
Any simulation study that relies on random numbers.

Multiple regression
Some variable, called the dependent variable or the response, is expressed as a linear combination of predictor variables plus random error. The coefficients of the variables in this combination are the unknown parameters of the model and are estimated from the data. The predictor variables can be non-linear functions of each other, for example $x1$, $x2$, $x1 \times x1$, $x2 \times x2$ and $x1 \times x2$ represent a quadratic surface. Other names for predictor variables include 'explanatory' variables.

Multiplicative rule
The probability of two events both occurring is the product of the probability that one occurs with the probability the other occurs conditional on the first occurring.

Multivariate distribution
Distribution of data m-tuples $(x1, x2, \ldots, xm)$.

Mutually exclusive
Two events are mutually exclusive if they cannot occur together. A set of events is mutually exclusive and exhaustive if exactly one must occur.

Negative binomial distribution
A generalisation of the Pascal distribution to non-integer r.

Normal distribution
A bell-shaped pdf which is a plausible model for random variation if it can be thought of as the sum of a large number of smaller components. It is also important as a sampling distribution, especially of the sample mean.

Null hypothesis
A supposition that is set up to be tested. If there is no evidence against the null hypothesis we will act as though it is true, i.e. accept the hypothesis. If there is evidence against the hypothesis, quantified by a small P-value, we claim evidence to reject the null hypothesis at the $100 \times P$-value level. We say the result is statistically significant at the $100 \times P\%$ level.

Odds
Ratio of the probability an event occurs to the probability that it does not, and more generally the odds on A against B is $\Pr(A)$ to $\Pr(B)$.

Or
In probability 'A or B' is conventionally taken to include both.

Order statistic
The ith-order statistic is the ith largest if the sample of n is sorted into ascending order.

Orthogonal
In a designed experiment the values of the control variables are usually chosen to be uncorrelated, when possible, or nearly so. If they are uncorrelated they are said to be orthogonal. (If the values are put in columns and thought of as 'vectors', the vectors are orthogonal.)

P-value
The probability of an event as extreme, or more extreme, than that observed conditional on the null hypothesis being true.

Paasche index
A relative cost index based on a current typical basket of goods.

Parameter
A constant which is a salient feature of a population, its value being usually unknown.

Pascal distribution
The distribution of the number of failures before the rth success in a sequence of Bernoulli trials.

Paver
A paving block. Modern ones are made from concrete in a variety of shapes and colours. Also called paviors.

Percentage point
The upper alpha percentage point of a pdf is the value beyond which a proportion alpha of the area under the pdf lies; a lower point is defined similarly.

Percentiles
The lower x percentile is the value of the variable below which $x\%$ of the distribution lies.

Point process
A process in which events occur at discrete time instants (points).

Poisson distribution
The number of occurrences in some length of continuum if occurrences are random and independent.

Poisson process
Occurrences in some continuum, often time, form a Poisson process if they are random, independent and occur at some constant average rate (in the standard case at least). The time from now until the next occurrence is independent of past occurrences.

Polynomial
An algebraic expression made up of a weighted sum of non-negative integer powers of the variable.

Power
The power of a statistical test is the probability of rejecting the null hypothesis as a function of specified alternative hypotheses being true.

Precision
The precision of an estimator is a measure of how close replicate estimates are to each other. Formally, the reciprocal of the variance of the sampling distribution.

Predictive distribution
A distribution of a variable which takes account of uncertainty about the parameters of the distribution.

Predictor variable
A variable used to predict the value of another variable in a regression.

Prior distribution
A distribution of some variable, based on our belief about it before an experiment to obtain further information.

Probability
A measure of how likely some event is to occur on a scale ranging from 0, representing impossibility, to 1, representing certainty. It can be thought of as the long-run proportion of times the event would occur if the scenario were to be repeated.

Probability density function (pdf)
A curve such that the area under it between any two values represents the probability that a continuous variable will be between them. The population analogue of a histogram.

Probability distribution
A rule which implies the probabilities that a discrete variable takes specific values or a continuous variable lies within a specific range. It can be expressed in various ways, e.g. pdf or cdf.

Probability mass function
A formula that gives the probabilities that a discrete variable takes any of its possible values. The population analogue of a line diagram.

Pseudo-random numbers
A sequence of numbers generated by a deterministic algorithm which appear to be random, and are indistinguishable from genuine random numbers by empirical tests. Computer-generated random numbers are actually pseudo-random.

Quartiles
The upper (lower) quartile (UQ, LQ) is the datum above (below) which one-quarter of the data lie.

Random digits
A sequence in which each one of the digits from 0 up to 9 is equally likely to occur next.

Random numbers
A sequence of numbers from a specified distribution, such that the next is independent of the existing sequence.

Random sample
A sample which has been selected so that every member of the population has a known, non-zero, probability of appearing.

Random variable
A rule which assigns a number to the outcome of an experiment.

Randomisation
Use of random numbers to allocate experimental units to treatments, or an order of runs of an experiment etc.

Range
Difference between the largest datum and the smallest datum when the data are put into ascending order.

Regression
A term that includes multiple regression and the special case of a simple linear regression in which Y is assumed to be linearly related to a single x.

Relative frequency
The ratio of the frequency of some event occurring to the number of scenarios in which it could potentially have occurred. That is, the proportion of times on which it occurred.

Relative frequency density
Relative frequency divided by the length of the class interval.

Reliability function
The reliability function, of t say, is the probability a lifetime exceeds t. It is the complement of the cdf.

Renewals accounting
Instead of allowing for depreciation, the cost of maintaining assets in continual good repair is budgeted for.

Residual
The observed value minus the fitted value in a regression analysis.

Response
The dependent variable in a regression.

R-squared
The proportion of the variability in a set of data that is accounted for by the model which is being fitted. This is usually defined as the ratio of the: (corrected sum of squares with no model − sum of residuals squared)/(corrected sum of squares with no model). It is also known as the coefficient of multiple determination, and the squared multiple correlation coefficient. In a linear regression with one predictor variable it is the square of the correlation coefficient.

Saddle point
A minimum when proceeding in one direction and a maximum if proceeding at right angles to the first direction. If the variables are continuous the shape of the surface is a saddle.

Sample space
A list of all possible outcomes of some operation which involves chance.

Sampling distribution
An estimate is thought of as a single value from the imaginary distribution of all possible estimates, known as the sampling distribution. The idea of a sampling distribution is necessary to measure the precision of an estimator. The term 'sampling distribution' refers to the context in which the distribution arises, rather than the form of the distribution. For example, the sampling distribution of the mean is approximately normal for large samples. The t, chi-square and F-distributions are usually introduced in the context of sampling.

Sampling frame
A list of members of a population in which each member is identified by a unique number. A sample is drawn by taking members whose numbers correspond to those selected by a random mechanism.

Scalar
A single number, rather than an array of numbers, in a matrix context.

Simple random sample
A sample chosen so that every possible choice of n from N has the same chance of occurring as the final sample. It implies all members of the population have the same probability of selection, but many other sampling schemes also have this property.

Skewness
A measure of asymmetry of a distribution. Positive values correspond to a tail to the right.

Standard deviation
Square root of the variance.

Standard error
The standard deviation of some estimator. Commonly used, without qualification, for the standard deviation of the sample mean.

Standard normal distribution
The normal distribution scaled to have a mean of 0 and a standard deviation of 1. Its cdf and percentage points are tabled.

Standardised residual
The residual divided by the estimated standard deviation of the errors in the regression.

State space
The range of values that can be taken by the state variable.

State variable
A variable which changes over time is often referred to as the state variable to distinguish it from the time variable.

Stationarity
A random process is stationary in the mean, if the mean does not change over time. That is, there are no trends or seasonal variations in the mean. It is second-order stationary if neither the mean nor the autocovariance structure change over time. In particular, this requires that the standard deviation does not change seasonally. It is strictly stationary if all moments are time invariant.

Stationary point
A point on a surface at which all partial derivatives are zero. In three dimensions these correspond to local maxima, minima or saddle points. The greatest value of a function defined over a finite domain may occur at a boundary rather than at a maximum.

Statistic
A summary number calculated from the sample, usually to estimate the corresponding population parameter.

Stochastic
A synonym for random.

Stratified sampling
The population is divided into sub-populations called strata. A random sample is then taken from each stratum.

Student's *t*-distribution
The sampling distribution of many statistics is normal, or at least approximately so, and can be scaled to standard normal. If the standard deviation of the statistic, which is used in the scaling, is replaced by its sample estimate, with v degrees of freedom, the normal distribution becomes a t-distribution with v degrees of freedom. If v exceeds about 30 there is little practical difference.

Studentised residual
A residual divided by its estimated standard deviation, which is not quite the same as the standard deviation of the errors. Hence, Studentised residuals are close to, but not identical with, standardised residuals.

Survivor function
An alternative name for the reliability function.

Taylor series
A polynomial approximation to a function, $f(\ \ )$, about a point at which the polynomial function equals the function to be approximated. If the point is $(0, f(0))$ the Taylor series is usually referred to as a Maclaurin series.

Test statistic
A statistic which has a sampling distribution specified by the null hypothesis and sensitive to any departures from the null hypothesis. Then unlikely values of the statistic, conditional on the null hypothesis being true, are evidence against the null hypothesis.

Time series
A sequence of values of a variable which changes over time, usually thought of as a realisation of some underlying random process.

t-ratio
The ratio of an estimate to its standard deviation. If the absolute value of the t-ratio is less than 1, a 66% confidence interval for the parameter will include 0. If the t-ratio exceeds about 1.7 the 90% confidence interval will exclude 0, and we can at least claim to be confident about the sign of the coefficient.

Type I and Type II errors
The Type I error is the probability (α) of rejecting the null hypothesis (H_0) when it is time. The Type II error is the probability (β) of not rejecting H_0 when it is false. The probability β depends on the specific alternative hypothesis which is being considered.

Unbiased estimator (estimate)
An estimator is unbiased, for some parameter, if the mean of its sampling distribution is equal to that parameter. That is, if the estimates are averaged over imagined replicates of the sampling procedure, the result is the parameter value. An unbiased estimate is a particular value of the estimator.

Uniform distribution
A variable has a uniform distribution between two limits, if the probability it lies within some interval, between those limits, is proportional to the length of the interval. Limits of 0 and 1 are by far the most usual, and arise in the context of random number generation.

Utility
The relative worth of some outcome, usually measured on a probability scale from $[0, 1]$. Often assumed to be the monetary gain or loss if the monetary amounts are small compared with a company's reserves.

Variable
A quantity that varies from one member of the population to the next. It can be measured on some continuous scale, be restricted to integer values (discrete), or be restricted to descriptive categories.

Variance
Average of the squared deviations from the mean. It has units equal to the square of the units of the variable. Independent variances are additive.

Variance–covariance matrix
If $(x1, \ldots, xm)$ is an m-tuple, the $m \times m$ variance–covariance matrix V has the variances down the leading diagonal and covariances in the off-diagonal positions.

Vector
In statistics, an array with 1 column is commonly referred to as a (column) vector, and similarly for rows.

Weighted mean
An average in which the data are multiplied by numbers called weights, added, and then divided by the sum of the weights. If the weights are all the same this is the usual mean.

References

Abdul-Aziz, Z. 2000: A comparison of the use of statistical methods and quality improvement techniques in Malaysia and UK. Department of Engineering Mathematics, University of Newcastle upon Tyne.

Abdul-Aziz, Z., Chan, J.F.L. and Metcalfe, A.V. 1998: Use of quality practices in manufacturing industries in Malaysia. *Total Quality Management*, 9(4), S13–S16.

ADA Decision Systems Staff. 1994: *DPL Decision Analysis Software for Microsoft Windows: Standard Version User Guide*. Wadsworth, Calif.

Aitchison, J. 1970: *Choices Against Chance*. Addison-Wesley, Reading, Mass.

Albert, J.H. 1996: *Bayesian Computation Using Minitab*. Duxbury, Belmont, Calif.

Allenby, R.B.J.T. 1995: *Linear Algebra*. Arnold, London.

Almond, R.G. 1995: *Graphical Belief Modeling*. Chapman & Hall, London.

Alwan, L.C. and Roberts, H.V. 1995: The problem of misplaced control limits. *Applied Statistics (Journal of the Royal Statistical Society Series C)*, 44(3), 269–78.

Anderson, V.L. and McLean, R.A. 1974: *Design of Experiments: A Realistic Approach*. Marcel Dekker, New York.

Appleby, J. 1998: BUSTLE – a bus simulation. *Teaching Statistics*, 20(3), 77–80.

Archibald, T.W., McKinnon, K.I.M. and Thomas, L.C. 1997: An aggregate stochastic dynamic programming model of multireservoir systems. *Water Resources Research*, 33(2), 333–40.

Axelrod, R. 1984: *The Evolution of Cooperation*. Basic Books, New York.

Barnett, S. and Cameron, R.G. 1985: *Introduction to Mathematical Control Theory* (2nd edition). Oxford University Press, Oxford.

Barnett, V. 1991: *Sample Survey Principles and Methods*. Arnold, London.

Barrett, H. and Weinstein, A. 1998: The effect of market orientation and organizational flexibility on corporate entrepreneurship. *Entrepreneurship Theory and Practice*, 23, 57–70.

Bartlett, M.S. 1937: The statistical conception of mental factors. *British Journal of Psychology*, 28, 97–104.

Bartlett, M.S. 1938: Methods of estimating mental factors. *Nature*, 141, 609–10.

Beal, S. 1995: Measuring your money. *Water Bulletin*, 638.

Bhattacharya, R.N. and Waymire, E.C. 1990: *Stochastic Processes with Applications*. Wiley, New York.

Black, F. and Scholes, M. 1973: The pricing of options and corporate liabilities. *Journal of Political Economy*, 81, 635–54.

Bowersox, D.J. and Closs, D.J. 1996: *Logistical Management*. McGraw-Hill, Singapore.

Box, G.E.P. and Cox, D.R. 1964: An analysis of transformations. *Journal of the Royal Statistical Society Series B*, 26, 211–43.

Box, G.E.P., Hunter, W.G. and Hunter, J.S. 1978: *Statistics for Experimenters*. Wiley, New York.

Box, G.E.P. and Kramer, T. 1992: Statistical process monitoring and feedback adjustment – a discussion. *Technometrics*, 34, 251–85.

Box, G.E.P. and Luceño, A. 1997: *Statistical Control by Monitoring and Feedback Adjustment*. Wiley, New York.

Brown, G., Draper, P. and McKenzie, E. 1997: Consistency of UK pension fund investment performance. *Journal of Business Finance and Accounting*, 24(2), 155–78.

Buchberger, S.G. and Wu, L. 1995: Model for instantaneous residential water demands. *ASCE Journal of Hydraulic Engineering*, 121(3), 232–46.

Bunday, B.D. 1996: *Queueing Theory*. Arnold, London.

Bunday, B.D. and Garside, G.R. 1987: *Linear Programming in Pascal*. Arnold, London.

Burghes, D.N. and Wood, A.D. 1980: *Mathematical Models in the Social, Management and Life Sciences*. Ellis Horwood, Chichester.

Casella, G. and George, E.I. 1992: Explaining the Gibbs sampler. *The American Statistician*, 6(3), 167–74.

Caulcutt, R. 1995: The rights and wrongs of control charts. *Applied Statistics (Journal of the Royal Statistical Society Series C)*, 44(3), 279–88.

Caulcutt, R. 1999: From measurement to action. *Proceedings of the Conference on the Control of Industrial Processes II*, 30–31 March 1999, University of Newcastle upon Tyne, UK.

Chatfield, C. 1989: *The Analysis of Time Series* (4th edition). Chapman & Hall, London.

Cheng, C.-L. and Van Ness, J.W. 1999: *Statistical Regression with Measurement Error*. Arnold, London.

Chib, S. and Greenberg, E. 1995: Understanding the Metropolis–Hastings algorithm. *The American Statistician*, 49(4), 327–35.

Clarke, G.M. and Kempson, R.E. 1997: *Introduction to the Design and Analysis of Experiments*. Arnold, London.

Cohen, M.A., Eliashberg, J. and Ho, T.-H. 1996: New product development: the performance and time-to-market tradeoff. *Management Science*, 42(2), 173–86.

Cook, R.D. 1977: Detection of influential observations in linear regression. *Technometrics*, 19, 15–18.

Cooke, D., Craven, A.H. and Clarke, G.M. 1990: *Basic Statistical Computing* (2nd edition). Arnold, London.

Cox, D.R. and Isham, V. 1980: *Point Processes*. Chapman & Hall, London.

Cox, T.F. and Cox, M.A.A. 1994: *Multidimensional Scaling*, Chapman & Hall, London.

Craven, B.D. 1995: *Control and Optimisation*. Chapman & Hall, London.

Danaher, P.J. 1997: Using conjoint analysis to determine the relative importance of service attributes measured in customer satisfaction surveys. *Journal of Retailing*, 73(2), 235–60.

Davies, O.L. 1954: *The Design and Analysis of Industrial Experiments*. Oliver & Boyd, London.

Delaney, R.V. 1995: Sixth Annual State of Logistics Report. Presented to the National Press Club, Washington, D.C., 5 June 1995.

Deming, W.E. 1986: *Out of the Crisis*. Cambridge University Press, Cambridge.

Dobbs, I.M. 1993(a): Individual travel cost method: estimation and benefit assessment with a discrete and possibly grouped dependent variable. *American Journal of Agricultural Economics*, 75, 84–94.

Dobbs, I.M. 1993(b): On adjusting for truncation and sample selection bias in the individual travel-cost method. *Journal of Agricultural Economics*, 44, 335–43.

Dunn, R.A. and Ramsing, K.D. 1981: *Management Science*. Macmillan, New York.

Eastaway, R., Wyndham, J., Eastaway, R. and Rice, T. 1998: *Why do Buses come in Threes? The Hidden Mathematics of Everyday Life*. Robson, London.

Efron, B. and Gong, G. 1983: A leisurely look at the bootstrap, the jacknife, and cross-validation. *American Statistician*, 37, 36–48.

Erickson, K.T. and Hedrick, J.L. 1999: *Plantwide Process Control*. Wiley, New York.

Erikson, W.J. and Hall, O.P. 1989: *Computer Models for Management Science* (3rd Edition). Addison-Wesley, Reading, Mass.

Everitt, B.S. 1987: *Introduction to Optimisation Methods and their Application in Statistics*. Chapman & Hall, London.

Everitt, B.S. 1992: *The Analysis of Contingency Tables* (2nd Edition). Chapman & Hall, London.

Everitt, B.S. and Dunn, G. 1991: *Applied Multivariate Data Analysis*. Arnold, London.

Farrow, M., Goldstein, M. and Spiropoulos, T. 1997: Developing a Bayes Linear decision support system for a brewery in *The Practice of Bayesian Analysis*, ed. S. French and J.Q. Smith, 71–106, Arnold, London.

Fearn, T. and Maris, P.I. 1991: An application of Box–Jenkins methodology to the control of gluten addition in a flour mill. *Applied Statistics* (*Journal of the Royal Statistical Society Series C*), 40(3), 477–84.

Ferguson, T.D. and Ketchen, D.J. Jr. 1999: Organizational configurations and performance: the role of statistical power in extant research. *Strategic Management Journal*, 20, 385–95.

Gallagher, C. and Metcalfe, A. 1982: How managers use numerate techniques. *Management Today*, December 1982, 33–8.

Gelman, A., Carlin, J.B., Stern, H.S. and Rubin, D.B. 1995: *Bayesian Data Analysis*. Chapman & Hall, London.

Ghosh, B.C. and Taylor, D. 1999: Switching advertising agency – a cross-country analysis. *Marketing Intelligence & Planning*, 17(3), 140–6.

Goodwin, R.M. 1990: *Chaotic Economic Dynamics*. Oxford University Press, Oxford.

Gorn, G.J. 1982: The effects of music in advertising on choice behaviour: a classical conditioning approach. *Journal of Marketing*, 46, 94–101.

Graham, G. and Martin, F.R. 1946: Heathrow. The construction of high-grade quality concrete paving for modern transport aircraft. *Journal of the Institution of Civil Engineers*, 26(6), 117–90.

Graybill, F.A. 1976: *Theory and Application of the Linear Model*. Duxbury, North Scituate, Mass.

Greenfield, T. and Savas, D. 1992: DEX. A program for the Design and Analysis of Experiments (Version 4.0).

Guttorp, P. 1995: *Stochastic Modelling of Scientific Data*. Chapman & Hall, London.

Hadi, A.S. and Ling, R.F. 1998: Some cautionary notes on the use of principal components regression. *The American Statistician*, 52(1), 15–19; response 52(4), 371.

Hand, D.J. and Crowder, M.J. 1996: *Practical Longitudinal Data Analysis*. Chapman & Hall, London.

Hand, D.J. and Jacka, S. 1998: *Statistics in Finance*. Arnold, London.

Hettmansperger, T.P. and McKean, J.W. 1998: *Kendall's Library of Statistics 5 – Robust Nonparametric Statistical Methods*. Arnold, London.

Hill, D. 1956: Modified control limits. *Applied Statistics*, 5, 12–19.

HMSO. 1985: *Code of Practical Guidance for Packers and Importers – Weights and Measures Act 1979*. HMSO, London.

Hogg, R.V. and Craig, A.T. 1978: *An Introduction to Mathematical Statistics* (4th edition). Macmillan, New York.

Howard, R.A. 1960: *Dynamic Programming and Markov Processes*. MIT University Press, Cambridge, Mass.

Hutchinson, P. 1996: Waiting for buses: size-weighted means. *Teaching Statistics*, 18(1), 9.

Ibrahim, K.B. and Metcalfe, A.V. 1993: Bayesian overview for evaluation of mini-roundabouts as a road safety measure. *The Statistician (Journal of the Royal Statistical Society Series D)*, 42, 525–40.

Isao, O. 1990: Improving productivity in a Japanese shipyard. *Journal of the North East Coast Institute of Engineers and Shipbuilders*, 196(4).

Jackson, A. and De Cormier, R. 1999: E-mail survey response rates: targeting increases response. *Marketing Intelligence & Planning*, 17(3), 135–9.

Jamieson, L.F. and Bass, F.M. 1989: Adjusting stated intention measures to predict trial purchase of new products – a comparison of models and methods. *Journal of Marketing Research*, 26(3), 336–45.

Johnson, R. and Wichern, D. 1988: *Applied Multivariate Statistical Methods* (2nd edition). Prentice Hall, Englewood Cliffs, N.J.

Johnson, R. and Wichern, D. 1992: *Applied Multivariate Statistical Methods* (3rd edition). Prentice Hall, Englewood Cliffs, N.J.

Joiner, B.L. and Campbell, C. 1976: Designing experiments when run order is important. *Technometrics*, 18, 249–59.

Jones, R. 1995: Control charts for monitoring two or more quality characteristics simultaneously. *Quality World Technical Supplement*, September 1995, 119–26.

Jones, R. 1999: The Cholesky decomposition of Hotelling's T^2. *Proceedings of the Conference on the Control of Industrial Processes II*, 30–31 March 1999, University of Newcastle upon Tyne, UK.

Jones, R.D., Kushler, R. and Winterbottom, A. 1996: The funnel experiment. *Proceedings of the Conference on the Control of Industrial Processes*, 19–20 September 1995, University of Hertfordshire. University of Hertfordshire Press, Hatfield.

Kendall, M. and Stuart, A. 1980: *The Advanced Theory of Statistics*. Griffin, London.

Kotz, S. and Lovelace, C.R. 1998: *Process Capability Indices in Theory and Practice*. Arnold, London.

Kourti, T. and MacGregor, J.F. 1996: Multivariate SPC methods for process and product monitoring. *Journal of Quality Technology*, 28(4), 409–28.

Kreyszig, E. 1993: *Advanced Engineering Mathematics* (7th edition). Wiley, New York.

Kushler, R. and Hurley, R. 1992: Confidence bounds for capability indices. *Journal of Quality Technology*, 24, 188–95.

Kwakernaak, H. and Sivan, R. 1972: *Linear Optimal Control Systems*. Wiley, New York.

Lamberton, D. and Lapeyre, B. 1996: *Introduction to Stochastic Calculus Applied to Finance*. Chapman & Hall, London.

Lambin, J.-J. 1997: *Strategic Marketing Management*. McGraw-Hill, London.

Lee, P.M. 1997: *Bayesian Statistics: An Introduction*. Arnold, London.

Lewis, C.D. 1981: *Scientific Inventory Control*. Butterworths, London.

Lindley, D. 1985: *Making Decisions* (2nd edition). Wiley, London.

Little, R.J.A. 1989: Testing the equality of two independent binomial proportions. *The American Statistician*, 43(4), 283–8.

Lochner, R.H. and Matar, J.E. 1990: *Designing for Quality*. ASQC Quality Press, Milwaukee, WI and Quality Resources, White Plains, N.Y.

Lockyer, K. and Oakland, J.S. 1981: How to sample success. *Management Today*, July.

London Economics. 1995: Capital costs have been steady for four years. *Water News*, Issue 59.

Loudon, D.L. and Della Bitta, A.J. 1993: *Consumer Behaviour* (4th edition). McGraw-Hill, New York.

Luangpaiboon, P., Metcalfe, A.V. and Rowlands, R.J. 1999: Effects of parameters of a genetic algorithm when optimising process yields. *Proceedings of the Conference on the Control of Industrial Processes II*, 30–31 March 1999, University of Newcastle upon Tyne, UK.

Luangpaiboon, P., Metcalfe, A.V. and Rowlands, R.J. 2000: Comparison of automatic optimisation algorithms on response surfaces with noisy measurements. Department of Engineering Mathematics, University of Newcastle upon Tyne.

McLeod, A.I. 1985: A remark on AS183. An efficient and portable pseudo-random number generator. *Applied Statistics*, 34, 198–200.

Makridakis, S., Wheelwright, S.C. and Hyndman, R.J. 1998: *Forecasting: Methods and Applications* (3rd edition). Wiley, New York.

Manly, B.F.J. 1994: *Multivariate Statistical Methods* (2nd edition). Chapman & Hall, London.

Marketing News. 1997: Consumers are just beginning. *Marketing News, Interactive briefs*, 3 March 1997, p. 35.

Massé, P. 1946: *Réserves et la Régulation de l'Avenir*. Hermann, Paris.

Mayr, O. 1970: *The Origins of Feedback Control*. MIT Press, Cambridge, Mass.

Metcalfe, A.V. 1991: Probabilistic modelling in the water industry. *Journal of the Institution of Water and Environmental Management*, 5(4), 439–49.

Metcalfe, A.V. 1992: The role of engineering control in quality management. *Total Quality Management*, 3(3), 331–4.

Metcalfe, A.V. 1994: *Statistics in Engineering*. Chapman & Hall, London.

Michell, P., Cataquet, H. and Hague, S. 1992: Establishing the causes of disaffection in agency–client relations. *Journal of Advertising Research*, March/April 1992, pp. 41–8.

Michie, D., Spiegelhalter, D.J. and Taylor, C.C. 1994: *Machine Learning, Neural and Statistical Classification*. Ellis Horwood, New York.

Mitrani, I. 1982: *Simulation Techniques for Discrete Event Systems*. Cambridge University Press, Cambridge.

Montgomery, D.C. 1991: *Design and Analysis of Experiments* (3rd edition). Wiley, New York.

Montgomery, D.C. 1997: *Design and Analysis of Experiments* (4th edition). Wiley, New York.

Montgomery, D.C. 1999: Experimental design for product and process design and development (with comments). *The Statistician (Journal of the Royal Statistical Society Series D)*, 48(2), 159–77.

Montgomery, D.C. and Peck, E.A. 1992: *Introduction to Linear Regression Analysis*. Wiley, New York.

Moore, D.S. 1979: *Concepts and Controversies*. W.H. Freeman, San Francisco.

Morrison, D.F. 1967: *Multivariate Statistical Methods*. McGraw-Hill, New York.

Muczyk, J.P. 1978: A controlled field experiment measuring the impact of MBO on performance data. *The Journal of Management Studies*, 15, 318–29.

Murray, P.M. and Murray, M.G. 1999: WatSup model – a new approach to demand modelling. *Proceedings of the Conference on the Control of Industrial Processes II*, 30–31 March 1999, University of Newcastle upon Tyne, UK.

Nelder, J.A. and Mead, R. 1965: A simplex method for function optimisation. *Computer Journal*, 7, 308–13.

Nicholson, N. 1998: Personality and entrepreneurial leadership. *European Management Journal*, 16(5), 529–39.

Norman, P. and Naveed, S. 1990: A comparison of expert system and human operator performance for cement kiln operation. *Journal of the Operational Research Society*, 41(11), 1007–19.

O'Hagan, A. 1994: *Kendall's Advanced Theory of Statistics – Volume 2B Bayesian Inference*. Arnold, London.

O'Hagan, A., Glennie, E.B. and Beardsall, R.E. 1992: Subjective modelling and Bayes' linear estimation in the UK water industry. *Journal of the Royal Statistical Society Series C*, 41, 563–77.

Olguín, J. and Fearn, T. 1997: A new look at half-normal plots for assessing the significance of contrasts for un-replicated factorials. *Applied Statistics*, 46(4), 449–62.

Pan, Y. and Chi, P.S.K. 1999: Financial performance and survival of multi-national corporations in China. *Strategic Management Journal*, 20, 359–74.

Pankratz, A. 1991: *Forecasting with Dynamic Regression Models*. Wiley, New York.

Papayannopoulos, A., Metcalfe, A.V. and Lowther, A. 1999: Elliptical control charts – a practical application. *Quality World*, 25(10), 32–34.

Pitt, L.F. and Abratt, R. 1988: Music in advertisements for unmentionable products – a classical conditioning experiment. *International Journal of Advertising*, 7, 130–7.

Pole, A., West, M. and Harrison, J. 1994: *Applied Bayesian Forecasting and Time Series Analysis*. Chapman & Hall, New York.

Poston, T. and Stewart, I. 1978: *Catastrophe Theory and its Applications*. Pitman, London.

Poundstone, W. 1992: *Prisoner's Dilemma*. Doubleday, New York.

Prabhu, V.B. 1999: Best practice in manufacturing: the N.E. experience. *Proceedings of the Conference on the Control of Industrial Processes II*, 30–31 March 1999, University of Newcastle upon Tyne, UK.

Press, W.H., Teukolsky, S.A., Vettering, W.T. and Flannery, B.P. 1992: *Numerical Recipes in Fortran* (2nd edition). Cambridge University Press, Cambridge.

Pritchard, D., Armstrong, R.J. and Metcalfe, A.V. 1993: SPC for hand assembly operations. *Proceedings of the Conference on Quality and its Applications*, 1–3 September 1993, University of Newcastle upon Tyne, pp. 611–14.

Raab, G.M. and Donnelly, C.A. 1999: Information on sexual behaviour when some data are missing. *Applied Statistics* (*Journal of the Royal Statistical Society Series C*), 48(1), 117–33.

Raj, S.P. 1985: Striking a balance between brand "popularity" and brand loyalty. *Journal of Marketing*, 49(Winter), 53–9.

Ranne, A. 1999: The investment models of a Finnish pension company. *Vector*, 15(4), 63–70.

Ray, W.H. 1981: *Advanced Process Control*. McGraw-Hill, New York.

Readman, M., Muldoon, M., Reynolds, L., Morris, R., Bayliss, M., Wood, D. and Stewart, I. 1998: Control of a spring coiling machine. *Wire Industry*, 65(780), 831–6.

Reedy, B.L.E.C., Metcalfe, A.V., De Roumanie, M. and Newell, D.J. 1980: The social and occupational characteristics of attached and employed nurses in general practice. *Journal of the Royal College of General Practitioners*, 30, 477–82.

Ross, S.M. 1997: *Simulation* (2nd edition). Academic Press, San Diego.

Salopek, D.M. 1997: *American Put Options*. London, Harlow, Essex.

Seddon, J. 1992: *I Want You to Cheat!* Vanguard Press, Buckingham.

Seddon, J. 1997: *In Pursuit of Quality: The Case Against ISO9000*. Oak Tree Press, Dublin.

Seddon, J., Davis, R., Loughran, M. and Murrell, R. 1993: *BS5750 Implementation and Value Added*. Vanguard Consulting Ltd, Buckingham.

Shewhart, W.A. 1931: *Economic Control of Quality of Manufactured Product*. Reinhold Company, Princeton, N.J.

Singh, S.N., Mishra, S., Bendapudi, N. and Linville, D. 1994: Enhancing memory of television commercials through message spacing. *Journal of Marketing Research*, 31, 384–92.

Spooner, H.A. and Lewis, C.D. 1995: *An Introduction to Statistics for Managers*. Prentice Hall, London.

Stewart, I. 1997: *Does God Play Dice?* (2nd edition). Penguin, London.

Taffe, J. and Garnham, N. 1996: Resampling, the bootstrap and Minitab. *Teaching Statistics*, 18(1), 24–5.

Taguchi, G. 1986: *Introduction to Quality Engineering*. Asian Productivity Organization, Tokyo.

Taguchi, G. 1987: *System of Experimental Design* (Volumes 1 and 2). American Supplier Institute, Dearborn, Mich. and Quality Resources, White Plains, N.Y.

Taguchi, G. and Wu, Y. 1980: *Introduction to Off-line Quality Control*. Central Japan Quality Control Association, Nagoya.

Taguchi, G., Elsayed, E. and Hsiang, T.C. 1988: *Quality Engineering in Production Systems*. McGraw-Hill, New York.

Taha, H.A. 1997: *Operations Research* (6th edition). Prentice Hall, Englewood Cliffs, N.J.

Target Group Index Report. 1976: Simmons Market Research Bureau Inc., New York.

Thom, R. 1972: *Stabilité Structurelle et Morphogénèse*. Benjamin, New York.

Tinley Oast Research Ltd. 1998: WatSup Model. Faculty of Technology, Kingston University, Kingston-upon-Thames, UK.

Tong, H. 1990: *Non-linear Time Series*. Oxford University Press, Oxford.

Tong, H. and Crowe, C.M. 1995: Detection of gross errors in data reconciliation by principal component analysis. *AICHE Journal*, 41(7), 1712–22.

Turgeon, A. 1980: Optimal operation of multireservoir power systems with stochastic inflows. *Water Resources Research*, 16(2), 275–83.

Turgeon, A. 1981: A decomposition method for the long-term scheduling of reservoirs in series. *Water Resources Research*, 17(6), 1565–70.

Volvo Car Components Corporation. 1992: Process capability. *VSB 92*, Volvo, Sweden, Issue 2.

Vonderembse, M.A. and White, G.P. 1996: *Operations Management* (3rd edition). West Publishing Company, St Paul, Minn.

Wahid, Z. and Metcalfe, A.V. 1996: Quality control of a welding process using robust design experiments with noise factors: a comparison of alternative analyses. *Proceedings of the Conference on the Control of Industrial Processes*, 19–20 September 1995, University of Hertfordshire. University of Hertfordshire Press, Hatfield.

Wardrop, R.L. 1995: Simpson's paradox and the hot hand in basketball. *The American Statistician*, 49(1), 24–8.

Wetherill, G.B. and Brown, D.W. 1991: *Statistical Process Control*. Chapman & Hall, London.

Wichmann, B.A. and Hill, I.D. 1982: An efficient and portable pseudo-random number generator. *Applied Statistics*, 31, 188–90.

Wicks, D. and Bradshaw, P. 1999: Gendered organizational cultures in Canadian work organizations: implications for creating an equitable workplace. *Management Decision*, 37(4), 372–80.

Wiener, N. 1950: *The Human Use of Human Beings*. Houghton Mifflin, Boston, Mass.

Young, M.R., De Sarbo, W.S. and Murwitz, V.G. 1998: The stochastic modeling of purchase intentions and behaviour. *Management Science*, 44(2), 188–202.

Zeeman, E.C., Hall, C., Harrison, P.J., Marriage, H. and Shapland, P. 1976: A model for institutional disturbances. *British Journal of Mathematical and Statistical Psychology*, 29, 66–80.

Index